Flow of Life in the Atmosphere

Flow of Life in the Atmosphere

An Airscape Approach to Understanding Invasive Organisms

Scott A. Isard

Stuart H. Gage

Michigan State University Press • *East Lansing*

∞ The paper used in this publication meets the minimum requirements
of ANSI/NISO Z39.48–1992 (R 1997) (Permanence of Paper).

Michigan State University Press
East Lansing, Michigan 48823-5202

Printed and bound in the United States of America.

07 06 05 04 03 02 01 00 1 2 3 4 5 6 7 8 9 10

LIBRARY OF CONGRESS CATALOGING-IN-PUBLICATION DATA
Isard, Scott A.
Flow of life in the atmosphere: an airscape approach to understanding
invasive organisms / Scott A. Isard, Stuart H. Gage.
p. m.
Includes bibliographical references (p.).
ISBN 0-87013-550-3 (alk. paper)
1. Air—Microbiology. I. Gage, S. H. II. Title.
QR101.I835 2000
579—dc21
00-064605

Book and cover design by Sharp Des!gns, Inc., Lansing, MI

Cover photo of *peronospora tabacina sporangiospores* provided by H. W. Spurr and C. E. Main, Department of Plant Pathology, North Carolina State University, Raleigh, N.C. Used by permission.

Cover photo of Mountain Cedar pollen release provided by Estelle Levitin, Department of Biological Science, The University of Tulsa. Used by permission.

Visit Michigan State University Press on the World Wide Web at:
www.msu.edu/unit/msupress

Contents

Figures

Tables

Acknowledgments

In the development of the ideas covered in this book, we wish to recognize the encouragement, stimulation, and knowledge gained from our mentors and colleagues including Dean Haynes, Mike Irwin, Pete Lingren, Charlie Main, Paolo Mandrioli, Bob Riley, and Bill Showers. The gratitude to our families for their infinite patience and encouragement while we pursued this and other academic activities, with at time excessive intensity, is immeasurable. Gail Kampmeier and Annalisa Ariatti must be acknowledged, not so much for their direct impact on this book, but for their organizational efforts and seemingly eternal dedication to aerobiology. Finally, we especially wish to thank Donna Isard for her many hours of editing that has made the text so easy and enjoyable to read.

Introduction

This book attempts to relay the importance of anticipating the consequences of the aerial flow of biota as scientists begin to develop new strategies to understand and manage the environment. A sound understanding of the biological and meteorological interactions that govern the movement of organisms in the atmosphere is a prerequisite to the development of successful management strategies for terrestrial ecosystems. Inflows and outflows of organisms to and from habitats can be as important to ecosystems as birth and death rates are in regulating the dynamics of populations.

The ultimate objective of environmental management programs should be to manipulate ecosystems so that they fulfill the needs of humans and at the same time maintain their integrity. Our approach focuses on predicting events that destabilize relationships among organisms and between populations and their environment. This preventive management strategy is based on the premise that the ability to understand and predict dynamics of populations in an ecosystem, allows for optimal and integrative use of a wide variety of methods to enhance human resource production and to reduce harmful impacts of diseases and organisms on humans. It is a paradigm that focuses on manipulating ecosystems to maintain the long-term stability of their diverse populations and the interactions among organisms and their environment. In many ecological systems, sudden and dramatic population fluctuations often result from movement of biota from one habitat to another. The design of grand plans to manage ecosystems, without concern for the inflow and outflow of organisms associated with those ecosystems, could be tragic.

The information in this book is intended to encourage and stimulate a broad perspective on aerobiology. Aerobiology is the study of factors and processes that influence the movement of biota in the atmosphere. Elements that control the aerial flow of organisms can be biological or meteorological. These factors usually interact to govern aerobiological processes and can have important influences on the development of organisms long before they move in the atmosphere.

We attempt to provide those who are interested in the concepts and actions associated with ecosystem manipulation, a window through which to view the aerial movement of biota and the airscapes they encounter. In the first third of the book, we aspire to provide a "systems thinking" approach to understanding the movement of organisms in the atmosphere, and the management of ecosystems impacted by these aerobiota. We explore the rationale for studying aerobiology, provide a conceptual framework for the science of aerobiology, and present a practical way to assemble aerobiology knowledge and methodologies into programs directed toward ecosystem management. A general model of the flow of biota through the atmosphere is developed to focus attention on the processes and scales of atmospheric motion systems that transport organisms. The movement process is divided into stages defined by interactions between the biology of organisms and ambient conditions in the atmosphere.

In the second portion of the book, atmospheric motion systems that commonly impact the movement of biota are described with respect to spatial and temporal scales of motion and the underlying landscape. Chapter 5 is designed as a "non-mathematical" primer for those who desire to learn about the basics of atmospheric motion. The four chapters that follow provide a systematic overview of atmospheric motion systems ranging from global wind belts to turbulent eddies. We attempt to provide readers with examples of how biota use motion systems that occur at each scale to assist their aerial movements among habitats.

In the final section of the book, we provide three examples of how programs to measure and study atmospheric movements of organisms among habitats might be designed. Two of these chapters are written as research proposals, in part because as scientists, we are conditioned to express our ideas in proposals, but more importantly, because the proposal format facilitates the expression of clear linkages among research problems, background literatures, scientific hypotheses, work objectives, and data collection methods. In the first example, we focus on the interactions of biological and meteorological factors that govern the ascent of weak-flying insects (aphids) in the atmosphere. A field measurement program to rigorously test hypotheses concerning the control of insect behavior and the stability of the lower atmosphere on aphid flight trajectories, is central to this research initiative. Next, we present a case study of an on-line decision support system that helps tobacco growers protect their crop from a tiny spore, *Peronospora tabacina* Adam, which commonly disperses long distances within the atmosphere. In some years, this spore causes blue mold epidemics that span the eastern half of North America, while in others, the plant disease occurs in relatively few, highly dispersed, local areas. The integration of a region-wide monitoring network into an Internet-based support system that enables producers to make informed management decisions with regard to this costly agricultural pest is emphasized. The third example focuses on the large-scale aerial flow of biota from a wide range of taxa between the subtropical region and the continental interior mid-latitude region of North America. This research initiative calls for an extensive and highly coordinated program of (1) field and satellite measurements of populations, vegetation productivity and phenology, and weather, (2) radar measurements of aerobiota and concurrent atmospheric motion systems, and (3) the development and implementation of a large-scale interactive forecast model of atmospheric bioflow.

Flow of Life in the Atmosphere: An Airscape Approach to Understanding Invasive Organisms is dedicated to the aerobiologists who have participated in the Alliance for Aerobiology Research (AFAR) over the past decade. Many of the ideas expressed in this book were formulated and refined in discussions among these scientists and outreach specialists during the AFAR Workshop and the numerous AFAR symposia and meetings over the ensuing years. The desire to continue these provocative dialogues and interactions was a primary motivation for our work. Consequently, we believe that it is appropriate to conclude this book with a brief presentation of the history and objectives of AFAR. The epilogue also provides an Internet address for the Movement and Dispersal homepage, so that interested readers may inquire about upcoming aerobiology meetings and symposia, learn more about ongoing movement and dispersal research projects, and, hopefully, join the alliance.

The diversity of the types of biota that move in the atmosphere is immense. Aerobiota include plant and animal viruses, fungi, bacteria, pollen and seeds of higher plants, soil nematodes, arthropods, and birds. Our perspective on how to treat this immense diversity is to consider the biota that move in the atmosphere, not as unique entities, but as organism complexes (e.g., passively transported microorganisms, weak-flying insects, or strong-flying birds) that move through a common medium. We strongly believe that successful manipulation of ecosystems to increase their value and use to humans, depends first and foremost on thoroughly understanding the biologies of their various inhabitants. However, this book is not about the biology of biota that flow in the atmosphere. Two recent publications, *Insect Migration: Tracking Resources Through Time and Space,* edited by Alistair Drake and Gavin Gatehouse (1995) and Hugh Dingle's monograph *Migration: The Biology of Life on the Move* (1996), cover that subject in truly magnificent fashion. The former grew out of a symposium entitled *Insect Migration: Physical Factors and Physiological Mechanisms* held as part of the 1992 XIX International Congress of Entomology in Beijing, China. The book focuses exclusively on migrating insects and covers three topics: migration in relation to weather and climate, adaptations for migration, and forecasting the migration of insect pests. Dingle (1996) provides a comprehensive synthesis of migration as a biological phenomenon. Using comparisons across an extensive range of taxa, he illuminates common physical characteristics and behaviors of organisms that migrate, and illustrates how natural selection may have molded these properties into successful life histories.

Two other recent books on individual taxa of aerobiota are *Atmospheric Microbial Aerosols: Theory and Applications* by Bruce Lighthart and Alan Jeff Mohr (1994) and *How Birds Migrate* by Paul Kerlinger (1995). The former is a technical treatment of the atmospheric physics and biochemistry of aerosol/microorganism complexes that move in the air. The edited volume provides a systematic treatment of the aerobiological processes experienced by these bioaerosols (including viruses, bacteria, fungi, protozoa, algae, and pollen grains) and the health, containment, and regulatory issues related to these aerobiota. Kerlinger's book is of a much lighter nature. It is for those curious about the behavior of birds during flight and their use of the atmosphere. *How Birds Migrate* contains a very delightful series of short chapters supported by lucid illustrations and interesting case studies.

Bioaerosols, edited by Harriet Burge (1995), explores the relationships between humans and the biological contaminants of the indoor atmosphere. It also examines some of the increasingly evident synergistic relationships between biological aerosols and other air pollutants. Although the book is primarily for practitioners, such as facilities managers, industrial hygienists, occupational physicians and health officials, *Bioaerosols* likely will be interesting reading for people who are concerned about the deteriorating quality of indoor air, in which we spend a large portion of our lives.

Finally, a recent textbook, *Methods in Aerobiology* (1998) written by the team of instructors for the biannual Advanced Aerobiology Course should be noted here. This volume, which focuses on aerobiological techniques, was edited by Paolo Mandrioli, Paul Comtois, and Vincenzo Levizzani, and contains chapters on the subjects covered in the international field course, including microphysics, sampling methods, indoor bioaerosols, and statistical analysis. The individual contributions provide technical summaries of aerobiological knowledge in topical areas, while the book as a whole affords a glimpse of the tremendous range of expertise of practitioners in the field.

In contrast to the above works, we do not attempt to characterize individual types, races, and species. Nor do we review the literature in a systematic and exhaustive manner. Our objective is to generalize in order to provide the reader with a vision of how plants and animals can use the atmosphere to move among habitats and how knowledge and information about this aerial movement phenomenon can help humans manipulate terrestrial ecosystems for their benefit. To maintain a focus on this objective and at the same time be succinct, we have tried to limit examples to those that elucidate important points and to provide references to those works that have had an important influence on our thinking. Truly believing that one picture is worth a thousand words, we endeavored to illustrate as many topics as feasible with diagrams, often redrawing and adapting figures from works of other scientists.

The temporal and spatial scales which are relevant for organisms that move in the atmosphere range from weeks to seconds and from continents to leaf parts. In the gaseous atmospheric medium, motions that occur across the entire range of scales are intimately linked in hierarchical structures. Although it is often difficult to understand the ways that atmospheric motion systems are nested, an indifference to the reciprocal connections between large- and small-scale atmospheric events invites failure in aerobiology research. In addition, the spatial and temporal scales of atmospheric motion systems are sometimes cross-dependent, in that small motions tend to be fleeting, while atmospheric systems that influence large areas tend to last from days to weeks. Consequently, short flights of insects among plants in a canopy occur quickly while seasonal migrations of birds throughout continents can last days to weeks. However, many equally important aerobiological events have mismatched spatial and temporal scales. For example, the release of spores from a single field during a few hours may impact large regions that are far downwind. The interwoven concepts of nested, cross- dependent, and mismatched spatial and temporal scales are fundamental to the airscape perspective on bioflow.

Issues of spatial and temporal scales also are among the most challenging aspects of ecology and ecosystem management today. Many biological studies focus on processes that operate at small geographic and short temporal scales. Recently, more holistic perspectives on applied problem-solving are beginning to emerge. Long-term biological and environmental measurement programs are leading to new discoveries and understanding of cycles that exist in biological, atmospheric, and Earth processes. Heightened awareness is also emerging about the need for broader perspectives on geographic scale. This new consideration for expanded thinking about space and time and their associated measurements can be related directly to the need for increased understanding of aerobiology processes. One cannot contemplate long-distance flow of biota from one geographic place, through the atmosphere, to colonize a different habitat, without considering the variety of spatial and temporal scales associated with these processes.

We have incorporated some of the emerging concepts about spatial and temporal scales into this book. We have also attempted to expose readers to some of the tools needed to apply these concepts to their own research issues. The electronic infrastructure provided by the Internet has revolutionized aerobiology by providing universal access to data, the easy transfer of information, rapid communications of ideas, a mechanism for publishing results, and the use of dynamic advanced visualization and audio technologies. It is our hope that each reader will realize the importance of learning these rapidly evolving concepts and technologies, to assist in their understanding and prediction of the flow of biota in the atmosphere.

CHAPTER 1

The Importance of Understanding the Atmospheric Flow of Biota

Many organisms utilize the atmosphere as a medium to move from one terrestrial habitat to another. Some accomplish this by drifting passively and allowing atmospheric motion systems to transport them. Others take a more active role, shifting their appendages to remain aloft while moving with the flow. Still others engage in directed flight and may navigate across or upwind when atmospheric conditions permit. Within the complex of organisms that utilize the atmosphere for transport, a multitude of distinct morphological, physiological, and behavioral characteristics that expedite movement between one geographical location and another have evolved. Despite this vast biological diversity, general principles that govern the aerial movement of biota exist, in large part, because the movement occurs within a common medium, the atmosphere.

Aerobiology and understanding the dynamics of populations in ecosystems.

The rates of movement of biota into and out of an ecosystem, along with birth and death rates, are among the fundamental processes that regulate the dynamics of populations. These rates are interactive, and, almost without exception, very little is known about the role that movement plays in the dynamics of local populations. Throughout history, humans have decreased the diversity of the biotic and abiotic components of many terrestrial habitats. Thus resources needed by their other inhabitants are becoming concentrated in areas that are becoming further and further separated. This is increasing the importance of long-distance movement in the life history of many biota. Organisms move among terrestrial habitats by floating, soaring, and flying in the air, using a variety of forms of terrestrial locomotion, and on occasion, by floating and swimming in water. It takes substantially less energy per unit body mass to float, soar, fly, or swim a given distance than to walk, and many organisms can increase the efficiency of their movement by taking advantage of air and water currents. For these reasons, it is not surprising that there is more movement among terrestrial habitats by organisms that use the air than by those that move over the land surface or through water, and that many of these "aerobiota" are highly mobile in that they have adapted to moving long distances in the atmosphere.

A large number of species important to humans move in the atmosphere, including plant and animal viruses, fungi, bacteria, pollen and seeds of higher plants, soil nematodes, arthropods, and birds. Yet empirical data on aerial movements of these organisms is sparse. From the perspective of science, these data are critically needed to increase our understanding of aerial movement processes. However, equally important, these data are necessary to provide the knowledge required to ensure human health and to efficiently and safely manage many terrestrial ecosystems.

Rabb (1985), in his treatment of conceptual frameworks for studying area-wide movements of populations, claims that the dearth of observational and experimental data on movements of organisms in general, and especially those that move in the air, is critical. This lack of information, he argues, currently limits progress in ecology. All too often scientists investigate single factors in laboratories or on small plots over short time intervals. This narrow spatial and temporal perspective is both a response to western scientific paradigms and the result of a myopic research support system. In addition, these data are sparse because scientifically sound information on movement is hard to measure in the field. It is not because ecologists discount the importance of movement and dispersal to population dynamics.

There are a number and variety of factors, both biological and atmospheric, which influence an organism's movements in the air. Because these different factors interact in many complex ways, it is difficult to gain a solid understanding of the processes that influence an organism's aerial movements. It is even more difficult to formulate scientifically acceptable generalizations about the movement of aerobiota that are applicable across a wide range of taxa. The diversity of species that move in the atmosphere, the difficulty of studying biota that move long distances far above our heads, and the lack of a scientific framework for studying the flow of aerobiota that is independent of a particular taxa all contribute to the paucity of aerobiology knowledge.

There exists great diversity in the cycles of growth, development, and reproduction among organisms that move in the atmosphere. Each species of organism also experiences its environment differently. The size, life span, and spatial arrangement of organisms in a landscape usually dictate the physical and temporal dimensions of the habitats that biologists investigate. Consequently, the scales at which population dynamics and aerial movement are studied vary among species. As a result of this diversity and complexity, few scientific studies, with the notable exception of Dingle (1996), have focused on the morphological, physiological, and behavioral characteristics that aerobiota have in common.

Another reason for this dearth of knowledge is because aerobiology, the study of the movement of biota in the atmosphere, is a multi-disciplinary endeavor. Institutions of higher education in North America do not have aerobiology departments, and the principles of aerobiology are rarely taught. Most scientists who study aerobiology are trained in either the biological or atmospheric sciences, but rarely both. Thus to obtain a comprehensive understanding of aerobiology, scientists from the biological disciplines must learn about the intricacies of meteorological processes and how they influence population dynamics, independent of their own formal training. Similarly, researchers trained in the atmospheric sciences who want to investigate aerial movements of biota need to comprehend the complexities of biological processes and how they influence the initiation of movement and the behavior of organisms in the atmosphere. Acquiring knowledge in an unfamiliar discipline is difficult. Many aspects of the biological and atmospheric sciences are dissimilar. For example, measurement technologies and the scales of measurement have traditionally been very different in these two branches of science. To characterize the dynamics of populations, a biologist is likely to take field measurements once a week, or less often. In contrast, a meteorologist must measure velocity and temperature fluctuations many times a minute to characterize the eddies and convection currents in which the same biota may move. Because of the interdisciplinary nature of the subject, successful aerobiology research programs are composed of teams of scientists that include participants from both the biological and atmospheric sciences.

Perhaps a greater discrepancy between the biological and meteorological sciences relates to their approaches to the standardization of field measurements. Throughout the 20th century there has been a concerted effort to measure atmospheric (weather) variables around the world using a single set of methods. Extensive measurement programs have been conducted whenever new and improved technologies became available to correlate the new and old methodologies and thus maintain a consistent time series. As a result, comparable daily (and some hourly) data on maximum and minimum temperatures, precipitation, and other meteorological variables are available for thousands of locations on Earth. With the advent of weather satellites and radars in the last 30 years, the types of atmospheric measurements and their spatial and temporal resolutions have increased dramatically and are now available in standardized formats for the entire Earth. In contrast, the standardization of biological sampling procedures has often occurred at the level of habitat type. This is especially true for arthropods and to a lesser degree for microbiota. Generally, the technologies biologists employ to sample populations, and the spatial and temporal resolution of their population measurements vary tremendously from one habitat type to another. For example, nets are often employed to catch moths and other insects in agricultural fields during the day while pheromone and light traps are commonly used to collect the same insects in the forest at night. In contrast to the efforts of national weather services, biological research agencies have not invested in spatially extensive, standardized data collection networks. As a result, the plethora of important ecological data that have been collected over the years are not very useful for studying the dynamics of populations that move among different types of habitats during their life cycles or for gaining insight into the regional dynamics of organisms. From a management perspective, the lack of historical biological data that are standardized among habitats makes it difficult to develop strategies for manipulating biota on a wide area basis. The paucity of population data comparable across space also makes it hard for aerobiologists to study the influence of atmospheric motion systems on long-distance movements of biota.

Yet as Rabb (1985) and Dingle (1996) clearly point out, ecologists have long recognized the importance of movement to the study of population dynamics. The dynamics of populations at a location do not depend on environmental and biological factors at that place alone. Populations are also strongly influenced by environmental and biological dynamics that occur at other locations, some far away from the specific place of interest. The numbers and diversity of organisms that move in the atmosphere are astounding. For example, airplane collections over Louisiana between 1926 and 1931 netted more than 28,000 insects, spiders, and mites, representing 216 families, 824 genera, and 700 species (Glick 1939). Perhaps 1200 species of bacteria, some 40,000 species of spore-producing fungi, mosses, liverworts, ferns, and their allies, and more than 10,000 species of pollen-producing flowering plants use the atmosphere to disperse (Gregory 1973). Many other biota use soil and water as media for movement. The magnitude and timing of their movements can have dramatic impacts on populations in both origin and destination habitats. In fact, our understanding of the dynamics of populations can depend as much on the knowledge of the movement processes as it can on the local factors that control the rates of birth and death.

Despite the importance of movement, the subject is frequently avoided in population studies. Moreover, the movement of organisms is often over-simplified and discussions are based on unrealistic assumptions, such as inflows are equivalent to outflows. In systems where biota move into and out of the study area, the true dynamics of populations may be misunderstood if movement is discounted, and thus result in erroneous predictions of population fluctuations. For most habitats, the inventory of organisms that move in the air, and information on the distance, frequency, and reasons for their movement, is incomplete. Because rates of birth, death, and movement are interactive, any improvement in knowledge about movement of biota will increase the understanding of factors that influence birth and death processes. Where movement is critical to population dynamics, the errors resulting from ignoring movement are likely to be great enough to make field measurements of factors that affect birth and death rates within the ecosystem irrelevant! Accurate information on the numbers and types of organisms that enter and depart from a system is fundamental to a scientifically sound understanding of dynamic ecosystem processes at all scales of organization. Similarly, this knowledge base should be an integral part of the foundation of any program to manage terrestrial ecosystems.

Managing ecosystems impacted by aerobiota.

Organisms that utilize the atmosphere as a medium for movement create some of the most challenging ecosystem management problems. This is because their populations can increase dramatically, often without warning and independent of local factors that operate within an ecosystem. Many of these biota are highly mobile, either because they have morphological, physiological, and behavioral characteristics that expedite long-distance movement, or because they are transported to new habitats within atmospheric motion systems. These biota can be pests or diseases of agriculture, forests, wildlife, and humans. However, the overwhelming majority of the individuals, the species, and the mass of biota that are transported in the atmosphere are considered to be beneficial by humans.

Information on the movement of both undesirable and beneficial organisms is essential to the success of ecosystem management programs. These programs are designed to implement strategies for manipulating biota and their environment to meet human needs. At the same time, flows of energy and materials and interrelationships among organisms, and their environment must be maintained to ensure ecosystem integrity and thus an ecosystem's potential to meet human needs in the future.

Throughout history, humans have tended to concentrate resources in space for a variety of reasons, including economies of scale. These actions have created ecosystems rich in the resources needed by plants and other animals. The life history strategies of many of the organisms that compete with humans for these resources (generally labeled pathogens, weeds, or insect pests) permit rapid population growth when resources are abundant. As a result, humans have had to manage resource-rich agricultural and forest ecosystems to limit the populations of these unwanted competitors. Often the strategies employed by humans to limit competitors disrupt the dynamics of the ecosystem by killing other important organisms, such as parasites and predators, leading to outbreaks of undesirable populations.

Pathogens, weeds, insects, and other human competitors are among the dominant life forms in terrestrial ecosystems, in terms of their abundance and diversity. They represent a variety of trophic levels and play an important role in regulating populations and thus maintaining ecosystem stability. Some of these competitors spend the majority of their life history in distant habitats and move through the atmosphere to find resources at specific stages in their life cycles, making the influxes of these aerobiota to ecosystems difficult to predict. Consequently, successful management of agricultural and forest ecosystems requires an understanding of how organisms use the atmosphere to move among habitats. Knowledge of factors that govern birth, death, and movement are all equally critical to developing management strategies that successfully limit populations of pathogens, weeds, and pests, while minimizing disruptive effects on beneficial organisms and the future value of these ecosystems to humans.

Plant pathogens, weeds, and insects destroy approximately 37% of the potential food and fiber crops in the United States. In 1986, about 45 billion kg (0.5 million tons) of 600 different types of pesticides were used to combat these organisms at a cost of $4.1 billion. An additional 180 billion kg (2 million tons) of pesticides at a purchase price of $16 billion was applied each year elsewhere on planet Earth. The direct return on each dollar invested in chemical control of these organisms in United States agriculture was approximately $4. Pesticides, however, do not always reduce crop losses. For example, in the United States, where the total value of agricultural crops grew approximately 5-fold between 1945 and 1989, the percentage of value of crop production lost to pests increased from 31.4 to 37%, even though the amount of pesticides (insecticides, herbicides, and fungicides) produced (about 90% of which is sold within the country) increased 33-fold (Pimentel et al. 1990, 1993).

Most estimates of the economic benefits of pesticides are based on direct crop returns. As such, they do not include the human health, environmental, and many of the indirect economic costs associated with the use of pesticides. A more comprehensive approach to estimating the costs associated with chemical control of undesirable organisms in agriculture should include: impacts on human health; livestock and livestock product losses; increased control expenses resulting from pesticide-related destruction of natural enemies and from the development of pesticide resistance; crop pollination problems and honeybee losses; crop and crop product losses; fish, wildlife and microorganism losses; and government expenditures to reduce the impacts of pesticide use on the environment and society (Pimentel et al. 1993). When these factors are considered, the annual environmental cost of pesticide use in the United States exceeds $8 billion, and is likely $100 billion per year worldwide (Pimentel 1997). The detrimental impact of agricultural pesticide use on human health alone is astounding. Between 1 and 3 million human pesticide poisoning events, including more than 200,000 deaths, occur world-wide each year (WHO/UNEP 1990, Pimentel 1997). In the United States there are on average 67,000 pesticide-related illnesses and 27 deaths annually (Litovitz et al. 1990). The estimated economic costs of pesticide poisonings and other related illnesses in the United States each year alone is approximately $800 million (Pimentel et al. 1993).

Traditionally, species that successfully compete with humans for food and fiber resources, or are detrimental to human health, are designated pests. Once so labeled, they frequently become the focus of eradication programs. Too often, organisms continue to be targeted for "control" long after their populations have declined and they are no longer a serious threat to humans (i.e., once a pest, always a pest). Currently, there is an important shift in thinking toward preventive management with the explicit goal of maintaining ecosystem integrity. Here we broadly define both ecosystem and its integrity. An ecosystem includes interacting biological, environmental, economic, and social components (see Slocombe 1993). Its boundaries are governed by population processes and patterns. The integrity of an ecosystem is its capability of maintaining a balanced, integrated, adaptive community of organisms (see Angermeier and Karr 1994). Important aspects of ecosystem integrity include species composition, diversity, and their functional organization.

In the preventive ecosystem management paradigm, all species, whether they are pests or beneficials, are considered, first and foremost, to be integral parts of the food web. From this perspective, species diversity is viewed as a critical indicator of ecosystem integrity. Pest outbreaks are viewed as periodic upswings in natural fluctuations of populations. Organisms, with the exception of those that are harmful to human health, are considered to be "beneficials" until their numbers begin to increase rapidly, at which time they become pests.

For example, there is an acute need to encourage a wide diversity of organisms in agricultural systems to restore and maintain ecological balance, so that production may be sustained without constant human intervention in the form of agrochemical inputs (see Altieri 1995). An herbivore or plant pathogen should only be considered a pest when its population grows large enough to reduce productivity below an acceptable economic level. In other human-constructed landscapes (e.g., suburban, urban, and recreational), organisms are pests if they are noxious or when they become overly abundant or limit human activity. Similarly, in more natural systems, all organisms are considered beneficial from the preventive management perspective, unless their numbers grow large enough to severely limit other important species or impact the functioning of the ecosystem so as to decrease its fulfillment of human needs.

We use this concept of pest throughout the following chapters: all plants and animals are considered to be beneficial in that they are integral to the food web and contribute to the diversity and stability of life on Earth; any organism may become a pest if it has a negative impact on human health and activities, if it reduces the productivity of a crop or other resource below an acceptable economic level, or it becomes numerous enough to destabilize an ecosystem and thus jeopardize an ecosystem's potential to satisfy human expectations for the future. Usually the status of pest is only temporary. Most pests become beneficial organisms after their populations decline.

The advent of preventive management programs aimed at maintaining ecosystem integrity has created an increased need to predict when, where, and which populations are likely to grow rapidly and achieve pest status. Where movement is critical to the dynamics of populations, the realization of this demand requires information on the flows of biota into and out of habitats. Tied to this is a need to understand the biological and atmospheric factors that interact to govern movements of biota among habitats. In short, aerobiological understanding must be part of the foundation of successful programs for managing terrestrial ecosystems.

Information on inflows and outflows of organisms is essential to establish parameters for predictive models of ecosystem dynamics. However, at a more fundamental level, this knowledge is necessary to establish reliable sampling practices for measuring and monitoring populations. For example, many organisms move among habitats during a single stage in their life cycles. Most of these organisms use eddies and convective currents during warm, sunny days to assist them in their travels. Yet large-scale weather systems that occur infrequently, at irregular intervals, and move rapidly across the landscape, can be equally important to the dynamics of a population. Many biota (e.g., active fliers such as insects and birds) avoid aerial movement during these intense weather events. Others, such as spores and pollen, use the energy from the wind and rain splashes associated with these systems to initiate movement. Still others that may have initiated movement during fair weather are overtaken by fast moving atmospheric motion systems. Flight behavior and converging wind systems can act together to concentrate biota in the atmosphere, while intense precipitation and atmospheric turbulence can deposit biota in high concentrations on the landscape. As a result, new biota may be introduced or dramatic increases in local populations may develop from existing widespread but low-level populations.

The success of management programs for maintaining the integrity of ecosystems with highly mobile biota requires a comprehensive understanding of the dynamics of populations based, in part, on the identification of the contribution to the total inflows and outflows of the movements of organisms that occur at different scales. Consequently, temporal and spatial scales associated with atmospheric motion systems may dictate the timing and placement of biological sampling and monitoring devices. However, only with a sound knowledge base about the biological and atmospheric interactions that govern aerial movements of organisms can we hope to develop reliable sampling and monitoring practices.

The effectiveness of other components of ecosystem management programs may be limited by insufficient knowledge of movement of important organisms as well. For example, aerial movements of biota are currently given little consideration in research programs aimed at developing food and fiber plants that can withstand the detrimental effects of herbivores and pathogens. This occurs even though a basic understanding of movement is fundamental to several aspects of host plant resistance breeding, such as exploring ways to disrupt the host-finding behaviors of harmful organisms. Field programs for evaluating resistant plants are hampered by unreliable methodologies for sampling aerobiota. Precise monitoring of the taxa, numbers, timing, and spatial variations of immigrants to test plots is a prerequisite to measuring the impact of herbivores, plant pathogens, and beneficial organisms on food and fiber plants.

Currently there is increasing emphasis on trying to moderate undesirable fluctuations in populations by manipulating the environment to enhance the numbers and effectiveness of their natural enemies. Pathogens and predators foreign to an ecosystem may also be employed to help dampen population fluctuations. These exotic organisms can be imported from other ecosystems and new enemies can be developed in laboratories. To date, programs to introduce biological enemies have been overwhelmingly reactive in that they have been instigated in response to a dramatic population increase of an undesirable organism. In some circumstances, there has been pressure to enhance or release these "beneficial" organisms in response to urgent problems before the full range of consequences can be evaluated. There is always the potential that some of these new and exotic biota may do irreparable harm to an ecosystem and may disperse to other ecosystems where they are undesirable. It takes many years to conduct extensive and thorough risk assessment of the likely spread of an organism and the long-term ecological and societal impacts of introducing new or exotic biota into an ecosystem. The paradigm of managing ecosystems to maintain populations of potential pests within the range that allows resource production at acceptable economic levels provides a framework for proactive planning for the use of biological enemies. However, effective research programs in this area also require improvement in systems for monitoring populations, predicting aerial movements of organisms, and modeling ecosystem dynamics.

Management practices can be used to alter the timing, spatial distribution, and architecture of plants in an ecosystem to maintain the stability of populations. For example, increasing the diversity of plants, and including natural refugia in human-constructed landscapes, usually increases the effectiveness of natural enemies in dampening fluctuations of pest populations. However, without basic knowledge about the exchanges of biota among different habitats, few management practices can be undertaken confidently to reduce undesirable populations and to increase the number and diversity of beneficial organisms.

Organisms can develop resistance to the chemicals used by humans to combat them. For example, almost 100 species of mosquitoes have developed resistance to one or more insecticides, usually after 2 to 10 years of uninterrupted use (Brown 1983). Between 1970 and 1980, the number of arthropods reported to have developed resistance to chemicals increased by 204 (91%), and by 1989 at least 504 species of arthropods, as well as plant pathogens, weeds, and nematodes were known to have developed resistance to pesticides (Georghiou and Lagunes-Tejeda 1991; Figure 1.1). Yet incidence of chemical resistance in many taxa are likely unrecorded, and resistance development among pathogens of plants, humans, and other animals likely far outstrip those of arthropods (Real 1996). Understanding gene flow within and among populations is essential for predicting resistance development. Gene flow can either retard or hasten the development of resistance to chemicals. The rate of resistance development depends on environmental conditions and the movements of organisms among and within habitats. Our current paucity of information about the principles of movement prevents us from generalizing about gene flow in different species and under different environmental conditions.

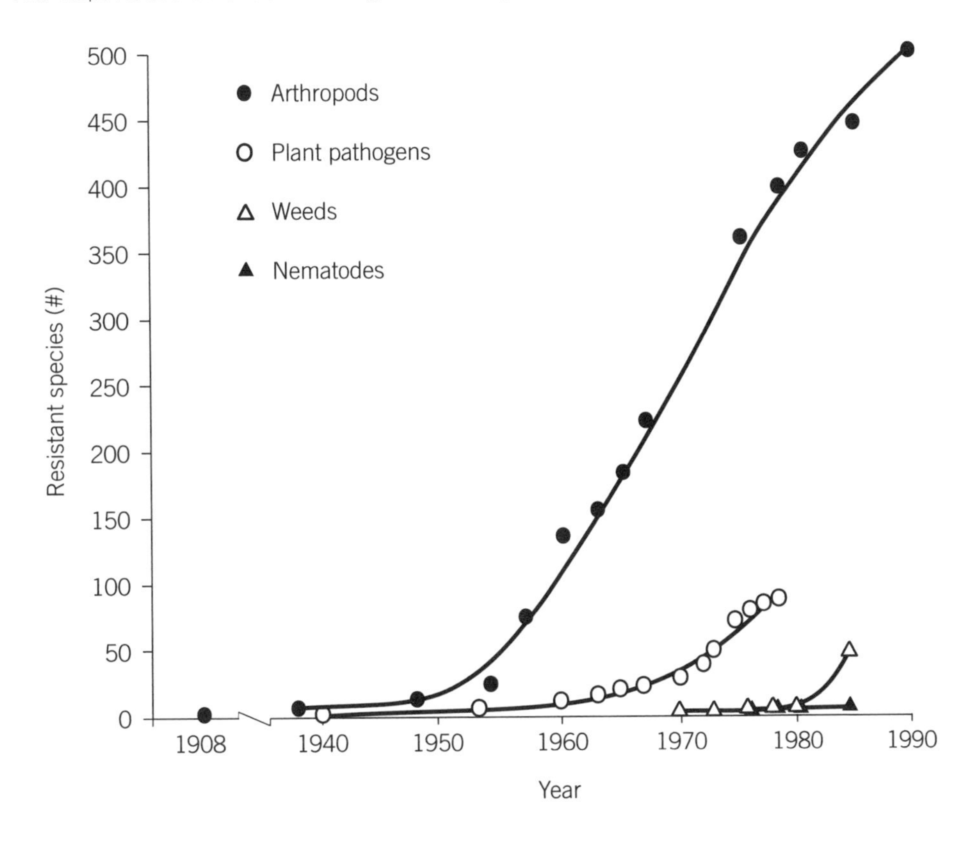

Figure 1.1. Recent increases in pesticide resistance among selected taxa.

"The bright future projected for crop protection and public health as a result of the introduction of synthetic organic pesticides is now open to serious question because of an alarming increase in the number of instances of resistance in insects, plant pathogens, and vertebrates, and to a lesser extent in weeds. There are no longer available any effective pesticides against some major crop pests, such as the Colorado potato beetle on Long Island and the diamond-back moth on cruciferous crops in much of the tropical world. Likewise, the malaria eradication programs of many countries are in disarray, in large part because vector mosquitoes are no longer adequately controlled with available insecticides. The incidence of malaria is resurging at an alarming rate. Because of the costs of bringing new pesticides to market, there are fewer new pesticides, and those produced are targeted only for major crops and pests. Resistance to pesticides, which first involved only insecticides, now exists for fungicides, bactericides, rodenticides, nematicides, and herbicides." (Preface from the *National Research Council Report on Pesticide Resistance: Strategies and Tactics for Management,* Glass 1986). Figure adapted from Georghiou (1986) and Georghiou and Lagunes-Tejeda (1991).

Other components of management programs also provide a platform for increased concern about organisms that move in the atmosphere. The bio-engineered crops that are rapidly extending our food supplies are providing new opportunities for organisms that are not currently classified as pests to become harmful. The technology to engineer beneficial organisms is also progressing rapidly. Because these advances are produced in the laboratory, little is understood about responses of bio-engineered organisms to natural fluctuations of weather and climate and their capacity and inclination to move within atmospheric motion systems. This knowledge is necessary to meet legal requirements for deploying genetically engineered organisms to enhance food production and protect human health. For many people, there are serious ethical considerations and responsibilities associated with the release of genetically engineered organisms.

Impacts of aerobiota on human health and society.

Airborne bacteria, fungi, viruses, and pollens are responsible for numerous human diseases (Table 1.1). Rapid urbanization during the past century has increased dramatically the importance of human disease transmission by aerobiota, because the airborne spread of disease is especially common within indoor environments. In addition, nonbiological particulates, gases, and other pollutants in urban environments are known to aggravate the effects of airborne diseases and allergenic agents.

The current concern about the quality of the air we breath indoors has renewed attention to airborne contagion. In the interest of energy efficiency and conservation as well as abatement of noise and outdoor pollutants, people in middle and high latitude regions and in cities worldwide have decreased the ventilation of their workplaces, schools, and homes. As a consequence, many humans experience prolonged exposure to high aerial concentrations of organisms, including bacteria, viruses, mites, and others that cause human disease and discomfort. An influenza outbreak involving 38 of 54 passengers (72%), contracted from a single individual on a grounded unventilated airplane, is a classic example of the new health risks humans face as a result of advanced technologies (Moser et al. 1979). The situation is especially serious in hospitals, where reduced ventilation has increased the risk for the transmission of contagious diseases. The reemergence of tuberculosis transmitted by airborne particles has also helped fuel renewed interest in indoor aerobiology. The development of antibiotic-resistant tuberculosis strains coupled with a dramatic increase in human susceptibility to the disease (due to the AIDS epidemic) is creating a major human health problem.

The outbreak of Legionnaires' disease in 1976 awoke humans to the threat of airborne contagion perhaps more than any other single event. During that summer, 182 of 4400 individuals associated with the Legionnaires' Convention in Philadelphia became ill, 29 of whom died (Fraser et al. 1977). The spread of the bacterium was airborne and exposure likely occurred in the lobby of the headquarters hotel. The quest to find, identify, treat, and subsequently control the pneumonic killer, was one of the most intriguing in medical and microbiological history and was followed closely by the press and public worldwide. Out of the tragedy came insights into the ecology of pathogenic microorganisms, mechanisms associated with their aerial dispersal, and the discovery of a previously unrecognized human pathogen, *Legionella.*

Many of the plants and fungi that coexist on Earth with humans produce pollen or spores which are part of the air that we breathe outdoors and may act as antigens in interactions with the human immune system. Some algae, bacteria, actinomycetes, and protozoa are also considered aeroallergens when they are transported in the atmosphere. Allergic responses to aerobiota impose major adverse effects on the physical and economic well-being of humans. They can trigger allergic rhinitis or hayfever (symptoms:

Table 1.1. Some important diseases and allergens caused by aerobiota.

TYPE OF DISEASE	CAUSAL AGENT OR SOURCE
Bacterial diseases	
Pulmonary tuberculosis	*Mycobacterium tuberculosis*
Pulmonary anthrax	*Bacillus anthracis*
Staphylococcal respiratory infection	*Staphylococcus aureus*
Streptococcal respiratory infection	*Streptococcus pyogenes*
Meningococcal infection	*Neisseria meningitidis*
Pneumococcal pneumonia	*Diplococcous pneumoniae*
Pneumonic plague	*Pasteurella pestis*
Whooping cough	*Bordetella pertussis*
Diphtheria	*Corynebacterium diphtheriae*
Klebsiella respiratory infection	*Klebsiella pneumoniae*
Staphylococcic wound infection	*Staphylococcus aureus*
Fungal diseases	
Aspergillosis	*Aspergillus fumigatus*
Blastomycosis	*Blastomyces dermatitdis*
Coccidioidomycosis	*Coccidioides immitis*
Cryptococcosis	*Cryptococcus neoformans*
Histoplasmosis	*Histoplasma capsulatum*
Nocardiosis	*Nocardia asteroides*
Sporotrichosis	*Sporotrichum schenckii*
Diseases caused by viral and related agents	
Influenza, febrile pharyngitis or tonsillitis, common cold, croup, bronchitis, bronchiolitis, pneumonia, febrile sore throat, pleurodynia, and psittacosis.	
Common aeroallergens and sources	
Pollens	Wind-pollinated plants including grasses, weeds, and trees.
Molds (common aeroallergenic fungi include *Alternaria, Aspergillus, Botrytis, Cladosporium, Curvularia, Espicoccum, Fusarium, Helminthosporium, Hormodendrum, Macrosporium, Penicillium, Phoma, Pullularia, Rhodotorula, Spondylocladium, Stemphylium, and Trichoderma*)	Usually saprophytic. Prevalence in an area depends upon the distribution of host plants, substrate, and atmospheric temperature and humidity.

Adapted from Table 1.2 in Edmonds 1979.

running nose and eyes), allergic asthma (symptoms: wheezing and shortness of breath), and atopic dermatitis (symptoms: inflammation of the skin).

A sinister aspect of aerobiology is the release of organisms that are transported by atmospheric motion systems as acts of terrorism. The potential hazards and risks from the airborne introduction of engineered animal and plant pathogens in biological warfare is considerable. Harmful biota also could be released inadvertently, disperse in the atmosphere, and subsequently cause human, other animal, and plant disease epidemics. Whether

Table 1.2. Status of major diseases transmitted to humans by aerobiota and predicted sensitivity of their geographic distribution to climate change.

DISEASE	VECTOR	POPULATION AT RISK (MILLIONS)	PREVALENCE OF INFECTION (MILLIONS)	PRESENT DISTRIBUTION	POSSIBLE CHANGE OF DISTRIBUTION DUE TO CLIMATE CHANGE
Malaria	mosquitoes	2100	270	tropics & subtropics	highly likely
Lymphatic filariases	arthropods	900	90.2	tropics & subtropics	likely
Onchocerciasis	black flies	90	17.6	Africa & Latin America	likely
African trypanosomiasis	tsetse flies	50	*	tropical Africa	likely
Dengue	mosquitoes	?	?	tropics & subtropics	very likely
Yellow fever	mosquitoes	?	?	Africa & Latin America	likely
Japanese encephalitis	mosquitoes	?	?	East & Southeast Asia	likely
Other arboviruses	arthropods	?	?	tropic/temperate zones	likely

*25,000 new cases/year.

Data from World Health Organization presented in Patz et al. 1996.

biota are considered harmful or beneficial by humans, the same knowledge base is required to understand and predict their aerial movements. The growing number of movies and novels about airborne pestilence attest to the public awareness and concern about these possibilities (see Krajick 1997).

Added to this is an increased urgency for knowledge on aerial movement spawned by the need to predict how the population dynamics of biota will be influenced by the changing global environment. Urbanization and suburbanization have created a variety of ecosystems with relatively homogeneous but abundant resources. The diversity of organisms that thrive in these new environments is increasing. Some of these organisms are transported by air currents, indoor and outdoor. The number of organisms that are considered noxious pests and dangerous pathogens in these environments is also increasing, simply because they are more often in contact with humans.

Rapid changes in global marketing have increased the transport of goods and the movement of people among continents. A concomitant increase in movements of organisms from one region of the world to another has occurred, both by human and atmospheric transport systems. Human-transported biota include an ever-increasing array of domesticated plants and animals, the pathogens and nematodes associated with these organisms, and other microorganisms on natural resources and manufactured goods.

Changes in the global climate potentially will alter the geographic range of many organisms as well. Vagrants, strays, and dispersing organisms will colonize environments that were previously unsuitable for them because they lacked one or more climate-related resources, such as food and shelter. This is especially true for highly mobile aerobiota. Some of these organisms may become devastating consumers and pathogens of food crops. Others may have detrimental effects on human health in their new environments. The possible expansion in geographic range of some major diseases transmitted to humans by aerobiota as a result of projected changes in climate are listed in Table 1.2. As human populations continue to grow, the potential for pest and disease epidemics will become greater, and the importance of issues associated with the spread of biota in the air will increase.

Shifts of atmospheric circulation patterns associated with changes in global climate will likely alter the atmospheric pathways that biota may use. These shifts, coupled with changing strategies of crop production, the introduction of new and exotic crops, and advances in crop genetics will likely make food and fiber production more reliant on chemical use. Simultaneously, western society is rapidly adopting new perspectives on promoting environmental protection, preventive ecosystem management, and regional management of pest systems. These paradigms have developed in response to a widespread desire to reduce the current use of potent chemicals that degrade our planet's biological processes. The trend toward increased reliance on chemical use for food and fiber production, coupled with heightened concern for our global environment, provides an immense challenge and opportunity to gain a better appreciation and understanding of aerobiology.

Summary.

In this chapter we have argued for a more in-depth understanding of aerobiology from three perspectives: first, as part of the expanding scientific knowledge base on intrinsic population processes in the life histories of biota, second, the essentialness of this knowledge base to the success of management programs aimed at maintaining the integrity of Earth's ecosystems, especially those with highly mobile organisms, and third, the impact of aerobiota on human health and society. The following two chapters build on these themes. Chapter 2 provides a conceptual framework for the science of aerobiology. In Chapter 3, we explore methodologies and technologies to incorporate aerobiology knowledge into programs to manage ecosystems.

CHAPTER 2

A Conceptual Framework for Understanding the Atmospheric Flow of Biota

Movement as part of the life cycle of organisms.

Life is a continuous cycle that has no real beginning or end. This continuum includes the processes of birth, growth, maturation, reproduction, movement, and death. The life cycles of organisms are often depicted as circles composed of growth stages and life processes. Population change may also be considered circular because it is not possible to determine a starting point unless we consider it only from a human perspective. Populations cycle in density between high and low over time. When plotted with time on the abscissa, these cycles appear wave-like. Fluctuations in populations are often irregular. This is because they are influenced by an extensive array of biological and environmental factors that interact to govern the rates of birth, death, and movement into and out of an ecosystem. Population change and life processes are intimately connected: cycles nested within cycles (Figure 2.1a–d).

Many populations build until their supply of food or another key resource becomes limiting. If resources are abundant and/or are highly valued by humans, they may become a pest. In time, competition within the species for resources, or a reduced resource supply, induces stress on the rapidly increasing population. Some organisms react to this stress by moving to new environments. Individuals that remain behind are more vulnerable to parasites, predators, and diseases that compete readily with the stressed organisms. As disease, parasitism, predation, and emigration continue to reduce a population, the number of natural enemies begins to decline. When the resources upon which the organisms feed begin to recover, the population will again increase to utilize the renewed resources, and the cycle will continue. At any time in this cycle, immigration of biota from other environments can augment an organism's population, their food or hosts, and their natural enemies.

Dramatic changes in rates of birth, movement, and death occur periodically during the life cycles of many organisms, and these can destabilize the structure of a population in an ecosystem. Often population outbreaks occur because the organisms involved can increase in density rapidly when environmental conditions are favorable for reproduction, although the factors, and interactions between factors, that cause pest outbreaks (e.g., genetics, trophic interactions, host changes, and reproduction strategies) vary across taxa

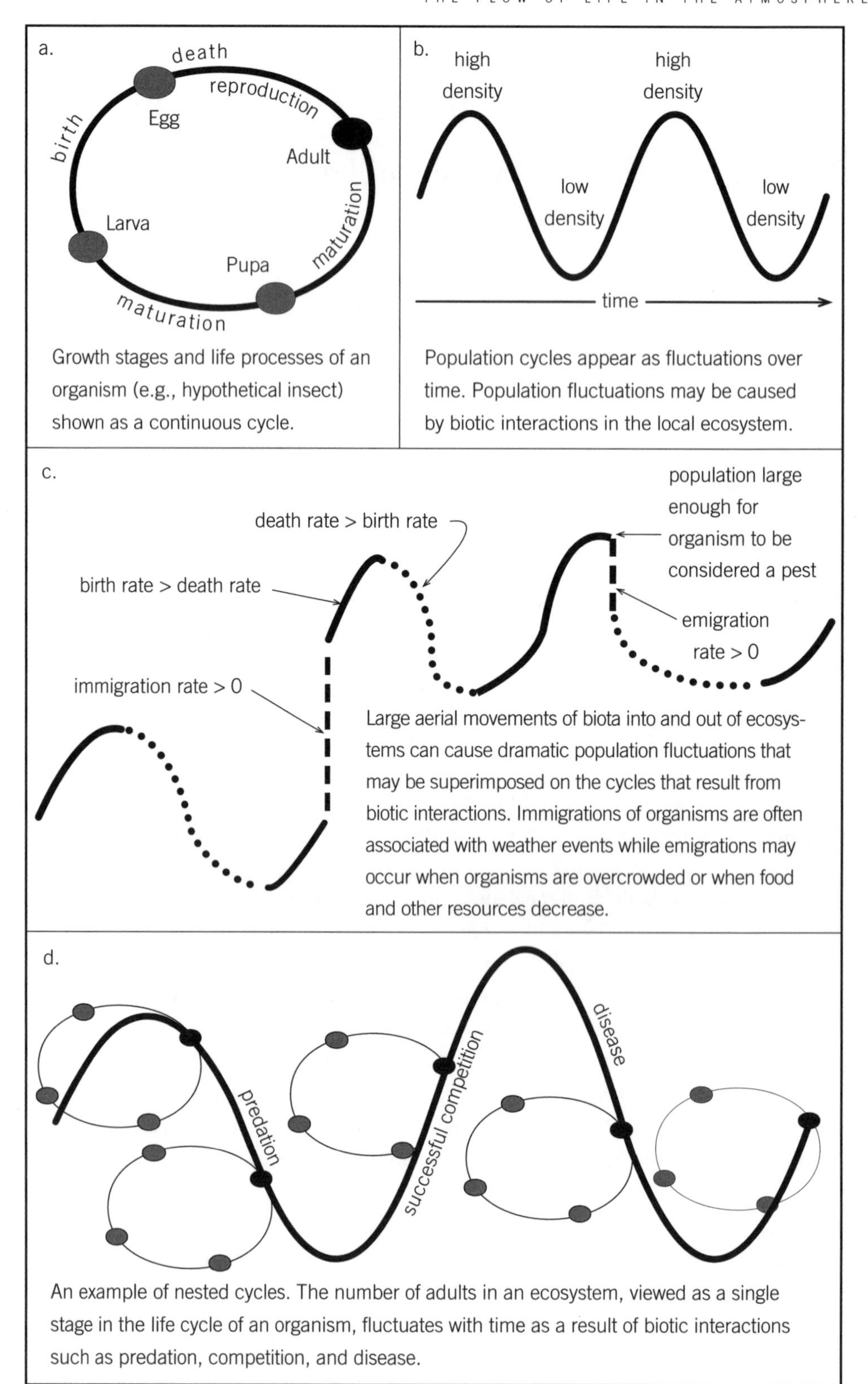

Figure 2.1. Examples of simple population cycles and how they can be nested.

Table 2.1. References to works that have influenced our thinking about the systems approach.

SYSTEMS APPROACH (GENERAL):
Watt 1966, Churchman 1968, Bertalanffy 1968, Holling 1978, Edmonds 1979, Checkland 1981.

SYSTEMS APPROACH TO ECOSYSTEM MANAGEMENT:
Rabb and Guthrie 1970, Haynes et al. 1973, Hollings 1973, Ruesink 1976, Haynes et al. 1980, Teng 1985, Brown 1991, Worner 1991, Teng and Savary 1992, Allen and Hoekstra, 1992, Gunderson et al. 1995, Barbosa 1998.

and ecosystems (see Barbosa and Schultz 1987). Massive influxes of biota also can contribute to population outbreaks. From a narrow perspective, dramatic movements of populations among distant habitats may appear random in time and space. However, there is generally a scale large enough to allow movements of these organisms to be viewed as cyclical, and thus to a large extent, predictable (Figure 2.2)

The same circularity applies to other Earth processes. Atmospheric and geochemical processes are as cyclical as biological processes. The Earth orbits the sun and rotates on its axis creating seasonal and diel cycles of energy. These drive atmospheric circulation systems, the hydrologic cycle, and biological and chemical processes on Earth. For many biota that live in midlatitude and polar environments, the resources necessary for life also cycle with the seasons and are only available during the warm months. In many tropical ecosystems, seasonal cycles of precipitation regulate resource availability. Population cycles in ecosystems that result from biotic interactions can be superimposed on the cycles of resources that are governed by the seasonal variations of heat and precipitation.

Some organisms have mechanisms to survive in an inactive state within their habitat during resource shortages. Others may die when resources become scarce or they may move through the atmosphere to more hospitable environments. Where immigration leads to the reemergence of a new population in a following season, the number of immigrants and the timing of their arrival may strongly influence their population density further in the life cycle. Potentially, the timing and strength of immigration can be just as important to determining if a population becomes an important pest as the availability of resources, natural enemies, and weather conditions. The plethora of complex interactions among organisms and their environments that cause populations to cycle make it difficult to understand and predict the population dynamics of ecosystems. This is especially true if important organisms periodically enter and leave the system (Figure 2.3).

Systems approach to aerobiology.

The capability of understanding these complex biotic and abiotic cycles and the intricate web of interactions among populations and their environment increased dramatically during the 1960s as scientists gained access to computers. At the same time, Watt (1961) began advocating the study of ecological relationships with mathematical computer simulation models using the systems approach. During the International Biological Program (IBP) in

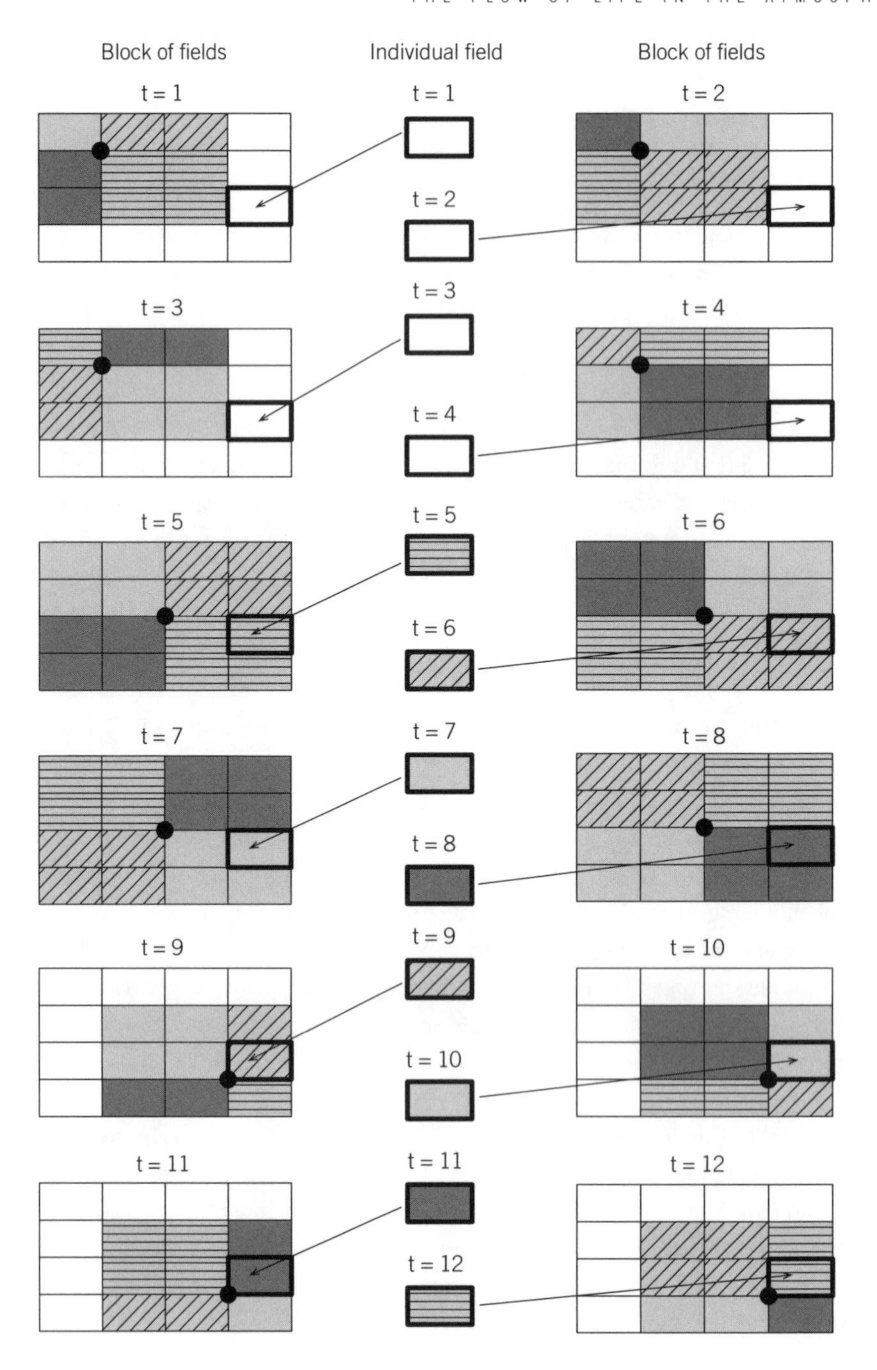

Figure 2.2. Changes in populations that are systematic at large spatial scales can appear random from a narrow perspective.

The levels of shading represent different population densities in this hypothetical example. At the larger scale, the pattern of population density rotates in a clockwise direction once every four time intervals. The center point of rotation (solid circle) shifts one cell downward and to the right with each rotation. From the perspective of a single field or cell (shown as the set of small highlighted rectangles in the middle column), the relatively short series (12 time steps) of hypothetical population density measurements appears random.

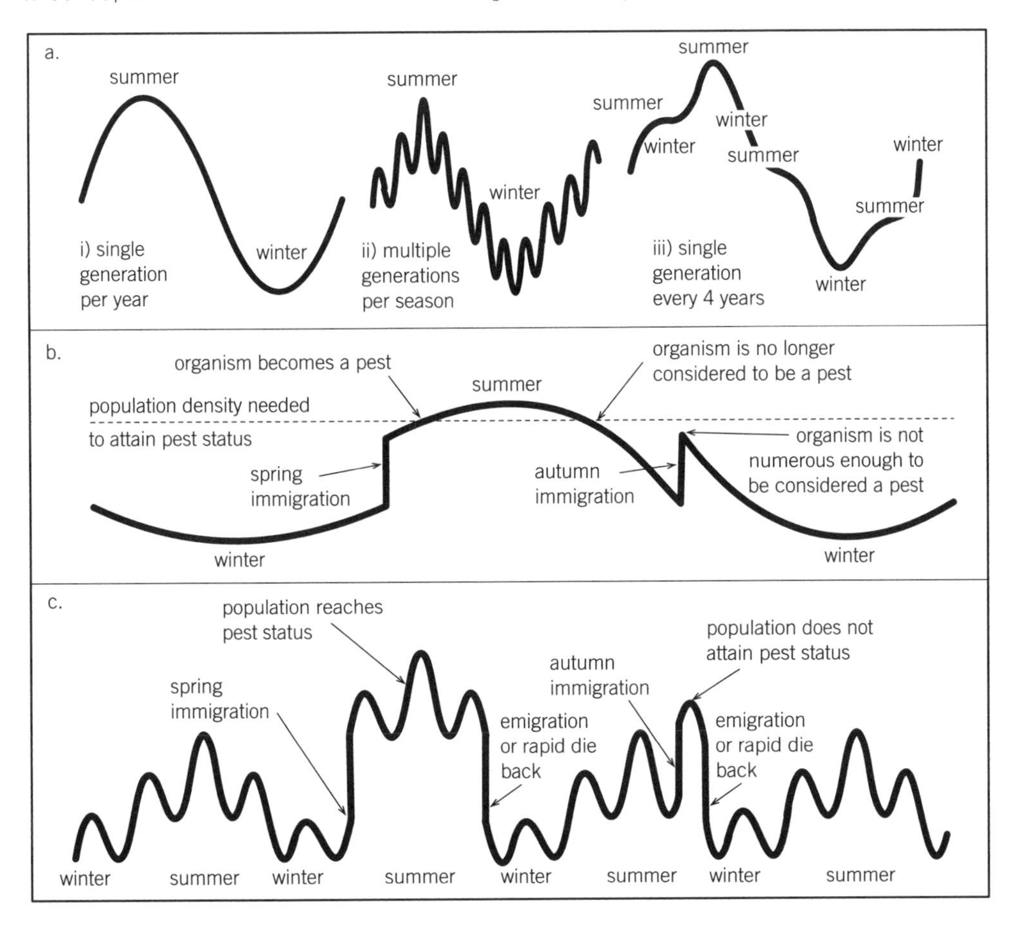

Figure 2.3. More examples of population cycles and how they can be nested.

(a) Fluctuations in adult populations (life cycles not shown) that result from biotic interations are superimposed on population fluctuations associated with seasonal cycles of resources in the ecosystem. The resulting fluctuations may appear simple when the biology is synchronized only with the seasons (i) and more complex when it is not (ii and iii). (b) The timing of an immigration can be important to determining whether or not an organism becomes a pest. In this example, the spring and autumn immigrations are of the same magnitude but they impact population dynamics differently. The spring immigration causes the population to reach pest status. The autumn immigration initially increases the organism's numbers, but its population subsequently decreases rapidly as resources decline with the approach of winter. Consequently, the autumn immigration does not cause the organism to become a pest. (c) Fluctuations in an adult population that cycle through multiple generations per year (life cycle not shown). In this example, fluctuations in the numbers of organisms that result from biotic interactions and cycles of resources are further compounded by immigration of organisms to the ecosystem from other habitats. Complex patterns of fluctuations such as the one shown here make it very difficult to forecast changes in populations.

the early 1970s, an aerobiology section was organized as part of the United States/IBP. An important outcome of the program was the formalization and application of the systems approach to aerobiology (Edmonds 1979). It provided a unifying approach to synthesize existing knowledge by which processes associated with the movement of biota in the atmosphere could be studied as an integral part of dynamic ecological systems. A few of the

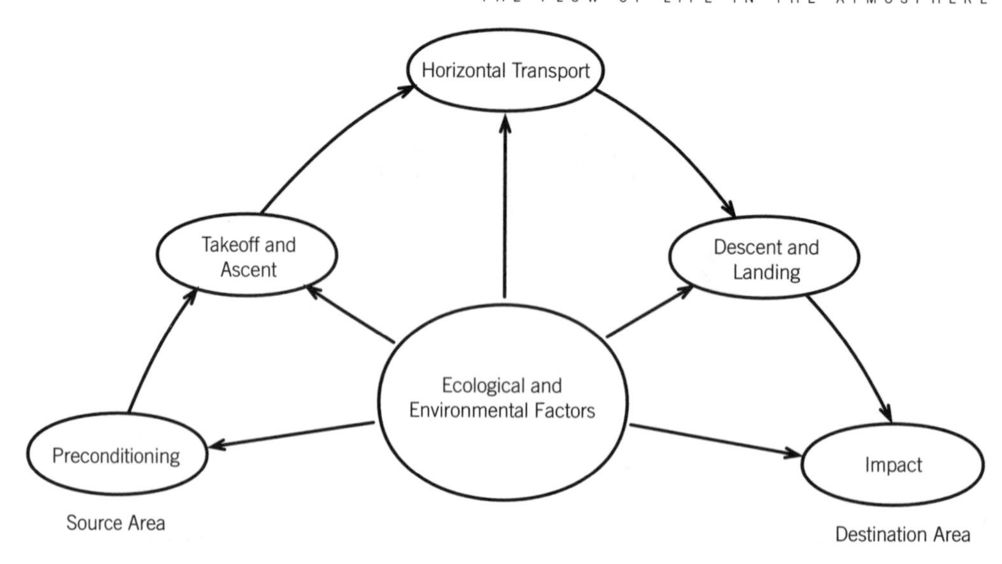

Figure 2.4. Aerobiology process model.

Each organism that uses the atmosphere as a medium for transport proceeds through these stages in order. The purpose of this systems model is to focus research on the environmental factors that affect biota in each stage and to enable the coupling of the various stages of the aerobiology process. The model was originally called the aerobiology pathway model, and the five stages were source, release, dispersion, deposition, and impact. The name aerobiology process model is used here because pathway denotes a route between geographic locations. The stages were renamed to increase their applicability across the entire range of aerobiota from microorganisms to birds. Figure adapted from Benninghoff and Edmonds 1972.

publications that have influenced our ideas are presented in Table 2.1 and represent a starting place for those interested in learning more about systems thinking as it relates to ecosystem management.

The systems approach is both a philosophy for acquiring knowledge and a methodology for learning about the structure and behavior of systems. At the philosophical level, it is a holistic approach to "thinking" about systems in order to understand complex phenomena and to make decisions about difficult problems. The essence of systems thinking is that there is a logical way to approach the acquisition of knowledge. In aerobiology, the systems philosophy provides a useful way of conceptualizing the complicated biological and meteorological interactions that govern the movement of biota in the atmosphere. This involves identifying both the total system in which the biota live, including source and destination ecoregions, as well as the atmospheric environment during transport, and all relevant types of knowledge about the system's structure and behavior.

The aerobiology process model (Figure 2.4), developed in the 1970s during the United States/IBP in aerobiology, encapsulates the systems philosophy as it relates to research on movement of biota in the atmosphere. This conceptual model is comprised of five components or stages: preconditioning in a source area, takeoff/ascent, horizontal transport, descent/landing, and impact at the destination. Each organism that moves in the atmosphere proceeds through this sequence of stages. The main purpose of the conceptual model

Table 2.2. Types of questions that should be asked when developing a systems approach to solving a complex problem.

- What are the objectives of the total system?
- What are the appropriate measures of performance for the total system?
- What entities or components are part of the system?
- What are the basic types of relationships among components, e.g., linear, non-linear?
- What constitutes the system's environment?
- What are the resources of the system and their importance?
- What are the goals, activities, and performance measures for the system's components?
- Who uses the system? What are their needs and expectations?
- Who manages the system? What are their goals?

These questions should be addressed regardless of whether or not the model of the system is conceptual or mathematical, analytical or numerical, stochastic or deterministic, or simple or complex.

Adapted from Churchman 1968.

is to focus research on the many factors that affect biota in each stage of the process. Division of the aerobiological system in this manner facilitates the coupling of the various stages in the life cycles of aerobiota. In Edmonds (1979), this model was called the aerobiology pathway model, and the five stages were source, release, dispersion, deposition, and impact. We have changed the name to aerobiology process model because "pathway" is typically associated with a route between geographic locations. The three middle stages are renamed "takeoff and ascent," "horizontal transport," and "descent and landing" because "release," "dispersion," and "deposition" apply to passively-transported biota. We believe that a general systems model for the field should be applicable across the entire range of aerobiota from microorganisms to birds.

As a methodology, the systems approach is a logical procedure or technique for building models to analyze the structure and predict the behavior of systems. Systems models range from relatively simple box and arrow diagrams to elaborate mathematical computer algorithms. At this level, a system is viewed as a set of related components that interact together in time and space. Cause-effect relationships link components in a system. Feedbacks allow the changes in one part of the system to affect other components both near and far. The particular set of components and cause-effect interrelationships that comprise a system are carefully chosen to represent the real world. Each component of the system is addressed separately by individuals who are knowledgeable in the area. On the one hand, if the goals of an ecological system are not clearly recognized and outcomes ambiguous or uncertain, the final model of the system may be a conceptual diagram, or "soft" systems model, showing flows and feedbacks among state variables and the interactive control of biotic, physical, and social factors on these flows. On the other hand, if the system has clear goals and/or predictable outcomes, the team of investigators, often aided by a systems analyst, may work from the conceptual system diagram to integrate the state variables and process controls of the system into a computer algorithm, often called a "hard" systems model, to simulate the behavior of the system. Table 2.2 provides examples of some of the important questions that should be asked during the conceptualization and construction of soft and hard systems models.

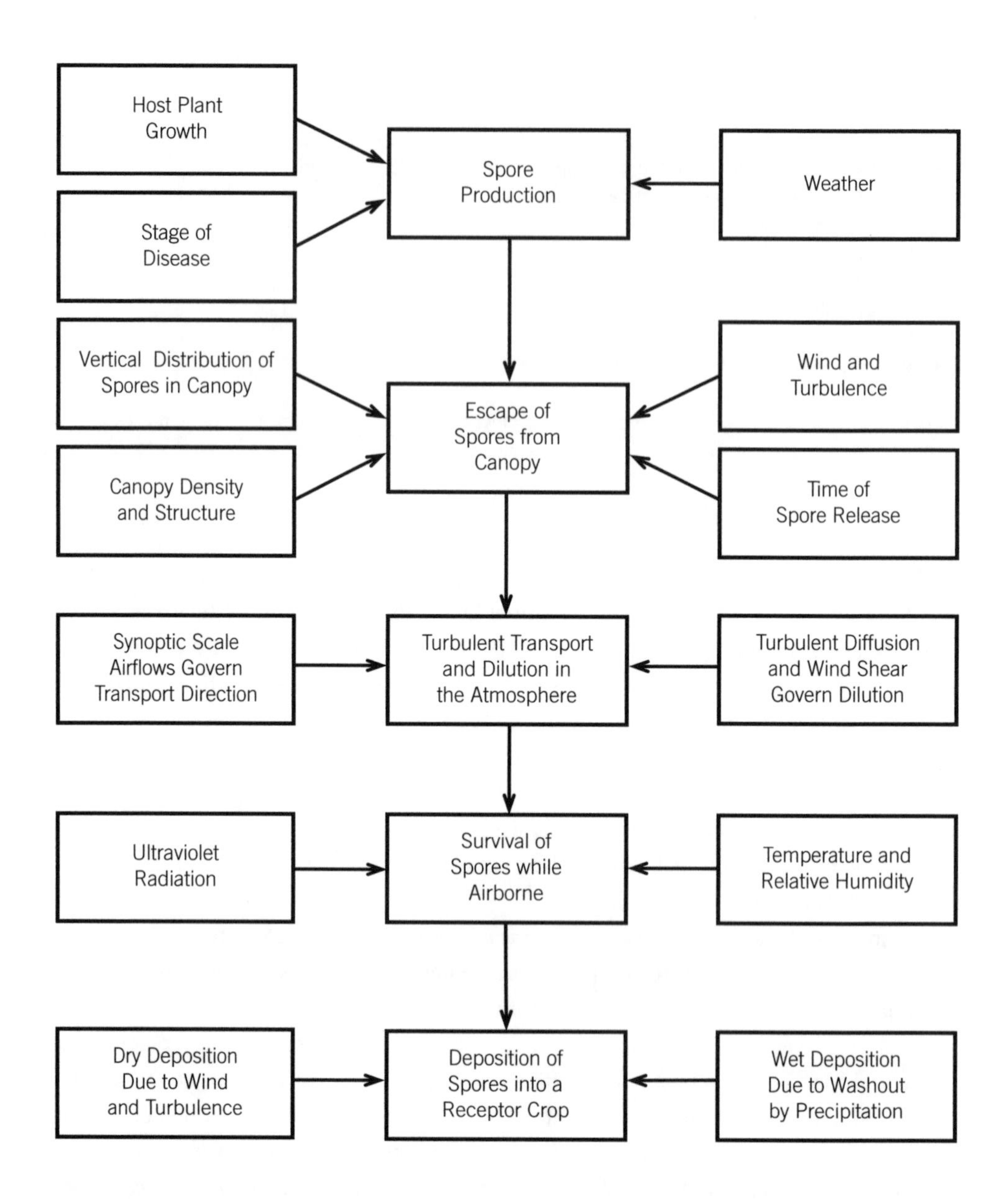

Figure 2.5. Model for understanding the atmospheric flow of fungal spores.

This framework was developed to study the blue mold of tobacco (*Peronospora tabacina* Adam), a foliar fungus disease disseminated by microscopic spores in the atmosphere. Blue mold has attacked the major tobacco production areas in North America repeatedly during the 1980s and 1990s. The spores that initiate these epidemics throughout the eastern United States and Canada are very likely transported on atmospheric motion systems from sources in the Gulf Coast states, Mexico, Cuba, and/or Jamaica. (Aylor et al. 1982, Davis and Main 1984, Davis et al. 1990.) Figure adapted from Aylor 1986.

Figure 2.5 is a conceptual diagram depicting a systems model that was constructed to study the atmospheric dispersal of fungal spores (Aylor 1986). The aerobiology process model was used as a template for coupling stages in the life history of this organism and for establishing biotic and environmental controls over its dispersal. This conceptual model was the precursor of a computer model currently used to forecast the long-distance atmospheric movement of tobacco blue mold spores (see Chapter 11).

Like any other methodology, systems models have their advantages and pitfalls. When built in a comprehensive manner, using the systems approach, simulation models can provide valuable information for assessing the consequences of manipulating a system without actually modifying it. The systems approach provides the framework for determining the interdependence between the object of control and its surrounding environment (Figure 2.6). The object of control—environment dichotomy helps researchers and managers identify the "real" problem, and enables them to be realistic about those aspects of the object of control that they wish to manage, as well as those aspects of the environment to which they want their control strategies to be responsive (Haynes et al. 1980).

Measuring the impact of altering ecosystem components on population fluctuations of target organisms, enhancing the number and effectiveness of their natural enemies, as well as the short- and long-term productivity of an ecosystem is complicated in the field. One must be able to compare important elements of a system that has been modified with the same elements in a "control" system. This is virtually impossible to achieve in a natural setting because the types and arrangement of elements in two landscapes are never the same. Similarly, in agricultural experiments, many pairs of matched plots must be used to overcome the effects of natural variations among plots. The validity of results from experiments on small plots is also questionable because they are seldom big enough to have the capability of maintaining a balanced, integrated, and adaptive community of organisms that are integral to the integrity of the larger agroecosystem. In addition, our understanding of the ecological and economic value of particular changes can be rigorously evaluated only after the accumulation of many years of experimental field data.

Computer simulation models can provide a useful tool for expanding our knowledge of ecological relationships. This is because once a computer model has been constructed and shown to provide reasonably accurate predictions, one can simulate the consequences of manipulating elements of an ecosystem on population fluctuations by changing input data and parameters in the model. However, there are a number of constraints that limit the success of simulation models for predicting ecosystem behavior. To represent the complexity of biological systems, simulation models must include many variables and flows of information with multiple feedback loops. In addition, these systems models must capture the effects of spatial variations in biotic and abiotic factors on populations if they are to be useful tools for understanding and predicting the behavior of real ecosystems. As a consequence, relatively high spatial resolution data are needed to drive simulation models. The spatial resolution can depend on the distances target populations move and the diversity of the landscape. The accuracy of predictions from simulation models can also suffer from gaps and uncertainty in our knowledge about relationships among important components

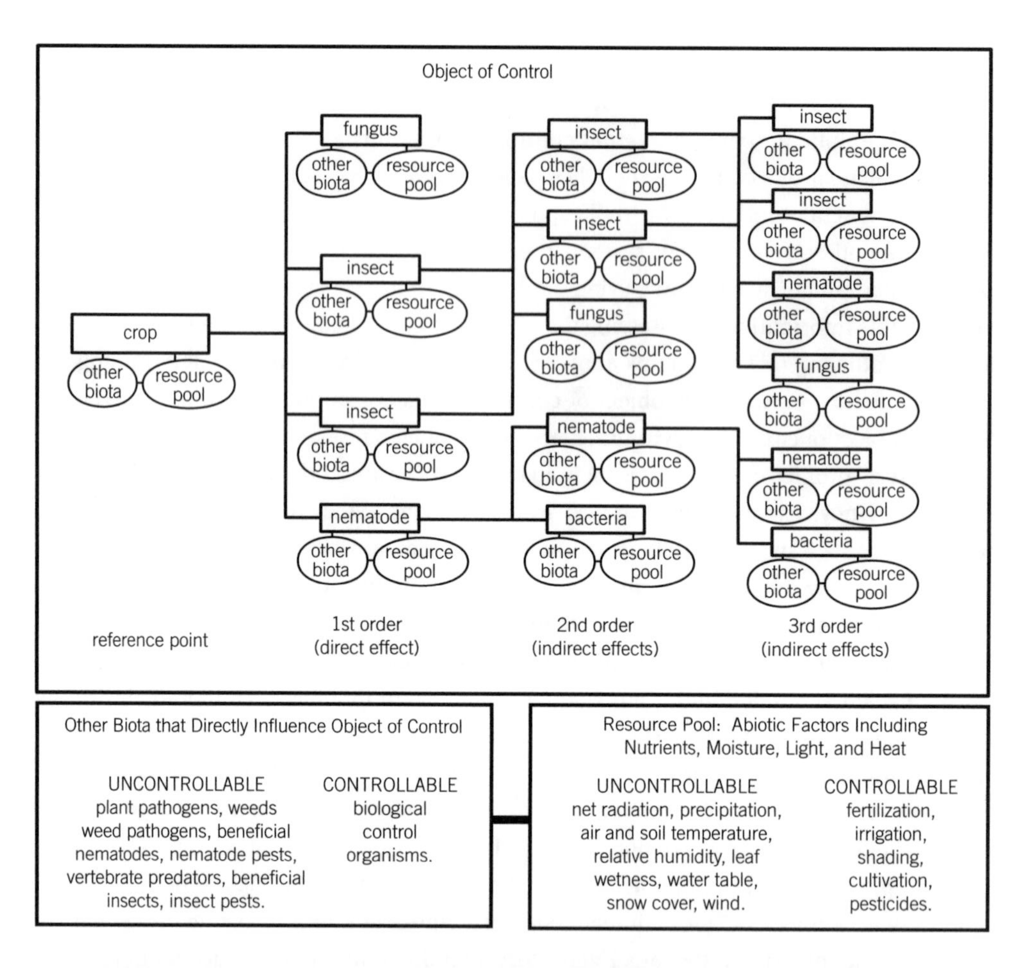

Figure 2.6. Separation of the object of control from its surroundings.

The separation of an agroecosystem into an object of control (set of important organisms and the web of their interactions) that one desires to manage and its surroundings (other important biota and abiotic factors that influence the object of control's resource pool) provides a framework for quantitative evaluation of the interdependence of organisms and their environment, and the construction of systems models. In this example, a crop is the reference point for the web of interacting organisms that comprise the object of control; however, the web could be arranged so that a pest is its focus. We show four primary pests (right of crop) and "other biota" (below crop) that directly impact the crop. To the right of these pests are organisms that impact them and thus have 2nd order or indirect effects on the crop. Other biota and resource pools represent the large array of important biotic and abiotic factors that directly influence an organism. Some factors can be controlled and some cannot. Usually relationships among organisms and between populations and their environments are too complex and our knowledge about them too limited to manipulate all of the parameters in an ecosystem in a practical way. The separation of an agroecosystem into an object of control and its surroundings allows managers and researchers to be realistic about those aspects of the ecosystem that they wish to manage (object of control), the surrounding biotic and abiotic factors that they can control, and those factors that they cannot but that they should monitor. Figure adapted from Haynes et al. 1980.

Table 2.3. Gaps in knowledge about the atmospheric flow of fungal spores.

PROCESS	RELATIVE UNCERTAINTY
Spore production	100–1000?
Escape of spores from canopy	2–5
Turbulent transport and dilution of spores in the atmosphere	
For the Planetary Boundary Layer (or Mixed Layer)	10–20
For escape from the Planetary Boundary Layer (or Mixed Layer)	?
Survival of spores while airborne	
High concentrations of spores	2–5
Low concentrations of spores	?
Deposition of spores into a receptor crop	
Dry deposition	2–5
Wet deposition	?

Small values of relative uncertainties indicate a high level of knowledge about the corresponding process. A question mark indicates that very little is known about the process. The modeling effort suggested that the production of spores, survival rates for low concentrations of spores in the atmosphere, and the rate of wet deposition of spores into a receptor crop were the primary areas where data were lacking. Although little was known about the escape of spores from the planetary boundary layer to the free atmosphere above, Aylor suggested that this flux is likely to be small, and that further research in this area was not as critical as in the other three areas where the uncertainty level of knowledge was high. Adapted from Aylor 1986.

of the ecosystem. There are also constraints associated with how well and at what scales microclimate and other environmental data needed for the models are measured.

Even so, models of ecological systems are important tools for increasing our knowledge about the complex interactions among organisms and their environments that cause populations to cycle and periodically result in organisms becoming pests. Simple conceptual box and arrow diagrams can be as useful as more complex mathematical computer simulation algorithms. Soft systems models are extremely valuable because they stimulate holistic thinking by bringing together people with different world views. Hard systems models can provide a capability for assessing both the intentional and inadvertent impacts of human activities on future environments. They are especially useful for finding gaps in our knowledge about relationships among populations and elements of their environment. In addition, they can identify where field data are needed to establish relationships among the components of an ecosystem. Table 2.3 shows some of the uncertainties that were discovered during construction of the fungal spore dispersal model.

Establishing aerobiological principles.

Because of the immense multitude of distinct morphological, physiological, and behavioral characteristics of biota that expedite their movement in the atmosphere, there is a critical need to focus aerobiological research on elucidating scientific principles of movement. In the physical and life sciences, principles are commonly developed through hypothesis testing. To be scientifically acceptable, aerobiological principles must pertain to groups of organisms that use the atmosphere as a medium for their travel. In other words, the generality (or value) of an aerobiological principle is related to the number and diversity of taxa to which it applies. Consequently, general principles in aerobiology can be developed only if hypotheses are tested across a wide range of biota and spatial scales.

Humans will always be discovering undesirable competitors and mismanaging Earth's ecosystems to create new diseases and pests. General aerobiological principles are needed to guide the development of research programs for understanding the population dynamics of many of these organisms. Without general scientific principles, much of the effort and time required to acquire aerobiological knowledge will be repeated each time a previously unstudied organism becomes an important disease or pest.

In 1992, a group of scientists from a large number of disciplines, institutions, and nations established the Alliance for Aerobiology Research (AFAR) to advance the understanding of atmospheric transport of organisms and biological particles (see Epilogue). At the founding workshop, they developed a set of fundamental aerobiological research hypotheses. These hypotheses (Table E.1) pertain to the maintenance of the movement process and the three stages in the conceptual aerobiology process model during which biota are airborne (see Figure 2.4). The biological and meteorological interactions that govern takeoff/ascent, horizontal transport, and descent/landing processes are central to the discipline of aerobiology. Although these hypotheses do not address the preconditioning and impact stages, the scientists acknowledged that to understand the long-distance aerial movement processes holistically, as an integral part of the dynamics of ecosystems, one must understand what happens at each end of the movement process. Dynamics of source area populations are crucial to the initiation of movement and the status of the biota during that movement. At termination, if organisms descend into suitable habitats, the dynamics of the receptor ecosystem largely determine the impact of the immigrants.

Dingle's (1996) book is a testament to the utility of seeking generalizations about the movement of organisms that span a wide range of taxa. His rigorously defined taxonomy of movement provides the framework for a comprehensive review and interpretation of the migrations of plants and animals of all types, in air, water, and on land. Freed from the constraint of studying a single taxon, Dingle discovered three fundamental generalizations about migration as a biological phenomenon: "(1) migration involves specialized behavior that is both qualitatively and quantitatively different from other types of movements; (2) migration is a syndrome integrating behavioral, physiological, morphological, and life-history traits; and (3) migration is an adaptation to shifting or patchy environments."

Summary.

There are four key aspects to the conceptual framework for understanding the atmospheric flow of biota. First, it is important to have a "non-linear" perspective. Change must be considered as a feedback process rather than a negative process. In western thinking, harmony is usually viewed as non-dynamic and devoid of change and thus not circular. A flat line or,

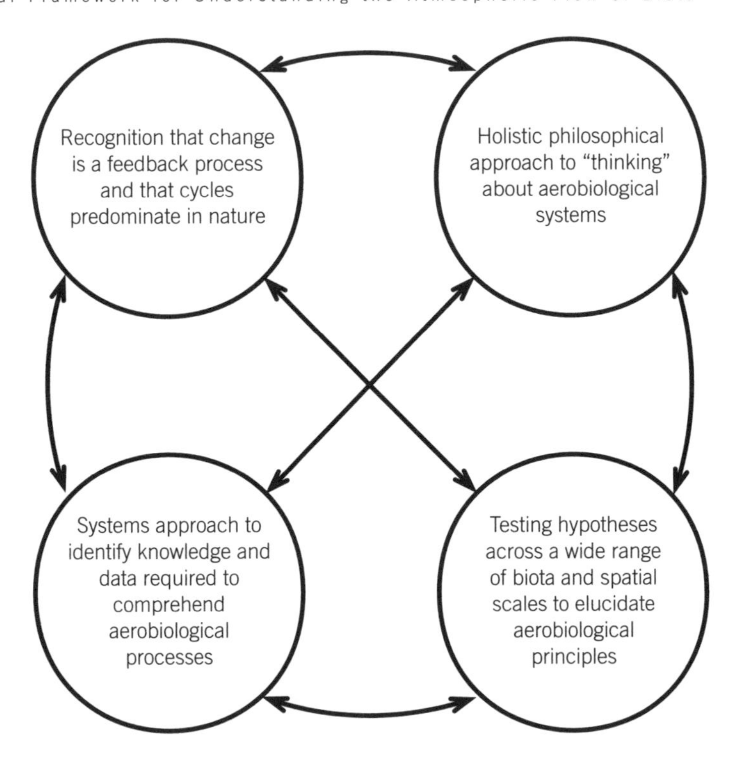

Figure 2.7. Conceptual framework for understanding the atmospheric flow of biota.

better yet, a line with a positive slope or an ascending arrow is the preferred way to describe progress and growth. However, in nature, cycles prevail and "change . . . is a natural tendency, innate to all things and situations" (Capra 1982). Negatives will eventually become positives if dynamics, feedbacks, and time are considered as part of a system. Second, it is important to have a holistic philosophical approach to "thinking" about aerobiological processes. Movement of biota in the atmosphere cannot be understood independent of other life processes. It can only be truly understood if it is studied as an integral part of dynamic ecological systems. Third, a systems approach to the study of movement of biota in the atmosphere can substantially increase the aerobiology knowledge base. Systems models may range from simple conceptual diagrams to complex computer simulation algorithms, and are especially useful for determining the interdependence between the object of control and its surrounding environment, and for identifying gaps in our knowledge about these relationships. They can identify where field programs should concentrate their efforts and resources so as to collect the data that are needed to establish these relationships. In addition, simulation models can provide valuable information for assessing the consequences of manipulating a system without actually modifying it. Finally, the ultimate goal of aerobiology, as a scientific discipline, is to discover generalizations about the movement of biota in the atmosphere. The development of scientific principles in aerobiology depends on the identification and testing of important hypotheses across a wide range of biological and meteorological systems and spatial and temporal scales (Figure 2.7).

CHAPTER 3

A Framework for Integrating Information to Support Decisions about Manipulating Ecosystems Impacted by Aerobiota

Managing ecosystems with the objective of maintaining the integrity of their many populations is extremely difficult. Not only does this require a sound understanding of the life histories of organisms and the relationships among organisms and their environments, but also it necessitates good programs to evaluate the current state of the ecosystem and to forecast its future states. To establish realistic goals for programs to manage ecosystems, it is essential to assess how humans value and use the ecosystem, not only today but also in the future. Very often human manipulations of ecosystems, regardless of whether they are well-intended or inadvertent, result in unforeseen consequences, many of which can be very detrimental.

Need for preventive strategies for managing ecosystems impacted by aerobiota.

The tide of new and resurgent human diseases rising around the world over the last few decades is perhaps the most serious consequence of the changes humans have wrought to our ecosystems. Many of these diseases are viruses transmitted by arthropods (arboviruses) that move in the atmosphere. At least 520 arthropod-borne viruses have been identified, of which more than 100 can produce human disease (Patz et al. 1996). Quite often these viruses have symptomless relationships with the insects or ticks that transmit them to birds, rodents, and ungulates. These hosts also have evolved to tolerate the viruses, and thus they constitute a reservoir for reinfecting arthropods. This triangle of virus, vector, and host is usually quite specialized and remains stable and unnoticed until humans disturb it. However, when humans break these ancient relationships, arboviruses can spread rapidly from one species and ecoregion to another, often as a result of aerial movements of their arthropod hosts. If the virus can survive in humans, widespread sickness or death may result. The plethora of human manipulations to ecosystems during the second half of the twentieth century has destroyed the integrity of many relationships involving arboviruses and other pathogens that are detrimental to human health. From a historical perspective, we are currently experiencing an epidemic of human disease epidemics. Since 1953 in South America alone, there have been four

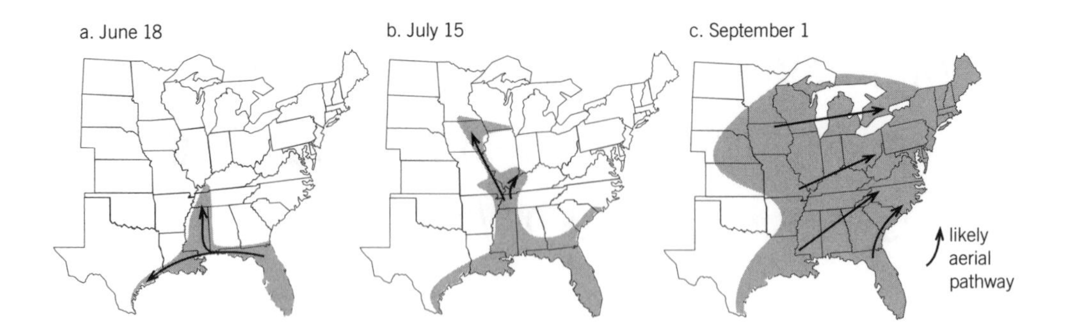

Figure 3.1. Aerial spread of southern corn leaf blight across North America in 1970.

"In 1970 an epidemic disease swept swiftly over the corn crop of the United States. A great agricultural resource of the country was threatened. In some sense, science and technology had been responsible. The yield of corn dropped an estimated 50 percent or more in some southern states and 15 percent nationwide; that even greater losses might occur in subsequent years seemed possible. Memories of earlier plant disease epidemics made the corn blight even more alarming. For example, at the turn of the century, a chestnut blight epidemic had moved down the mountain spine of eastern North America, leaving it bereft of chestnuts. Innumerable 'Chestnut Hills' remained on the maps, but the chestnuts themselves were gone. [History has recorded] the wheatless days of 1917, the great Bengal famine that killed thousands in India in 1943, and the Irish famine of the 1840s. Each had come about from the destruction of a staple food crop by [airborne] plant disease.

"No one really felt that the corn blight epidemic would cause famine in the United States; there were too many other high-carbohydrate crops for that. It did, however, prompt numerous questions: How serious? What about the next year? What happened? What caused the epidemic? Why was it not foreseen? Where did the technology go awry?" (Preface from the National Academy of Sciences Report on Genetic Vulnerability of Major Crops, Horsfall 1972). Figure adapted from Schumann 1991.

devastating epidemics of new hemorrhagic fevers caused by arboviruses, all linked to ecological disruptions (Karlen 1995).

The shift in modern agriculture, from small farms with mixed cropping systems to large-scale monocultures with increasing genetic homogeneity, has increased the vulnerability of human food and fiber crops to disease and pest epidemics as well. These are exactly the sorts of ephemeral habitats to which many of the biota that move in the atmosphere are adapted (see Dingle 1996). The southern corn leaf blight epidemic in 1970 was a dramatic example of how dangerous it is to manage ecosystems to maximize crop production, using high yield cultivars with narrow genetic bases, without consideration of the consequences of aerobiota. In a single growing season, this corn leaf blight transformed from a disease of minor status to an epidemic of catastrophic dimensions in North America. Southern corn leaf blight was noticed in the United States in 1968 and became common in seed and hybrid test fields in the Gulf Coast states during 1969. In early 1970, it spread into the main corn crops of Florida, Georgia, Alabama, and Mississippi and reached epidemic proportions by June (Figure 3.1a). Crop losses of 50% were common in fields throughout the southern states. During July, the blight was carried by atmospheric motion systems up the Mississippi River Valley into Kentucky, Ohio, Indiana, and Illinois (Figure 3.1b). By September, it had spread locally in these states and to Wisconsin, Minnesota, and southern Canada (Figure

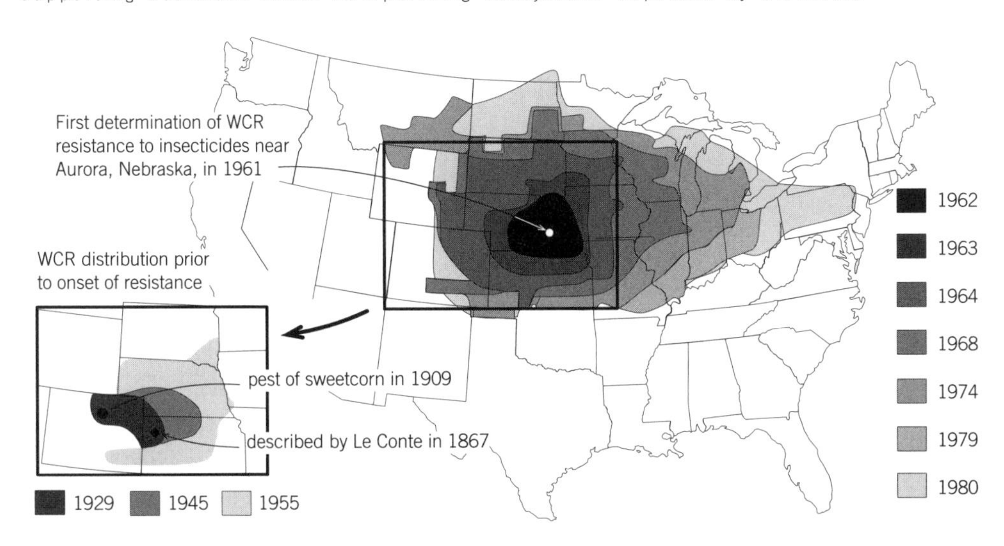

Figure 3.2. Spread of western corn rootworm beetle across the Central Plains into the eastern United States.

In 1864, western corn rootworm (WCR) beetle was first recorded in Kansas. In 1909, it became a pest of sweetcorn near Fort Collins, Colorado. During the first half of the twentieth century, the beetle inhabited the western prairies from South Dakota to Colorado. In the early 1950s, heavy investment in irrigation technology by producers within the western Corn Belt resulted in the widespread practice of growing corn in the same field year after year. This corn monculture created near optimum conditions for the increase of the WCR beetle population and established a pathway of continuous corn over which the beetle spread rapidly eastward (Ruppel, 1975). In 1955, WCR inhabited Nebraska, half of Colorado and Kansas, and small portions of South Dakota and Iowa. The use of soil insecticides to control WCR was begun in 1948. By 1961, WCR adults in central Nebraska had developed resistance to these pesticides. Over the next few years, the area inhabited by resistant WCR expanded rapidly. By 1964, the geographic range of WCR extended into North Dakota and east across the Mississippi River. In 1968, WCR was reported in northwestern Indiana, and by 1980, the beetle had crossed the Appalachian Mountains into southeastern Pennsylvania and northwestern New Jersey. Figure adapted from Chiang 1973, Metcalf 1983, and Krysan 1986.

3.1c). Estimates of losses in the United States for 1970 were over 700 million bushels, approximately 15% of the corn harvest, at a cost of about $1 billion (Tatum 1971). Had the United States relied on corn as a food staple, the effect of the corn leaf blight likely would have been catastrophic.

Organisms can also change their behavior to adjust to human manipulations of the environment, making some of the hardest earned gains of management programs fleeting. The changing behavior of the western corn rootworm beetle, from laying eggs almost exclusively in cornfields to laying them in both corn and soybean fields (and perhaps in other crops), is an example of an adaptation that is likely to have catastrophic consequences over the next few years. Prior to the 1950s, this strong flying beetle was a local pest along the western edge of the corn belt in Colorado, Nebraska, and Kansas (Figure 3.2). Once irrigation technology was adopted in the 1950s and 1960s in this region, the practice of planting corn year after year in the same fields became prevalent, and the western corn rootworm beetle spread from a single location in Nebraska to envelop the entire corn belt in little over a decade (Metcalf 1983).

Until the last few years, this insect laid its eggs almost exclusively in the soil near the base of corn plants. After diapausing for a single winter, western corn rootworm larvae would emerge in the spring or early summer to eat the roots of young corn plants. The roots of the plants can be protected from these larvae by applying soil insecticides to cornfields when the plants are young. Due to increasing concerns over widespread use of agricultural chemicals, the United States Department of Agriculture (USDA) Cooperative Extension Service (CES) embarked on an extensive educational program to convince producers who rotated corn with other crops that using a soil insecticide was unnecessary. Until the mid-1990s, this strategy worked where crops were rotated on an annual basis because beetle larvae hatched from eggs laid in cornfields would emerge the following year into fields planted in soybeans or other crops and quickly starve. Because both the environmental and economic benefits to adopting this strategy were substantial, the percentage of corn acres treated with soil insecticides decreased dramatically throughout the central United States (e.g., in Illinois it fell from 65% to 33% between 1978 and 1990). However, western corn rootworm beetle continues to be one of the most serious insect pests on non-rotated corn in this region. Between 80% and 90% of these fields are treated prophylactically with soil insecticides (Pike and Gray 1992).

It now appears that the effectiveness of the two-crop rotation strategy for checking the population of western corn rootworm beetles is decreasing rapidly. It is likely that a small number of beetles always laid their eggs in soybean fields. Over the last few decades, the intense two-crop (corn and soybean) rotation in east-central Illinois and west-central Indiana has selected for these beetles and has resulted in the development of a new strain of western corn rootworm beetle that lay their eggs in soybean fields (Onstad et al. 1999, Isard et al. 1999). If this is indeed the case, then farmers in the central the United States will have to quickly change their management practices if they want to produce corn profitably. If crop rotation fails as a management tool for western corn rootworm beetle, the cost of reverting to using soil insecticide use would likely exceed $100 million per year for Illinois producers alone. In light of increasing concerns about the detrimental effect of pesticides on human health, beneficial organisms, and the environment, reversing the hard-earned gains of the last two decades is unacceptable.

These examples highlight the need for proactive preventive strategies to manage Earth's ecosystems. Fundamental to this approach of manipulating ecosystems to maintain the stability of their populations is the view that the dynamic processes associated with biological systems are interwoven and cyclical. As humans continue to change the face of planet Earth, putting their own interests first, the fluctuations in the numbers of other organisms will likely increase in both frequency and magnitude. The practice of waiting until a population reaches pest status before reacting is inefficient, short-sighted, and ill-advised. Sometimes the best option is to do nothing and let a population cycle naturally. However, if potential human health and economic consequences are severe, intervention at earlier stages in the life cycle of the organism, its natural enemies, and/or its hosts, may prevent the organism from becoming a pest.

All too often programs for managing ecosystems fail to fulfill expectations. Sometimes the long-term goals and objectives of the programs are unrealistic or ill-conceived. In other cases, the perspective and management goals are too narrowly defined (e.g., eradication of a pest population). Often the understanding of the dynamics of life cycles of the key organisms and how they are interwoven with the cycles of other biota and the environment is incomplete. In these cases, research programs may need to be initiated first, and the development of management programs may have to wait until our knowledge about the system is more extensive. It is also essential to understand the human social and cultural components of an ecosystem. To be successful over the long run, strategies must be compatible with how humans value, use, and make policies for the ecosystems we manage (see Grumbine 1994).

Perhaps the most challenging issue in ecosystem management today revolves around developing and implementing methodologies to integrate different types of information into decision support systems to help managers make the best decisions possible about manipulating populations and their environments. Where atmospheric movements of biota into and out of an ecosystem have important impacts on the stability of its populations, its productivity, or human health, many issues, including those associated with how, when, and where to measure the populations of these organisms, are especially difficult.

Often decision-makers need information on the timing and magnitude of an immigration before the organism arrives in order to manage an ecosystem in a manner that is both effective and compatible with human values and use. Consequently, decision support systems designed to help manage ecosystems need good predictions of the timing and magnitude of the important movements of biota among habitats. In principle, predictions for decision-making may be generated by comprehensive ecosystem models that simulate populations and their movements at the habitat, ecosystem, landscape, and/or continental scales. However, even where the important causal relationships in dynamic ecosystems are fairly well-understood, the spatial and temporal scales in the resulting simulation models are so large that their primary use is limited to developing long-term management strategies. The predictions from complex simulation models of ecosystems are notoriously inaccurate—generally much too imprecise to be used to provide reliable tactical advice relevant to current conditions in a specific habitat.

On-line decision support systems and their application in aerobiology.

One useful alternative is to construct an ecosystem management decision support system around a program which continuously monitors changes in populations and their environments. The idea of parameterizing an ecological systems model with a near-continuous flow of information on the state of the system was pioneered by Dean Haynes and colleagues, and is referred to as an "on-line" (Haynes et al. 1973) decision support system (on-line DSS) or an "adaptive" simulation model (Wilson 1989). The components and flows of information within an on-line DSS are depicted in Figure 3.3, arranged by type of task and time frame. An effective program for measuring relevant environmental and economic factors and for sampling important populations on a regular basis in the field is the key component of an on-line DSS. A well-designed comprehensive monitoring program can provide managers with current information about the general health and productivity of natural and

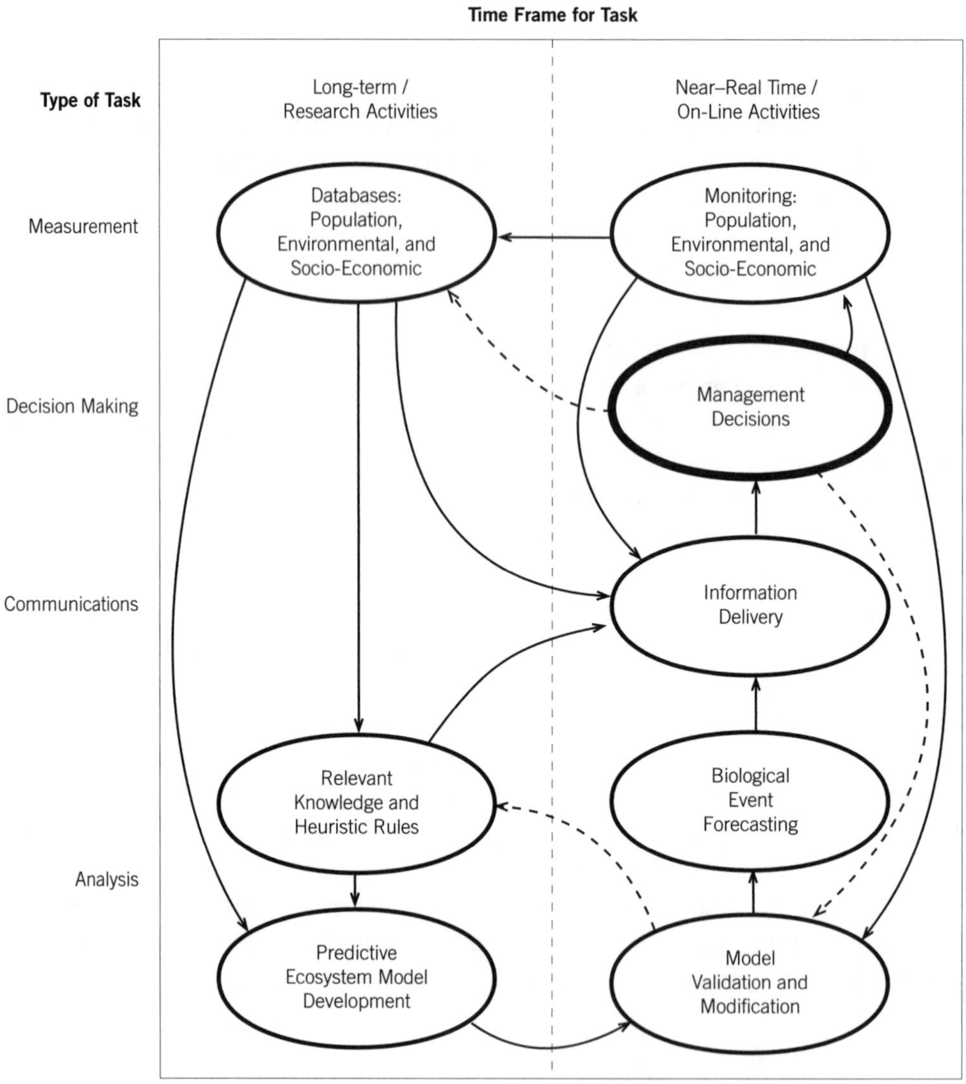

Figure 3.3. Component tasks of an on-line decision support system for ecosystem management and their time frames.

A conceptual model of an on-line decision support system for ecosystem management using preventive strategies showing the various component tasks arranged by type and time frame. Solid lines indicate flows of information and dashed lines represent feedbacks. Near real-time monitoring of populations and their environments is the key component, but it must be supported by research activities aimed at developing databases and ecosystem models. The monitoring program provides the basis for validating and modifying the ecosystem models to give managers accurate forecasts of populations and biological events. All relevant and available information must be delivered to decision-makers rapidly. A structure for making management decisions is provided in Figure 3.4.

agricultural ecosystems. Temporal and spatial data series to monitor and study the life cycles of biota, including their movements among other habitats near and far, would also result. Accurate forecasts of pests and pathogens of plants, humans, and other animals can be generated so that management tactics can be implemented before these undesirable organisms become threats. An ecosystem monitoring program must be designed to detect new pests and diseases as well as those biota endemic to the ecosystem. Quantitative time series data about new problems are needed to convince government officials to legalize the use of alternative tactics (e.g., new or previously banned chemicals) for pest control. Finally, an ecosystem monitoring program can provide an historical database to assess the impacts of climate change on populations and ecosystems.

Equally important to the foundation of an on-line DSS are on-going research programs for constructing historical population, environmental, and socio-economic databases and for developing predictive ecosystem models. Potentially, the learning processes associated with developing these databases and models can be as valuable as their products. If the level of knowledge about an organism's life cycle is limited, its population and the factors that influence its birth and death rates, as well as its movement into and out of the ecosystem, must be measured. As the understanding of the ecosystem increases, relationships between variables are identified and quantified. Over time, increased knowledge and databases are used to construct pragmatic indices (e.g., degree-days) and analytical, statistical, and simulation models to predict the dynamics of populations. In turn, indices and models will reduce the number of variables which need to be continuously monitored.

In the near real-time frame (right hand column of Figure 3.3), measurements of important factors in the ecosystem as well as changes in their spatial patterns are used to evaluate model predictions. Frequent analyses of differences between predicted and observed populations increase the knowledge about relationships among system components and provide a means for validating models as well as a basis for modifying them. Then, prior to the next round of biological forecasts, these data are incorporated into the models to update environmental and economic parameters as well as the densities of populations and their rates of birth, movement, and death. If the time interval between measurements is relatively short, simple linear models or extrapolations are used, in place of complex systems models, to project populations into the near future.

Rapid communications is also a key ingredient to an effective on-line DSS. Historical information from population and environmental databases, near real-time population, environmental, and economic measurements, forecasts of biological events, and other relevant information must be integrated and delivered to users quickly to enable them to make well-informed management decisions (Figure 3.3). Because so many of the Earth's processes are cyclical, human experiences are also an invaluable knowledge base. The heuristics of practitioners ("experts"), captured in relatively simple "rules of thumb," are extremely useful for making decisions about complicated management problems. A "systems thinking" approach to an on-line DSS interfaces heuristic knowledge with analyses of current and historical databases, technical information, and the understanding of system function and behavior gained through the development and use of models. As the knowledge base about the ecosystem increases, the monitoring programs, databases, and models can be refined,

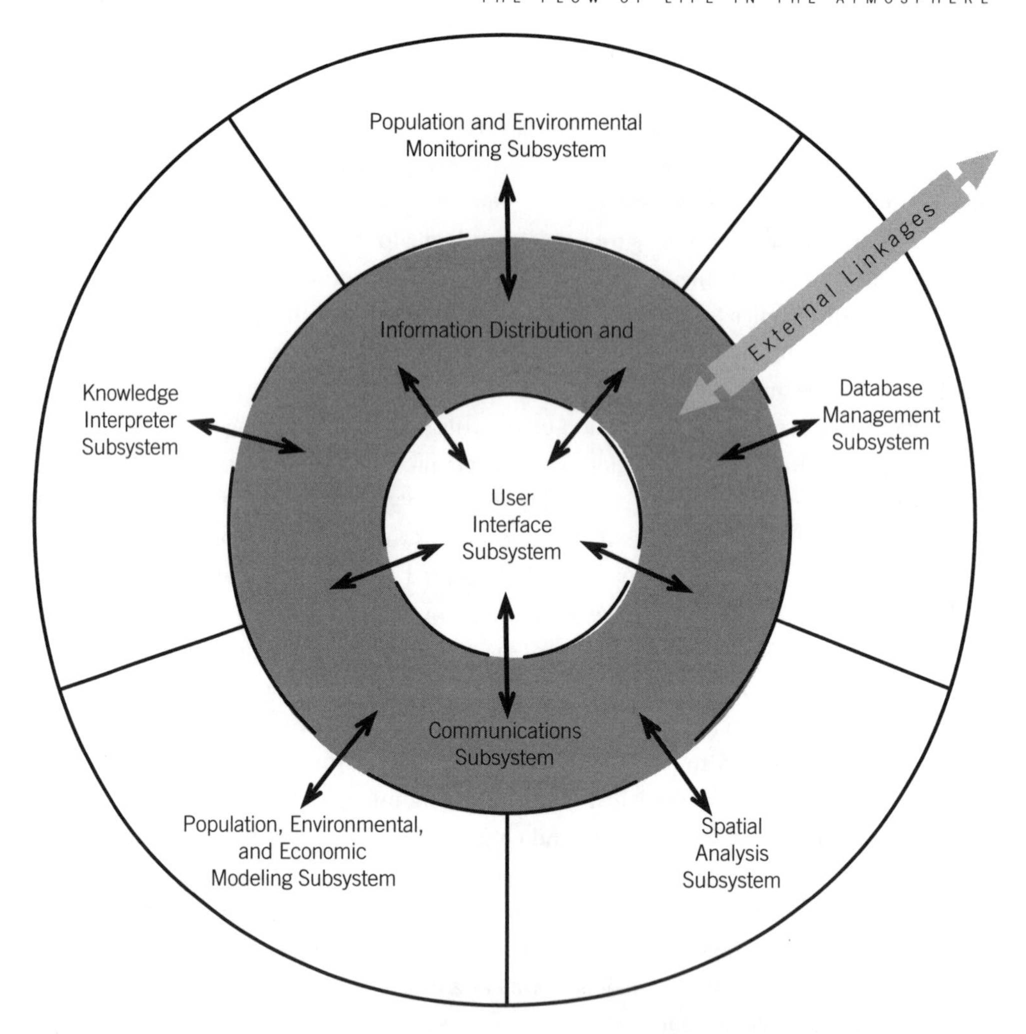

Figure 3.4. Component subsystems of an on-line decision support system for ecosystem management.

Central to an on-line decision support system is the user interface. The Information Distribution and Communications Subsystem provides the infrastructure that enables the other subsystems to interact with each other and users. The arrows represent the flows of information among subsystems. Figure 3.3 shows other component tasks of an on-line decision support system and their time frames while Figure 3.10 depicts many of the internal and external linkages.

making them more accurate and thus reducing their use of rules of thumb in decision-making.

Figure 3.4 shows an on-line DSS arranged into function-based subsystems. This figure progresses from Figure 3.3 by identifying specific subsystems appropriate for implementing the decision support process. They include Population and Environment (Ecosystem) Monitoring, Database Management, Spatial Analysis, Population, Environment, and Economic Modeling, and Knowledge Interpreter subsystems. In addition, there must be a User Interface and a component that controls the flow of information among the subsystems and can communicate across networks among a variety of computers and users. Many of the ideas

Table 3.1. References to works that have influenced our thinking about on-line DSSs.

BIOLOGICAL MONITORING AND ON-LINE DSSS:
Haynes et al. 1973, Tummala and Haynes 1977, Gage and Mispagel 1979, Brown et al. 1980, Gage et al. 1982, Gage and Russell 1987, Pickering et al. 1990, Wilson 1989.

DSSS WITH A FOCUS ON LONG-DISTANCE MOVEMENT OF BIOTA:
Norton 1991, Knight and Cammell 1994, Tang et al. 1994.

DSSS FOR ECOSYSTEM MANAGEMENT (GENERAL):
Coulson and Saunders 1987, Coulson et al. 1989, Stone and Schaub 1990, Plant and Loomis 1991, Plant and Stone 1991, Coulson 1992, Loh and Rykiel 1992, Liebhold et al. 1993, Edwards-Jones 1993, Warwick et al. 1993, Starfield et al. 1993, Power 1993, Stone 1994, Greer et al. 1994.

about on-line DSSs that are presented in this chapter were influenced by the publications listed in Table 3.1.

Population and Environmental Monitoring Subsystem.

The establishment of a good field data collection program is key to an effective on-line ecosystem management DSS (Figure 3.4). Regular quantitative monitoring of the occurrence and abundance of organisms in time and space is essential to predict population fluctuations and thus to manage ecosystems to maintain their stability. The data collection system needs to be designed to answer useful questions including: where are the important organisms?, how long have they been there?, how many are there?, how are their numbers changing?, are they causing human suffering or economic damage?, and, are parasites, predators, diseases, and other natural controls present?

For a biological monitoring program to be effective, it is critical that standardized measurements be taken at prescribed time intervals and at fixed sites located across habitats, ecosystems, landscapes, and/or continents. Only by establishing permanent collection sites can spatial analysis be achieved with integrity. Standardized data from permanent sites enable estimation of species distribution and variability at seasonal and annual time scales and the analysis of population fluctuations in ecosystems. Where possible, biological sampling should occur near meteorological observation sites (e.g., climate stations, airports, and weather radar installations). This is because atmospheric conditions strongly influence most aspects of our terrestrial ecosystems, and both atmospheric conditions and populations can vary substantially from one location to another. If meteorological and population measurements are synchronized, the analysis of relationships among concurrent atmospheric and biological factors can contribute tremendously to our understanding of the dynamics of ecosystems. Finally, a wide range of biota must be measured at each site and in an efficient format to make a biological monitoring program cost-effective.

An on-line DSS for ecosystem management with important biota that immigrate must also monitor the organisms' population and environment in the source area to enable prediction of the timing and magnitude of their flow to the destination area. For many biota, this information includes measurements of the size of the source area as well as the abundance and phenology of the organism and its hosts in the source area. Where organisms move long distances in the atmosphere (for example, from overwintering habitats in the subtropics to the interior of midlatitude continents), large-scale monitoring programs are required. In the past, frequent monitoring of populations and their environments at regional and continental scales was perceived as too labor-intensive and/or costly to be feasible in most areas of the world. Today, however, a large variety of data on ecosystems can be obtained relatively cheaply using remote sensing technology and can be communicated rapidly across computer networks. The types of measurements and their resolution will only increase with time, making large-scale biological monitoring more practical in the future. A concomitant increase in our understanding of the phenology of pests and pathogens, their natural enemies, and their hosts will allow targeting of specific time periods for sampling and thus further increase the feasibility of spatially extensive population and environmental monitoring programs. These monitoring programs should include human field observations, because many measurements of important behavioral and physiological characteristics that occur at fine resolution spatial and temporal scales are either impractical for, or beyond, the capability of remote sensing technology.

Perhaps the most serious roadblock to effective large-scale population and environmental monitoring (see Figure 3.4) programs is the lack of standardized procedures for sampling populations of biota among different habitat types. The dearth of regular and standardized measurements from different types of habitats dispersed across ecoregions and continents makes quantitative comparisons of populations and their relationships with environmental factors difficult and often unreliable. This is an especially important constraint to managing ecosystems with biota that move long distances in the atmosphere.

Large-scale biological monitoring programs for highly mobile pests have been successfully deployed in many instances (Figure 3.5, also see examples in Drake and Gatehouse 1995). Unfortunately these networks usually collect data on only a single or a narrow range of taxa. One such biological monitoring program has been gathering information on the distribution and aerial abundance of aphids for more than 30 years. This suction trap network was established in the United Kingdom by the Rothamsted Insect Survey (Taylor 1973) and by 1990 had expanded to include 55 traps in 12 European countries (Tatchell 1991). The network of standardized suction traps has provided a continuous record of aphid abundance, at fixed locations, that is extremely valuable for investigating population fluctuations and the ecology of these pests. Current measurements of aphid populations are also integrated into an on-line DSS to provide timely tactical advice to farmers and their advisors (Knight et al. 1992, Knight and Cammell 1994).

The management of agricultural ecosystems, using precision farming technologies, represents an intensive application of standardized biological monitoring programs at the local scale. The capability to simultaneously measure a variety of environmental factors and organisms, which contribute to the many different potential problems that producers encounter, is necessary for these technologies to be truly effective. Management strategies and tactics that rely on precision agricultural technologies need to discriminate between undesirable biota and beneficial organisms, and to anticipate within-field spatial variations due to the immigration of pests and plant diseases.

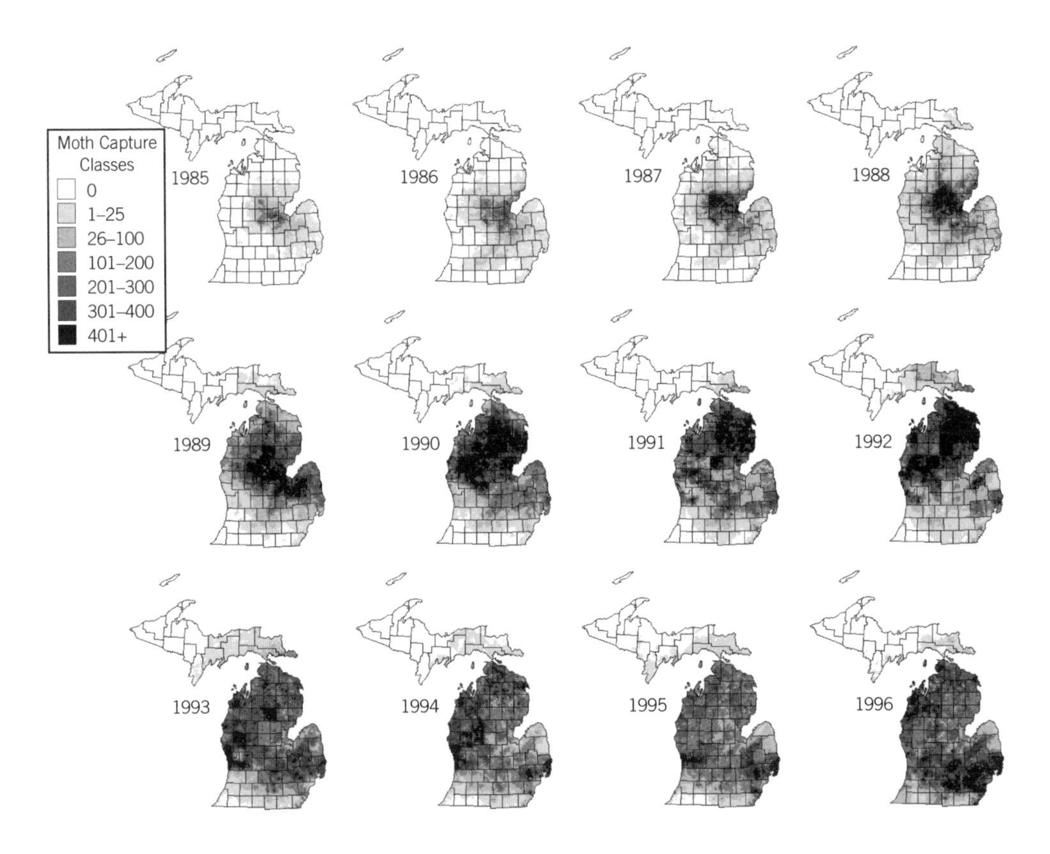

Figure 3.5. Distribution and abundance of male gypsy moth in Michigan from 1985 to 1996.

In the 1860s, the gypsy moth (*Lymantra dispar* L.) was introduced into North America and since has become an important forest and urban pest (Leonard 1981). Its larvae forage on deciduous trees causing defoliation where populations are large. If the trees are stressed by drought or disease, they may die; however, trees usually refoliate in late June after the insects pupate into adults. The gypsy moth has a high reproductive capacity; populations often increase 10- to 100-fold in consecutive years. The moth extends its range locally through wind dispersal of newly hatched caterpillars and long distance by the inadvertent transport of egg masses by humans (Mason and McManus 1981). Unlike many native forest insects, the gypsy moth also affects urban environments where it causes defoliation on shade and ornamental trees in residential and recreational areas, and its frass can cause human health problems (White and Schneeberger 1981).

The gypsy moth was first observed in Michigan during 1954, and defoliation due to larvae became common by 1984. In 1992, 280,000 hectares (700,000 acres) of defoliation was recorded, a 10-fold increase over the amount for 1986. In 1985, the Michigan Department of Agriculture began a statewide gypsy moth monitoring program. More than 3,000 pheromone traps were established in a grid with 2 permanent locations in each township. Data from the monitoring program have been integrated within a GIS and used to understand gypsy moth population dynamics, forecast the spread of this pest, and assess risk to trees in urban and rural environments (Gage et al. 1990).

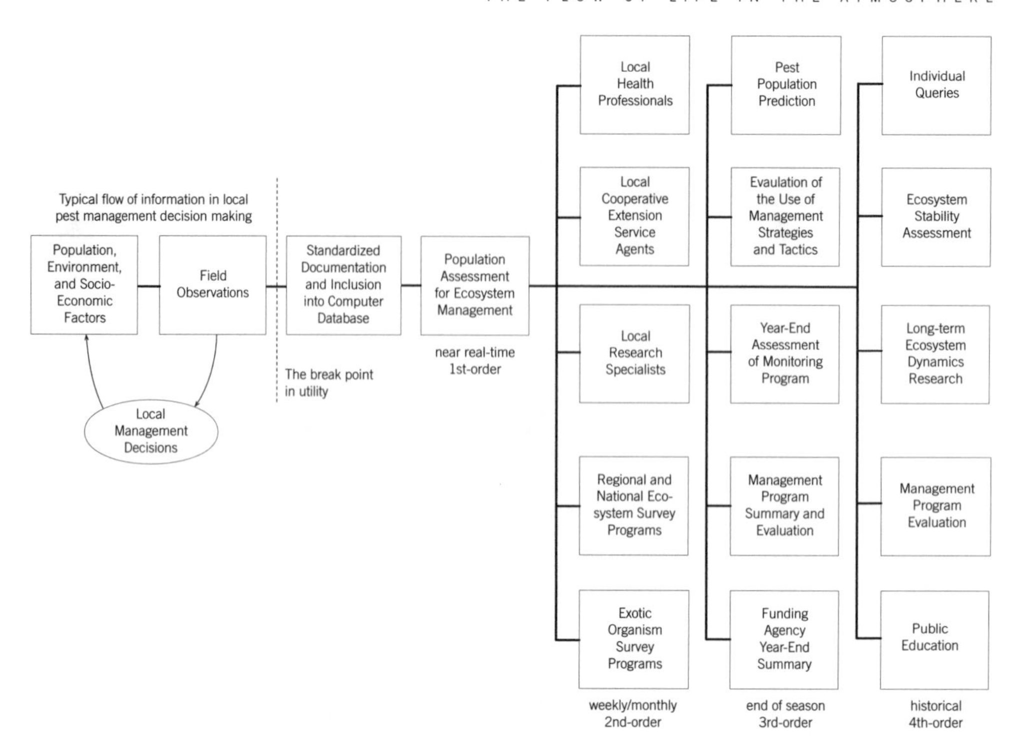

Figure 3.6. Flow of data from ecosystem monitoring program and its uses over different time scales.

Standardized collection, reporting, and storage in a relational computer database system allow continued use of and benefits from information. The dashed line represents the usual break in the flow of information. If the data flow is broken because it is not collected in a standard format and incorporated into a computer data base, 2nd-, 3rd-, and 4th-order uses cannot occur efficiently. Easily retrievable standardized data are especially needed by scientific researchers to expand our understanding of the dynamics of ecosystems. Other important users include practitioners, government administrators, politicians, teachers, and the press. Figure adapted from Gage and Russell 1987.

Database Management Subsystem.

Without comprehensive and efficient Database Management Subsystems (Figure 3.4) for storing and retrieving information, knowledge of the factors that cause populations to fluctuate are likely to remain poorly understood, even if regular and standardized measurements of important populations and environmental factors are obtained. In addition, it is only when management actions and their subsequent effects on the ecosystem are preserved in a retrievable manner that we can ascertain whether a specific strategy or tactic had a positive or negative impact. The value of a good database management system is that once data is collected, it can provide for these and other information needs, not only for ecosystem managers, but also for scientists determined to understand the dynamics of ecosystems, for government administrators and politicians assessing ecosystem management programs, and for teachers and the press who educate the public about planet Earth. Figure 3.6 diagrams these information flows and their time frames. Too often important biological observations are obtained but not recorded in a format that is useful and available to many potential users of the information. Unless regular and consistent field observations of populations and their environments are recorded in an easily accessible computerized database management system, the full potential for the analysis and use of this information cannot be realized.

The data used to compute the maps of gypsy moth abundance and distribution in Michigan presented in Figure 3.5 were simply male gypsy moth catches in pheromone-baited traps, deployed in early July and collected in August. The traps were placed in the same location (sections 8 and 26 of each township) throughout the state each year. The information was used for local pest management decision support (e.g., predicting a high probability of defoliation for a year following a trap catch of 200 or more moths). However, because the same trapping locations and methodologies were employed each year across the entire state, and the data were managed in a relational database, it was possible to extend the value of the information far beyond local management decision making (see "break point in utility" line in Figure 3.6). For example, the Michigan gypsy moth abundance and distribution data set has been used to: (1) develop relationships between male moth catch in pheromone traps and defoliation events (Gage et al. 1990), (2) determine where in the state high infestations would occur the following year to plan for defoliation assessment by the Michigan Department of Natural Resources, (3) target education programs for communities infested with the gypsy moth, (4) support federal government efforts to determine national infestation levels and the rate of change in gypsy moth population, (5) provide a basis for determining the leading edge of the dispersing moth population, resulting in new national program development (see United States Forest Service "Slow the Spread Project" *http://www.ento.vt.edu/st/*), (6) forecast long-term trends in gypsy moth populations, (7) support development of new analytical methods for analyzing population data in space and time (Lele et al. 1998), and (8) provide examples for classroom teaching on the analysis and mapping of animal populations (see *www.cevl.msu.edu/*).

It is important to store all data for an on-line DSS, regardless of their types, in a relational database subsystem, rather than dispersing them among the subsystems. Relational databases are flexible for finding relationships among data that are not physically linked. Perhaps more important, they can accommodate new relationships and data elements easily. Metadata are used to describe the data's availability and their location. This results in an adaptable on-line DSS; adding new subsystems, such as a model or even a knowledge interpreter, simply requires inserting metadata into the database. This arrangement also minimizes data redundancy and thus ensures better data integrity.

The databases that are useful for managing an ecosystem containing biota that immigrate are usually large and diverse. Also, the time frames associated with these data span the range from historical to nearly continuous. Hence the database management subsystem for an on-line DSS needs to be flexible and must have the capacity to upload, store, and download a variety of data types from surveys, maps, aerial photographs, satellite sensors, weather radars, and other sources while maintaining their integrity and security. Shown in Table 3.2 are some of the important geographical databases that are appropriate for use with on-line decision support systems. Historical data often include measurements from both source and impact habitats on ecosystem characteristics such as physical features and soils,

Table 3.2. North American information resources and databases appropriate for on-line decision support systems.

NAME OF FACTORS, MEASUREMENTS, SENSORS OR SYSTEMS	TEMPORAL RESOLUTION	SPATIAL RESOLUTION	SOURCE
Topography			
Digital Elevation Model (DEM)	static	7.3 min	USGS
Topographic maps	static	1:24,000 to 1:250,000	USGS
Soils			
State Soils Geographic Data Base (STATSGO)	static	1:12,000 to 1:63,360	NRCS
Land Use/Land Cover			
Omernik's Map	static	1:750,000	USGS
Land Use & Land Cover Map	static	1:250,000	USGS
Bailey's Ecoregion Map	static	1:750,000	USFS
Advanced Very High Resolution Radiometer (AVHRR) Composite Map	static	4 km	EROS
Hydrology			
Hydrologic Network	static	1:500,000	USGS
Human Populations			
Human population characteristics (measured)	10 yr.	fixed points, counties, cities, states, and regions	USCB
Human population characteristics (estimated)	1 yr.	fixed points, counties, cities, states, and regions	USCB
Agricultural population characteristics	5 yr. discontinued	fixed points, counties, and states	USCB
Crop Productivity			
Crop yield	1 yr.	fixed points (counties)	NASS

vegetation, and human land use, and time series of the local climatology as well as the life cycles and abundance of organisms. The database management subsystem must have the capacity to easily update these historical records on an annual or seasonal basis using survey and remotely sensed data on populations and their environments. It needs to be able to manage weekly population surveys, perhaps only for the warm season, and daily weather information from local and regional climate networks. The subsystem must also be able to store and retrieve high spatial and temporal resolution data from relatively short, intensive field measurement programs during important periods in the life cycle of key biota. These might be daily population data on the abundance and spatial extent of an organism in its source area during a stage in its life cycle when it is likely to emigrate. In turn, the intensive population monitoring program in the source area might trigger the need to manage a vast array of almost continuous and near real-time information on local weather conditions in the destination area and wind patterns over entire continents. It may be important to use meteorological forecasts available over the Internet to identify atmospheric motion systems with high potential to transport organisms from their source to the destination area. Hourly

Seasonal Dynamics of Vegetation			
AVHRR Composite Map	14 d	1 km	EROS
Thematic Mapper	16 d	30 multi-spectral (MS)	EOSC
RADARSAT	6–24 d	10 m–100 m	RIC
SPOT	3–26 d	10 m pan or 20 m MS	SPOT
Air Temperature—Earth's Surface			
Climate Network	24 hr	>5,000 sites	NOAA
Federal Aviation Administration (FAA) Network	1 hr	airports	FAA/NOAA
Air Temperature—Troposphere			
Radiosonde Network	12 hr	≈70 sites	NOAA
Precipitation			
Climate Network	1 d	>8,000 sites	NOAA
FAA Network	1 hr	airports	FAA/NOAA
Air Pressure			
Radiosonde Network	12 hr	≈70 sites	NOAA
Winds			
FAA Network	1 hr	airports	FAA/NOAA
Radiosonde Network	12 hr	≈70 sites	NOAA
WSR-88D (NEXRAD) Radar	continuous	≈100 sites	NOAA/DOD

A more complete list of GIS and remote sensing information resources for the United States is available at *http://www.gis.umn.edu.*

USGS = United State Geological Survey; NRCC = National Resource Conservation Service; USFS = United States Forest Service; EROS = Earth Resources Observation System Data Center; USCB = United States Census Bureau; NASS = National Agricultural Statistics Service; EOSC = Earth Orbiting Satellite Corporation; RIC = Radarsat International Corpotation; SPOT = Spot Corporation; NOAA = National Ocean and Atmosphere Administration; DOD = Department of Defense

precipitation measurements from regional microclimate networks may be used to identify where population surveillance should occur because organisms likely would be washed out of the atmosphere. In most cases, these high resolution data are only stored temporarily within the database management subsystem.

Spatial Analysis Subsystem.

An on-line DSS needs a set of geostatistical tools for analyzing spatially referenced data to characterize patterns (Figure 3.4). The Spatial Analysis Subsystem should be a fully configured geographic information system (GIS). Generally, GIS software contains database management systems of their own and facilities to capture and display data. These functions need to be housed in the database manager and user interface subsystems to make an on-line DSS efficient and adaptable. Geostatistical procedures quantify spatial correlations between map layers or themes, each composed of a single type of data. They can be employed over a spectrum of spatial scales. Geostatistics can be used to interpolate between, and extrapolate beyond, sample points. Using such tools, landscape units with similar sets

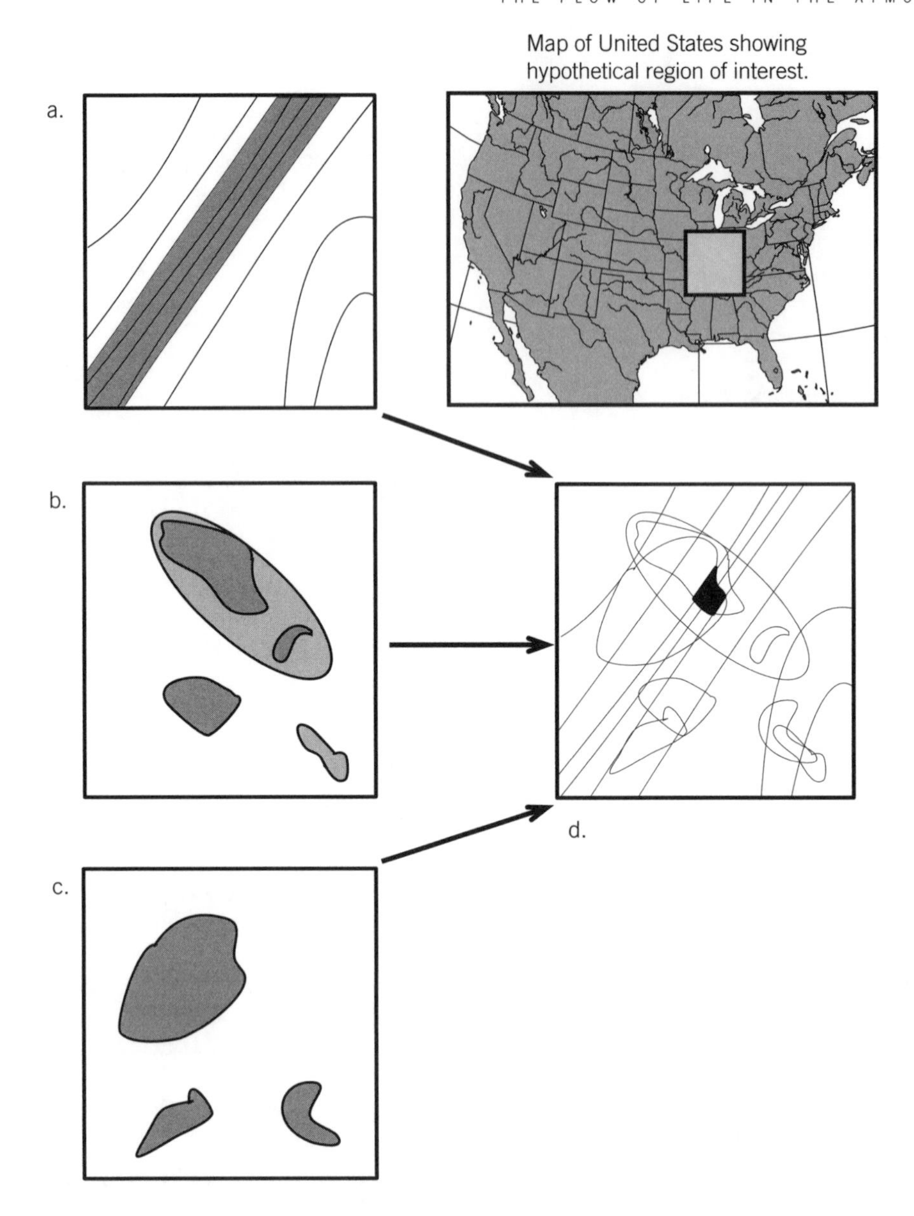

Figure 3.7. Hypothetical field surveillance map generated by a Spatial Analysis Subsystem.

(a) Real-time map of the 850 mb pressure surface depicting the wind direction and strength near the top of the planetary boundary layer above the region. The shaded area represents the most likely path of the target organism across the region. (b) Precipitation map from radar data of the region for the current 24-hour period. Dark shading represents areas of high intensity rainfall where target organisms may have been washed out of the atmosphere. (c) Distribution of vegetation communities that are potential hosts for target organisms. (d) Predictive map created by Spatial Analysis Subsystem (see Figure 3.4) showing the area that should be surveyed for the target organism to determine if it has immigrated to the region.

The Spatial Analysis Subsystem of an on-line DSS can guide intensive field surveillance for an organism in the destination region when population forecasts from its overwintering habitat suggest that it is likely to move through the atmosphere. A manager can use a GIS to overlay digital maps to view the spatial coincidence of populations and static and dynamic elements and conditions in their environments. This example only applies if overwintering of the target organism does not occur in the ecosystem.

of physical, biotic, and climatic characteristics can be identified and categorized. Through analyses of spatial and temporal relationships among and within elements of the terrestrial environment, populations, and atmospheric motion systems, the spatial analysis subsystem provides information about patterns of elements and events to the Knowledge Interpreter. It can furnish maps to guide field surveys for immigrants in the destination area once model forecasts for the source area indicate that atmospheric dispersal of the organism is imminent. Data and maps generated by the Spatial Analysis Subsystem are also important for parameterizing and initializing biological and socioeconomic models. The concept of building such an overlay map to guide sampling is depicted in Figure 3.7. A GIS is used to overlay dynamic elements and conditions in the environment, such as the upper air winds (Figure 3.7a) and precipitation (Figure 3.7b), on maps of static conditions, such as the distribution of habitats with potential host plants (Figure 3.7c). The spatial coincidence of these and other important factors (Figure 3.7d) can be used to target specific areas for intensive field surveillance for an immigrant.

Modeling Subsystem.

An on-line DSS produces quantitative output using predictive models and inputs from the Database Management and Spatial Analysis Subsystems (see Figure 3.4). The Modeling Subsystem combines the use of phenological models, simulation models of population dynamics and disease epidemics, mathematical models of airflows, and evaluation functions for cost/return and risk/benefit analyses. It is critical that human social factors be included in the models where appropriate. The models may be complex systems models that require extensive knowledge and input data and/or simple linear models and pragmatic indices that use only one or two input variables. As a general rule, the complexity of the types of models needed to effectively manage ecosystems is inversely related to the temporal and spatial resolution of the population, environmental, and socioeconomic measurements that are used to drive them. Current information about these inputs facilitates the use of simple methods for predicting future changes in populations. The modeling process must be streamlined as much as possible so that decisions can be made in a timely fashion.

The models supply the Knowledge Interpreter Subsystem, or "reasoning" system, with information about the current state of the system and provide forecasts of future system states. Figure 3.8 is an extension of Figure 3.7 showing an example of a population forecast based on survey results from the target area (identified by black-filled polygon in Figure 3.7). The number of females at sampling sites throughout the polygon provides an indication of the spatial distribution of the immigrant population. A population model is used to predict the trend in population density over a five-week period after the immigration (see population forecast plots bordering the polygon in Figure 3.8).

Near real-time forecasts of the long distance atmospheric movement of organisms are becoming increasingly available over the Internet. Two real examples of Population, Environmental, and Economic Modeling Subsystems (see Figure 3.4) include the North American Plant Disease Forecast Center (see Chapter 11 and *http://www.ces.ncsu.edu/depts/pp/bluemold/*) and the Cooperative Research Centre for Tropical Pest Management Helicoverpa

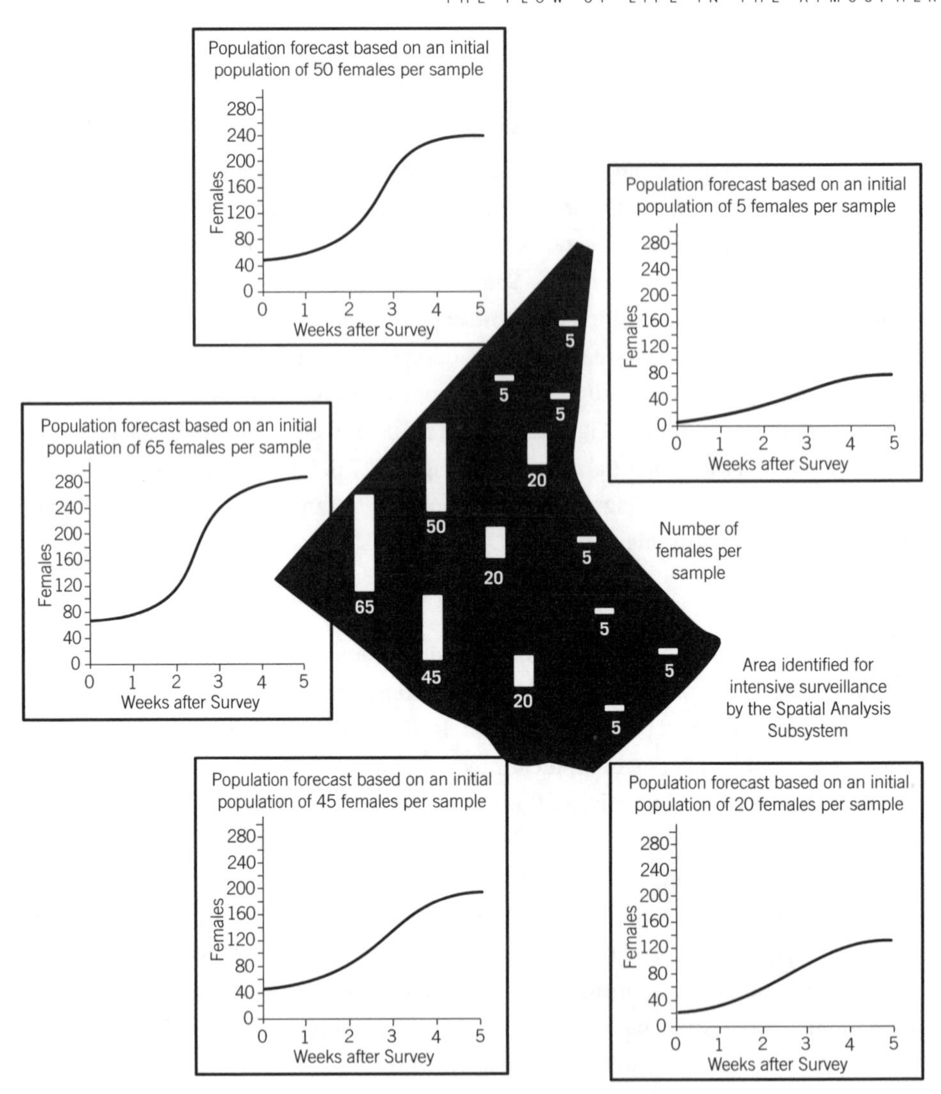

Figure 3.8. Hypothetical survey results and model forecasts of immigrant population growth.

An example of survey results and forecasts of the immigrant population growth generated by the Modeling Subsystem for the target area within the destination ecosystem shown in Figure 3.7. The on-line DSS has a "what if?" facility to allow managers to vary model inputs to view the results of possible alternative tactics for manipulating population growth that are compatible with how humans value and use the ecosystem.

Forecasting Project (*http://www.ento.etpm.uq.edu.au/forecasting.html*). These forecasting systems include a variety of information such as current maps of target organisms in their source areas, predictions of growth, development, and abundance in source areas, forecasts of probable atmospheric pathways, and information on the likelihood that a target organism will initiate aerial movement, encounter mechanisms that might wash them out of the atmosphere, and survive the journey. The Helicoverpa forecasting system is constructed so that a user can easily change the initial conditions and parameters in an atmospheric trajectory model and rerun it over the Internet in about 1 minute. This capability is important for an on-line DSS because it facilitates the evaluation of ecological and socioeconomic impacts of changes in the variables that managers can manipulate in the real ecosystem.

Knowledge Interpreter Subsystem.

If users are numerous, perhaps with varying levels of education and knowledge as well as different values and uses for the ecosystem, then it is important that an on-line DSS have the capacity to present alternatives and advice. This is especially true for agricultural applications of on-line DSSs. A Knowledge Interpreter Subsystem (see Figure 3.4) can provide this function by integrating data in order to present users with strategies and tactics for managing an ecosystem. This subsystem evaluates quantitative and qualitative data and information from the other subsystems using established rules. The Knowledge Interpreter Subsystem is composed of an "inference engine" and a knowledge base. The inference engine is the set of commands that control the reasoning process. Either forward or backward chaining can be used to reach solutions to problems. Forward chaining involves the use of data to evaluate a hierarchy of rules to reach a conclusion. In contrast, when a specific goal is desired, the reasoning process can proceed backward through the rule hierarchy to find those paths from current conditions that establish the goal.

The knowledge base is composed of rules and facts. Rules are usually if-then statements. The set of rules can incorporate both research-based knowledge and heuristic rules of thumb. The facts are the data supplied by the other subsystems, including technical information, knowledge of spatial patterns of biotic and abiotic factors in the ecosystem, information and forecasts of atmospheric flows and socioeconomic factors, and the current and projected phenology and density of important populations in the system. Figure 3.9 shows a continuation of the example (Figures 3.7 and 3.8) using the same polygon and a simple set of decision rules and recommendations. The inference engine associates probabilities with specific rules and facts and thus will calculate the degree of uncertainty associated with its advice. A Knowledge Interpreter has the capacity to supply the user with an explanation of the reasoning associated with a recommended management strategy or tactic.

User Interface Subsystem.

Central to an on-line DSS is the User Interface Subsystem that links managers (users) with the computer software (see Figure 3.4). The primary functions of the User Interface are to receive user input, translate inputs into software commands, send commands to the information distribution and communications subsystem, and export such information as displays, files, and hardcopy. It is extremely important that this component accommodate the problem-solving styles of the users. The User Interface must be easy to use, flexible, so that managers can apply decision support to their specific situations and provide answers in displays, reports, and maps that are easy to read and interpret. The user interface should have a "what if?" facility to allow managers to examine, and when appropriate change, data, model parameters, and decision rules to evaluate alternative scenarios quickly (e.g., the Helicoverpa Forecasting Project cited above).

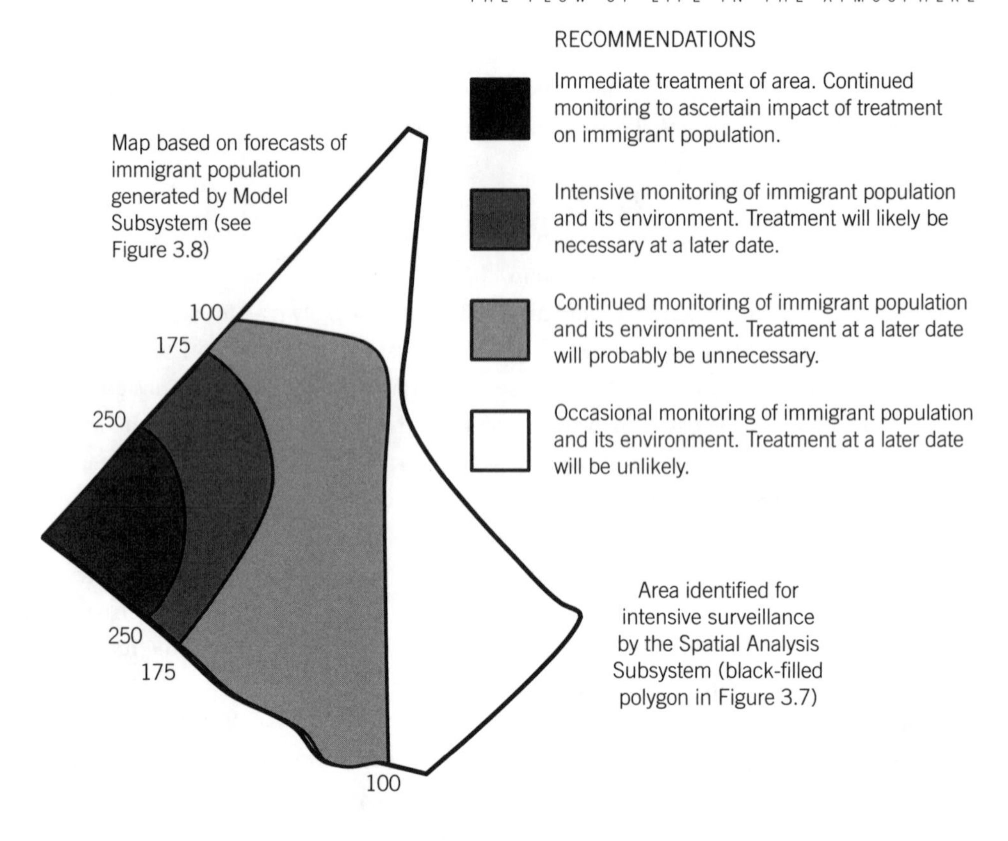

DECISION RULES

- If the number of females > 250 per unit area then the damage due to immigration will be high regardless of other biotic and environmental factors.
- If the number of females per unit area is between 100 and 250 then the impact of the immigrant will depend on other biotic and environmental factors.
- If the number of females < 100 per unit area then the damage from the immigration will likely be minimal.

Figure 3.9. Hypothetical management recommendations of a Knowledge Interpreter Subsystem.

The recommendations of the Knowledge Interpreter Subsystem of the on-line decision support system are based on field survey and model forecasts of population growth shown in Figure 3.8.

Information Distribution and Communications Subsystem.

Internal linkages among components of an on-line DSS (Figure 3.10) are provided by the Information Distribution and Communications Subsystem. It is the gateway through which data and information flow among the other subsystems as needed. Essentially, it is an interconnected set of managers, one for each subsystem. Each manager controls the bi-directional exchange of data for its respective subsystem. The Information Distribution and Communications Subsystem provides the protocols that permit different computers to communicate so that the decision support system can be implemented on a group of machines using networks.

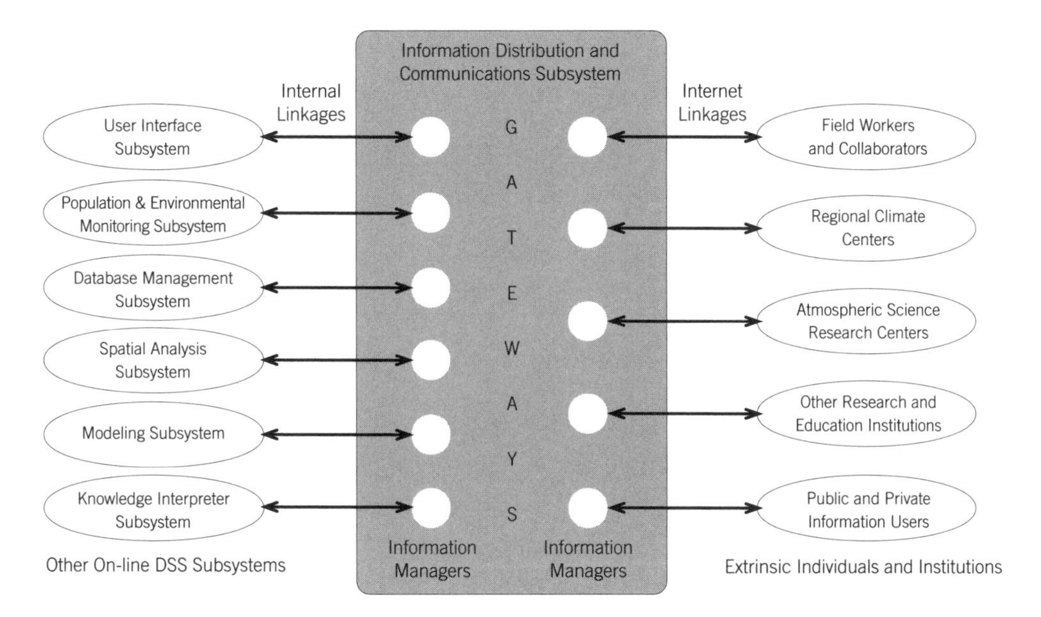

Figure 3.10. Flows within an Information Distribution and Communications Subsystem.

The internal managers of an on-line DSS Information Distribution and Communications Subsystem provide gateways for flows among subsystems and with extrinsic individuals and institutions over the Internet System. This depiction of on-line DSS information flows represents a realignment of Figure 3.4 with external linkages shown in more detail.

The Information Distribution and Communications Subsystem must be able to access and capture information over the Internet, such as current and forecast maps of atmospheric flows from the National Weather Service and atmospheric science research centers and daily weather and microclimate data from state climatology offices and regional climate centers (see External Linkages arrow in Figure 3.4). This capacity is critical because it allows for near real-time access to a large variety of data; it greatly reduces the time and effort needed to incorporate this information into databases; and it allows the system to access the results of models built and operated by other research and management programs and institutions. The Internet also can be used to disseminate tactical instructions to workers and collaborators in the field and to communicate field data to the data base management subsystem.

The validation of on-line DSS output for managing pests and diseases is difficult. This is especially true if the system contains a Knowledge Interpreter that provides users with advice. Most standard validation techniques are not appropriate for systems that deliver qualitative advice rather than quantitative results. Subjective methods for evaluating a decision support system include determining whether or not its advice seems reasonable, comparing its recommendations with those of human experts, deciding if the system achieves some minimum standard of performance, and prolonged field testing to ascertain the impact of its advice on the health and productivity of an ecosystem. It is also important to ascertain whether or not an on-line DSS provides information and advice not readily available to managers in other forms.

Adoption of an on-line DSS is contingent on the users' judgement of its value to them. This depends on many factors, some personal, and others related to whether or not users perceive that it increases the profitability or decreases the cost of managing the system. A decision support system must provide credible advice and convincing justification for recommendations using language and logic understood by the users. Good explanations are the key to user acceptance. It is best if users are actively involved in the design, construction, and evaluation of an on-line DSS so that they "own" the system. This can be accomplished through surveys to define user needs and attitudes, iterative modification of the system based on user feedback, incorporation of the users' "expert opinions" into the rule-base of the knowledge interpreter, and field evaluations to test the ease of use and the acceptability of advice given by the system.

Summary.

In this chapter we have argued that the development of successful strategies for managing ecosystems requires information in three important areas: (1) the ecological relationships of organisms and their environments, (2) monitoring populations and environmental conditions, and (3) the benefits and costs of manipulating the ecosystem. A solid understanding of the ecological basis of a population fluctuation will most often suggest ways and means of altering the environment to reduce an organism's numbers. Successful management of an ecosystem requires current and historical information on the types and numbers of inhabitants and the environmental conditions within which they live. Finally, a decision to manipulate an ecosystem should be made only after careful analysis of the short- and long-term environmental and social benefits and costs of the changes. We provide an outline of the on-line DSS approach to preventive ecosystem management as a tool that can be used to integrate these types of information. We argue that research programs designed to increase our understanding about ecological relationships among organisms and their environments, as well as the benefits and costs to humans of manipulating ecosystems, must be the basis of any attempt to manage Earth's ecosystems. Decision support systems must be built around biological monitoring networks to measure the current state of the system and to predict its likely future states.

Based on his experience with desert locusts in Africa and Asia, Pedgley (1982) provides a useful list of the necessary components for an effective forecasting service involving organisms that move long distances in the atmosphere: (1) a sampling system that spans the entire area, from source to destination, throughout which the movement takes place; (2) an analysis and forecasting center; (3) a network for rapidly communicating measurements from the many sampling sites to the analysis and forecasting center; (4) a model that relates the distribution of the target organisms to the changes (harmful and/or beneficial) they cause at their destinations and the impacts on the integrity of the ecosystems; and (5) a system for rapidly disseminating pertinent information to users, on the distribution and potential impact of the aerobiota. With the addition of information on available strategies and tactics for managing ecosystems impacted by aerobiota and the benefits and costs of using the tools, these are also the basic components of an on-line DSS.

From the day-to-day management perspective, the most important attribute of an on-line DSS is that it provides updated forecasts of populations and biological events on a frequent basis, accounting for current conditions in the area being managed. Frequent monitoring of populations and their environments is the key component to an ecosystem decision support system because it greatly simplifies the modeling process and generally increases the accuracy of population forecasts. The monitoring component is particularly useful for managing ecosystems with immigrants because it is often difficult to accurately forecast the timing and exact location of their arrival.

It is especially critical that population, environmental, and socioeconomic measurements are standardized and are obtained at prescribed time intervals and at fixed locations. They must also be incorporated into computerized databases that are easily accessible to potential users. Measurements that are standardized and easily retrievable are extremely valuable to many users in addition to those managing ecosystems. They are especially needed by scientific researchers to better understand the dynamics of ecosystems. Other important users include government administrators, politicians, teachers, and the press.

An on-line DSS has the capability of integrating and synchronizing different types of information, including those available from remote platforms in the atmosphere and over the Internet. It can be built to interpret information as well. Where appropriate, an on-line DSS also should serve as an implementation coach after decisions have been made. It is the capacity to adjust populations and parameters in the models using near real-time information from ecosystem monitoring programs that makes the on-line DSS approach useful for day-to-day tactical pest and disease management of ecosystems impacted by aerobiota.

CHAPTER 4

An Airscape Model of Biota Flow in the Atmosphere

Conceptual models for studying the aerial movement of biota.

Organisms exploit winds in the atmospheric medium to move within and among habitats for a variety of reasons. Their movements have been classified differently by authors depending on their perspective and focus of study. For example, Dingle (1996), using the pioneering works of John S. Kennedy (synthesized in Kennedy 1985) as a foundation, constructed an insightful taxonomy of movement as a behavior, relating it to the life histories of organisms (Table 4.1). He grouped movement behaviors into three types: (1) those that are home range or resource directed, (2) those in which cessation is not directly responsive to resources or home range, and (3) those that are accidental and not under the control of the organism. Movements that meet the criteria of the second category are considered migratory behaviors. This classification is extremely useful for characterizing migration behavior as a biological phenomena because it facilitates the identification of migratory behaviors in a wide range of biota and the commonalties in migratory behaviors among the different taxa.

Drake et al. (1995) developed a very useful model for comprehending the development and function of migration among insects. First they compiled a comprehensive list of important generalizations about insect migration that have advanced our understanding of the process (Table 4.2). They then used this knowledge base to construct a soft systems model for insects that undertake long-distance windborne migrations. The model, presented in Figure 4.1, includes four interrelated components: migration arena, population trajectory, migration syndrome, and genetic complex. A major emphasis is placed on the processes that connect these components. The model provides a useful template for understanding how insect migration systems function and evolve within highly variable environments.

It can be equally valuable to view the aerial flow of biota from an atmospheric perspective, independent of the reason for movement. This view facilitates the formation of generalizations about how a wide range of plants and animals use the atmosphere to move within and among habitats. Taylor (1958, 1974) coined the term "insect boundary layer" (also called flight boundary layer) to refer to the zone in the lower atmosphere within which an insect can control its flight direction and consequently its destination. Insects that ascend through

Table 4.1. Taxonomy of movement behavior.

MOVEMENT	CHARACTERISTICS
Movement that is home range or resource directed	
Stasis	Organism is stationary
Station keeping	Movements keep organism in home range
Kineses	Change in rate of movement or turning
Foraging	Movement in search of resources; stops when resources are encountered
Commuting	Movement in search of resources on a regular short-term basis, usually daily; ceases when resources are encountered
Territorial behavior	Movement and antagonistic behavior directed toward neighbors and/or intruders; stops when intruder leaves
Ranging	Movement over an area so as to explore it; ceases when suitable habitat/territory is located
Movement that is not directly responsive to resources or home range	
Migration	Undistracted movement; cessation primed by movement itself; responses to resources/home range suspended or suppressed
Movement that is not under control of organism	
Accidental displacement	Organism does not initiate movement; stops when organism leaves transporting vehicle

From Dingle 1996.

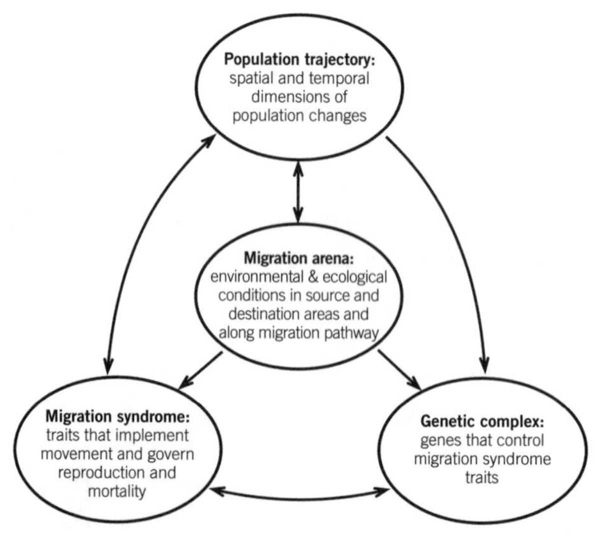

Figure 4.1. Conceptual model of the insect migration system.

The migration arena includes the biotic and abiotic factors that influence the system throughout the entire region in which migration occurs. The spatial and temporal dimensions of the demographic characteristics of the population is its trajectory. These include: the times and locations within the life history, when and where key events occur, and the development and movement of organisms between these events and locations. The migration syndrome is the set of co-evolved behavioral, physiological, biochemical, and morphological traits that together implement migration and govern the migrants' schedules of reproduction and mortality. The genes that code for the migration syndrome traits constitute the genetic complex. The arrows represent the processes that link components. These are the large array of abiotic and biotic processes that occur within the migration system. They influence both the environmental and ecological factors that govern the states within the migration arena and the migrants' responses to these factors and states (Drake et al. 1995). Adopted from Drake et al. 1995.

Table 4.2. Important advances in our knowledge about insect migration systems.

1. Migration is actively initiated, maintained, and terminated; it is not a 'passive' process or in any sense 'accidental'.
2. Migration can be distinguished behaviorally from other types of movement; migratory behavior is characterized by persistent, straightened-out movement and some temporary inhibition of station-keeping responses. Its outcome is that the insect is relocated to some new habitat beyond its previous foraging range.
3. Migration is an adaptation to temporary habitats that vary in availability and quality both in space and in time; seasonal variations may be quite regular, but others are often unpredictable.
4. By determining where and when reproduction occurs, migration has a major influence on fitness; it is thus a fundamental component of the life histories of migratory species and plays a central role in population processes.
5. Migration may be obligatory but more usually occurs facultatively in response to environmental cues. Induction of migration may involve developmental as well as physiological and behavioral responses, most obviously in wing dimorphic species where environmental cues initiate production of macropters as well as regulates their migratory behavior.
6. In females, migration usually occurs early in the adult stage, before reproduction begins; males exhibit more varied patterns of integration of migration and reproduction. Migration by larval stages occurs less commonly in insects.
7. Long-distance migrations usually involve flight above the 'flight boundary layer' (the zone near the ground in which the wind is slow enough for the insect to be able to make headway against it), frequently at altitudes up to 1 km (and occasionally higher), and often at night. Such migrations are achieved almost entirely through transport on the wind. However, some long-distance migrants fly exclusively or predominantly within their flight boundary layer.
8. Some migrations occur in regions and at seasons where the wind blows predominantly from one particular direction; others may be subject to winds that are highly variable. Correlations between wind direction and other environmental cues that regulate migration (e.g., temperature) may reduce the extent of variability in migration outcomes. Winds are, however, never constant and there is always an element of unpredictability about the destination of wind-assisted migrations.
9. The migrant is initially unresponsive to stimuli associated with favorable habitats, but this initial phase of migration is terminated spontaneously and followed by flight (or flights) with the migrant in responsive mode.
10. Migrants are characterized by a suite of co-adapted biochemical, physiological, behavioral and demographic traits, the 'migration syndrome'.
11. The phenotypic variance observed in many of the migration-syndrome traits has been shown to have a substantial additive genetic component, indicative of polygenic inheritance. Some (but not all) of the traits are associated by a system of genetic correlations.
12. The observed values and variances of the syndrome traits are likely to be the outcome of contemporary natural selection.

From Drake et al. 1995. The authors provide citations for each of the insights.

this boundary layer are carried along by strong winds in the atmosphere above (Johnson 1969). This concept, when broadened to apply to biota in general, is very useful for studying the interactions between biological and meteorological factors that govern the flow of organisms in the atmosphere. For example, the depth of the "biota boundary layer" (BBL) varies among organisms, weather conditions, and with time and place. Although biological factors are paramount to the pathways of organisms within this boundary layer, winds and atmospheric turbulence can also influence the trajectory of organisms. Atmospheric scientists often use the term planetary boundary layer (PBL) for the air layer that extends from the Earth's surface upward to an altitude that varies from approximately 300 m at night to 1–3 km during the day (see discussion in Chapter 5). In that part of the PBL above the BBL, winds and large-scale atmospheric structures govern the direction of flow. However, biological factors are also important to movement within this part of the PBL. They influence the time that an organism remains airborne and the levels in the PBL in which they flow. Biological and atmospheric factors can also interact to govern the concentration of biota in the PBL. It follows that aerial movements of biota can be divided into two categories: those that take an organism above its BBL into zones where atmospheric factors primarily dictate its pathway and those movements that occur entirely within an organism's BBL and thus are chiefly controlled by biological factors. An important utility of this classification is that it can be employed to distinguish between the takeoff/ascent (within the BBL), horizontal transport (above the BBL), and descent/landing (within the BBL) stages in the aerobiology process model (see discussion below).

The BBL concept also can be used to discriminate between local and long-distance aerial movements of organisms. This distinction is based on the potential distance of movement and the motion systems that organisms encounter during their travels rather than the actual distances between the points of initiation and termination of aerial movement. Any organism that ascends through its own BBL into the PBL above has the potential to be dispersed "long distance" by strong atmospheric motion systems to distant habitats. In contrast, aerial movement during times when organisms are unable or disinclined to escape their BBL is very likely to result in "local" journeys within the source area. It should be noted that the utility of this concept is limited for a number of cases. For example, organisms can escape their own boundary layer, only to return to their source area due to downwardly directed flight, gravity in the absence of strong winds, downdrafts and subsiding air parcels, and washout from precipitation. By this classification, these trips are considered long-distance movement. Also, as noted in Table 4.2, most birds and some strong-flying insects travel long distances within their biota boundary layer (which at night may extend to the top of the PBL).

A geographical framework for categorizing aerial flows of biota as either local or long-distance is very useful for understanding relationships among populations and their environments, within discrete spatial units. Changes in location within a habitat, regardless of the reason and route, are considered local movements. Conversely, aerial flows of biota among habitats constitute long-distance movement, regardless of the other circumstances associated with their journeys. Here we take an anthropocentric view of a habitat, defining it as an assemblage of plant and animal communities that by human standards differs in species structure or composition from the landscape elements in its surroundings. This classification allows quantification of the fluxes of organisms into and out of habitats and thus, as is argued in Chapter 3, is critical for interpreting data from biological monitoring programs and for understanding changes in populations for ecosystem management purposes.

Table 4.3. References to major works that have influenced our thinking about aerobiology.

AERIAL MOVEMENTS OF SPORES AND OTHER MICRO-ORGANISMS:
Gregory and Monteith 1967, Dimmick and Akers 1969, Ingold 1971, Gregory 1973, Edmonds 1979, Pedgley 1986, Cox 1987, Madden 1992, Lighthart and Mohr 1994.

AERIAL MOVEMENTS OF INSECTS AND OTHER ARTHROPODS:
Williams 1958, Johnson 1969, Rabb and Guthrie 1970, Rabb and Kennedy 1979, Huffaker and Rabb 1984, Rankin 1985, Sparks 1986, Danthanarayana 1986, Goldsworthy and Wheeler 1989, Rainey 1989, Drake and Gatehouse 1995.

AERIAL MOVEMENTS OF BIRDS:
Hochbaum 1955, Dorst 1962, Griffin 1964, Alerstram 1990, Gwinner 1990, Berthold 1993, Kerlinger 1995.

AERIAL MOVEMENTS OF BIOTA (NOT RESTRICTED BY TAXA):
Rainey 1976, Pedgley 1982, MacKenzie et al 1985, Rainey et al 1990, Dingle 1996.

For a current list of references to the plethora of refereed journal articles on the aerial movement of biota the interested reader should see Lighthart and Mohr 1994, Drake and Gatehouse 1995, Kerlinger 1995, and Dingle 1996.

The airscape perspective on the flow of biota in the atmosphere.

The assumptions that (1) aerial movements of biota that take place entirely within the BBL terminate in the same habitat as they begin, and conversely, (2) that movements above the BBL terminate in a different habitat from which they begin, make the atmospheric and geographic perspectives on movement compatible. As noted above, these assumptions may not be valid during very calm weather conditions or for strong-flying insects and birds. However, in the rest of this book, we combine the geographic and atmospheric view of movement into what we call an airscape perspective on the flow of biota within the atmosphere. We proceed with this approach as the foundation for refining the aerobiology process model (Figure 2.4) proposed during the IBP in Aerobiology (Edmonds 1979). Our objective in this book is to provide the reader with a vision of how plants and animals can utilize the atmosphere to flow among habitats and how knowledge and information about the movement phenomena can help humans manipulate terrestrial ecosystems in appropriate ways. To this end, we intentionally ignore many important aspects of movement behavior that are both common and varying among taxa. The intention is not to argue that meteorological or geographic perspectives on movement are more useful to understanding aerial movement or relationships between populations and their environments than are biological perspectives. On the contrary, as stated earlier, we strongly contend that successful manipulation of an ecosystem depends first and foremost on thoroughly understanding the biologies of its various inhabitants, and this includes understanding movement as a biological phenomenon.

The number and diversity of factors that impact the life history of biota that move in the atmosphere are immense. In some cases, the influence of these factors on the life cycles of specific aerobiota taxa have been studied in great detail (Table 4.3).

Table 4.4. Relative contribution of atmospheric motion systems to the control of aerial movement in major taxa.

MAJOR TAXA	CONTRIBUTION OF BIOLOGICAL FACTORS	CONTRIBUTION OF ATMOSPHERIC MOTION SYSTEMS	RATIO	COMMENTS
Viruses	negligible	high	very small	Primarily *passive transport*—biota exert little or no control over destination once airborne
Bacteria	negligible	high	very small	
Fungi	low	high	small	
Nematodes (soil)	low	high	small	
Higher plants (pollen & seeds)	moderate	high	small/moderate	
Mites & spiders	high	high	moderate	Primarily *active transport*—actions during transport effect destination
Insects	high	moderate	large	
Vertebrates	high	slight	very large	

Adapted from Rabb 1985.

Rabb (1985) provides an insightful treatment of many of these factors and classifies them into intrinsic (genotypic and phenotypic) and extrinsic (environmental) components. The intrinsic components include size, weight, morphology, physiology, age, and nutrition. Temperature, light and photoperiod, moisture, pressure, transport systems, resources for nutrition, reproduction and protection, predators, parasites and pathogens, and mutualists and commensals are considered elements of the extrinsic component that influence movement. Rabb's original classification pertains to organisms that move on land and in water and air. Here it has been modified (Table 4.4) to rank just aerobiota along a gradient of movement from passive to active. To some extent, all biota that flow in the atmosphere are active participants in the process in that they initiate movement when atmospheric conditions are conducive to transporting them to favorable habitats. At the other extreme, even the most robust aerobiota (e.g., waterfowl) take advantage of favorable winds whenever feasible and thus strive to conserve energy and be as passive as they possibly can.

The contribution of atmospheric motion systems to the control of aerial movement in viruses, bacteria, fungi, soil nematodes, and higher plants (pollen & seeds) is generally high. In these taxa, the ratio of the contribution to the control of aerial movement of intrinsic factors to that of atmospheric motion systems ranges from very small to small. Consequently, these biota are primarily transported passively, in that once they are airborne, they exert little control over their destination. In contrast, the behavior of mites, insects, arachnids, and vertebrates during aerial transport often influences their destination to an equal or greater degree than atmospheric motion systems. Many of these biota remain entirely within their BBL during most of their flights. The transport of these organisms is thus considered primarily active. The aerial movement of biota (e.g., viruses, bacteria, and nematodes) by vectors is considered passive with respect to atmospheric motion systems, regardless of whether the transport of the vector is primarily active or passive.

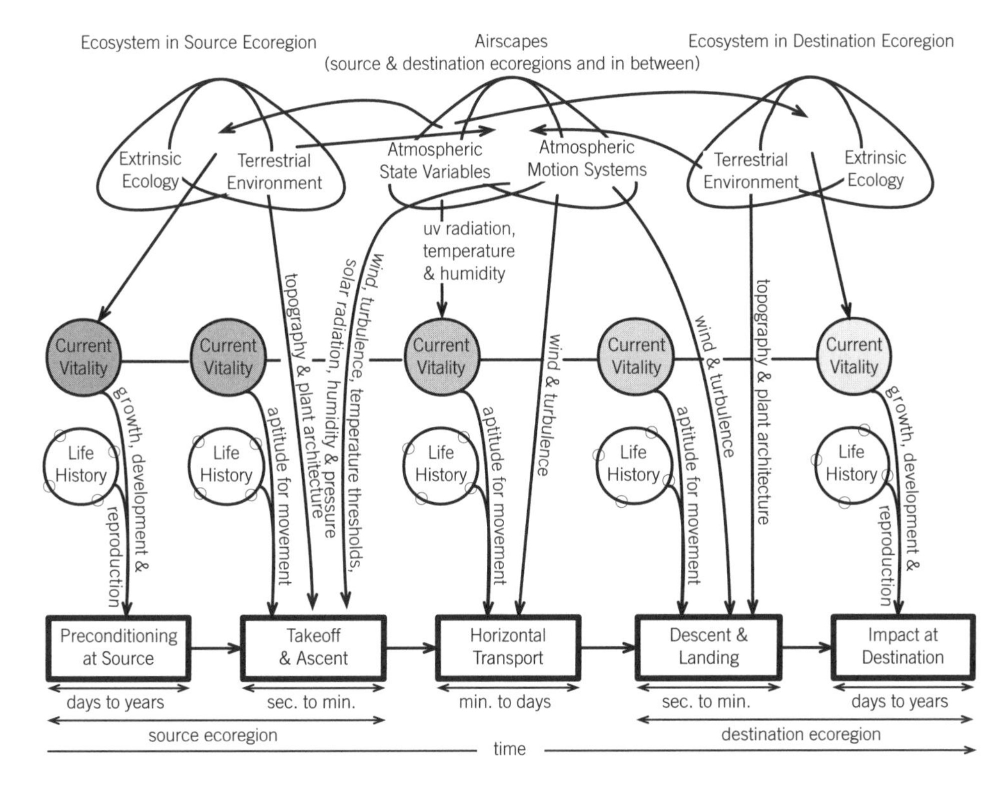

Figure 4.2. General aerobiology process model showing primary relationships among components that influence movement of biota in the atmosphere.

The factors affecting biota that move in the air are grouped into four components: (1) life history, (2) current vitality, (3) ecologies and environments in source and destination ecoregions, and (4) the airscapes encountered during movement. Life history is the sequence of stages and processes in the lives of organisms. Vitality relates to the current condition of individuals in populations and is highly variable. In source and destination areas, vitality is governed by extrinsic ecology (resources, predators, parasites) and terrestrial environment (soil, water, biota, air). The airscape is composed of atmoshperic state variations (temperature, pressure, moisture, solar radiation) and motion systems (tiny eddies to global wind belts) that impact biota as they move.

Relationships among components that influence organisms during aerial movement.

Figure 4.2 depicts relationships among components (groups of factors) that have a major influence on organisms that proceed through the stages of the aerobiology process model. The preconditioning and takeoff/ascent stages occur in the source area, and the time scales associated with these processes typically range from days to years and seconds to minutes, respectively. The descent/landing and impact stages transpire in the destination area, and their time scales usually span from seconds to minutes and days to years, respectively. The horizontal transport stage occurs above landscapes in the source and destination areas as well as those in between. This stage generally lasts from minutes to days.

When combined, the three middle stages of the aerobiology process model constitute the "flow" of biota in the atmosphere. The pathway of an organism through these stages is depicted in Figure 4.3. These flow stages have been given many different names over the

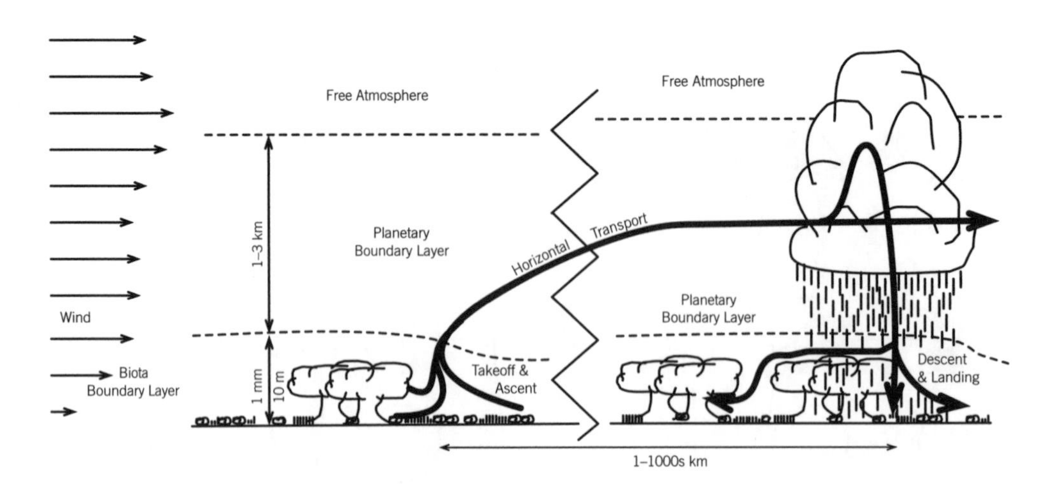

Figure 4.3. Stages of flow of biota in the atmosphere.

Takeoff/ascent, horizontal transport, and descent/landing, the three middle stages of the aerobiology process model, constitute the flow of biota in the atmosphere. The takeoff and ascent stage occurs after a period of development and preconditioning in the source ecoregion. Takeoff is influenced by the biological and environmental factors that dictate the timing and other activities associated with flight initiation. Ascent is the period of upwardly directed flight through the BBL that carries organisms to an altitude where strong winds govern their movements. Horizontal transport takes place in the PBL above the organisms' BBL and is the component that dictates flight pathways and consequently destinations. Descent through the BBL and landing is the final stage of the biota flow model. It is often assisted by rainfall. If flight is terminated in an unsuitable habitat, the biota may reinitiate flight. Otherwise, the dynamics of the receptor ecosystem strongly influence the impact of the immigrants.

years. The terminology that is used for the individual stages of flow depends in large part on the proportion of the energy expended by the organism to achieve the movement stage relative to that which is derived from the environment. For example, the terms takeoff, flight, and landing are frequently associated with relatively strong flyers such as insects and birds. In contrast, liberation, drift, and deposition are commonly used in conjunction with the aerial movements of more passively transported taxa such as pollen and spores.

Takeoff and ascent.

Takeoff is the process of commencing aerial movement. Takeoff is considered active when it involves the use of energy by the organism and passive when the energy comes from the environment, although this distinction is not always clear. Ascent is the upwardly directed movement through the BBL in which an organism may exert considerable control over its flight direction and/or speed. The notion of ascent through a BBL is seemingly less appropriate for relatively passively transported biota than for winged organisms. However, even for many passively transported organisms, the BBL concept has important meaning. In principle, these biota must cross a thin laminar boundary layer of slowly moving air before they can exploit the energy of turbulent air for dispersal. The airflow near a leaf during sunny and moderately windy conditions is depicted in Figure 4.4. Under these conditions,

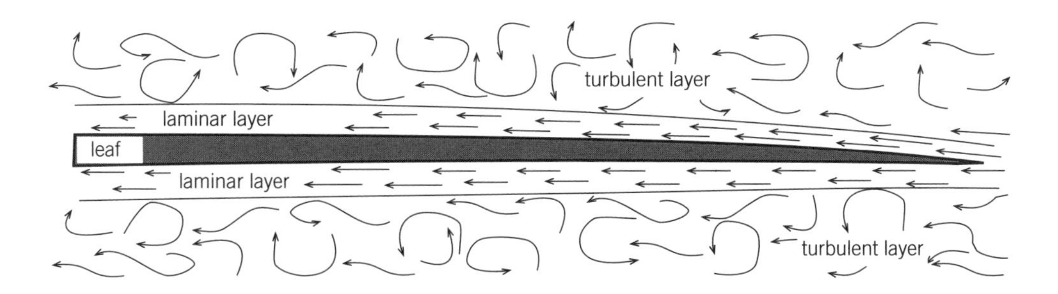

Figure 4.4. Laminar layer of air surrounding a leaf during moderately windy conditions.

Many microbiota have ejection mechanisms and/or use gravity to move through the thin laminar layers that surround plant parts before they can use the turbulent energy in the atmosphere for movement (see Figure 4.5). These takeoff mechanisms are often triggered by sunlight or the shaking of plant parts by wind because these conditions generally indicate that the laminar layer is thin. Conversely, during nights and overcast days when winds are calm, the laminar layer is generally much thicker and may present an impenetrable barrier to passively transported aerobiota. Figure adapted from Avery 1966.

the laminar layer around plant parts is generally quite thin, and thus conditions are conducive to the takeoff of weak and non-flying aerobiota. At other times, especially at night when it often extends upward as much as a few meters and on overcast days if winds are calm, the laminar layer near the Earth's surface can present a formidable barrier for passively transported biota to traverse. Plants have developed an amazing variety of mechanisms to use the turbulent energy of moving air for dispersal. Many spores, pollens, seeds, and other small aerobiota are launched into the atmosphere when potential energy engendered in the parent plant is suddenly released, while the unique presentation of others on exposed plant parts allows them to rely on gravity, the kinetic energy from raindrops, and windblown motions of plants, to propel them through their biota boundary layer. Still others, most notably many seeds, have hair-like structures that form parachutes and wings that lower their rate of fall, thus increasing their chance of encountering a wind gust that will carry them afar. A few of the more interesting liberation mechanisms of fungi redrawn from Ingold (1967) are shown in Figure 4.5.

The depth of the BBL is highly variable, as shown in Figure 4.6, for hypothetical spore (passive dispersal) and insect (active flier) movement. Not only does it vary among organisms but also with time and space. This is because the structure of laminar and turbulent airflows near the Earth's surface depends on both the weather and the roughness of the surface over which the wind blows (see discussion on turbulent eddies in Chapter 9). Organisms that ascend through their BBL into the faster airstreams above enter the horizontal transport stage of the aerobiology process.

Horizontal transport.

The horizontal transport stage of the aerobiology process dictates the direction of flow and consequently the organism's destination. This phenomenon is almost completely confined

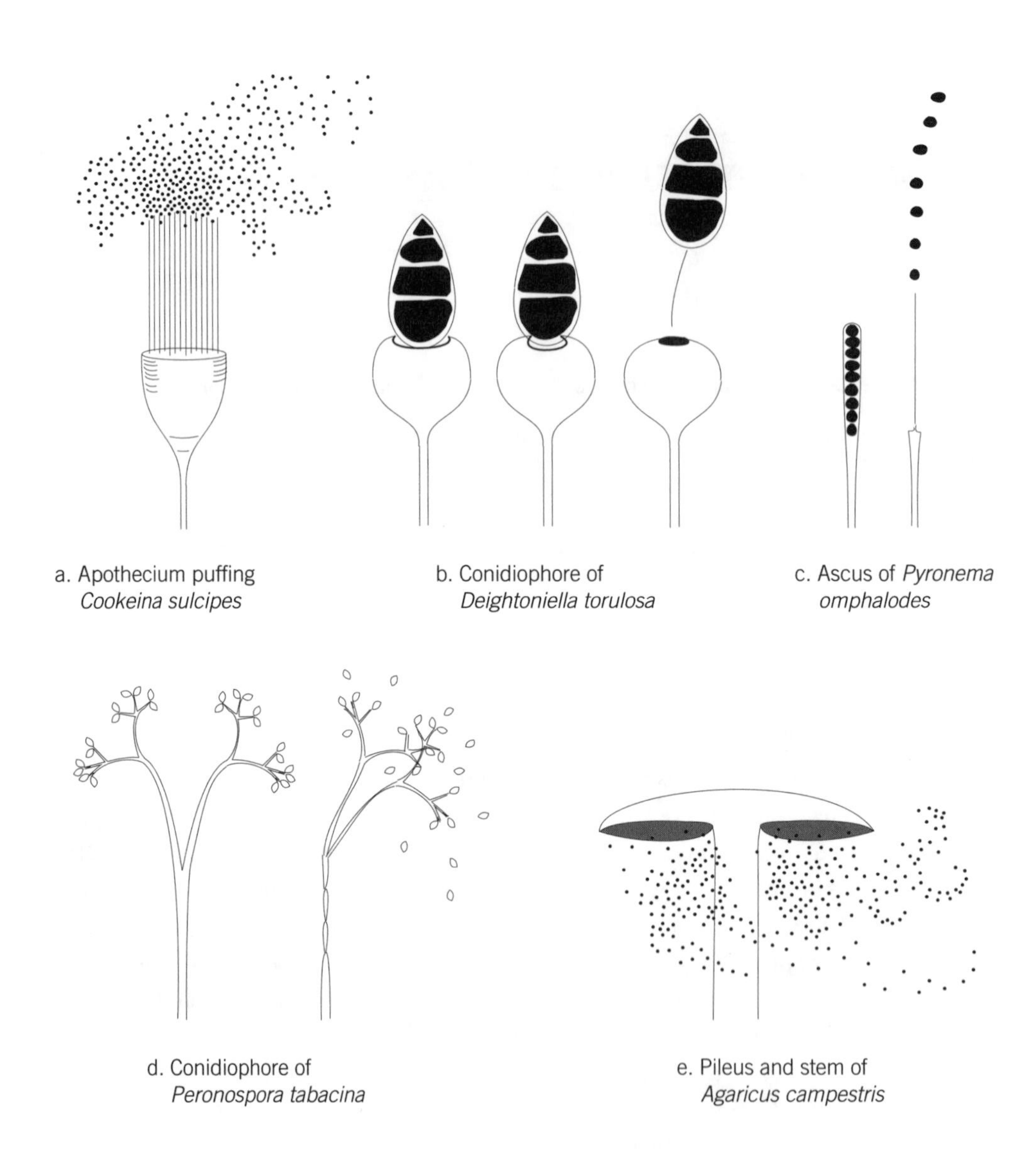

Figure 4.5. Liberation mechanisms of fungi.

Active takeoff of fungal spores usually involves either ejection by turgid living cells or is associated with drying. It generally occurs under conditions when mechanical or thermal turbulence is conducive to aerial transport. For example, in the cup-fungi (a), puffing is usually triggered by a gust of wind or by sudden illumination and heating by a shaft of sunlight. This fungi can produce 215,000 asci per sq. cm which can liberate more than 1.7 million ascospores (Kendrick 1992). In *Deightoniella torulosa* (b), discharge of conidium occurs due to rupture following evaporation of water from the terminal cell of the condiophore. Osmotically induced hydrostatic pressure ruptures the apex of the ascus of ascomycetes (c) causing their contents to squirt out into the air above. The loosely attached spores of tobacco blue mold (d) are thrown into the air when drying causes the conidiophore to twirl, while mushrooms (e) can drop a half million spores per minute from their gills into the airflow below (Ingold 1967). Figure adapted from Ingold 1967.

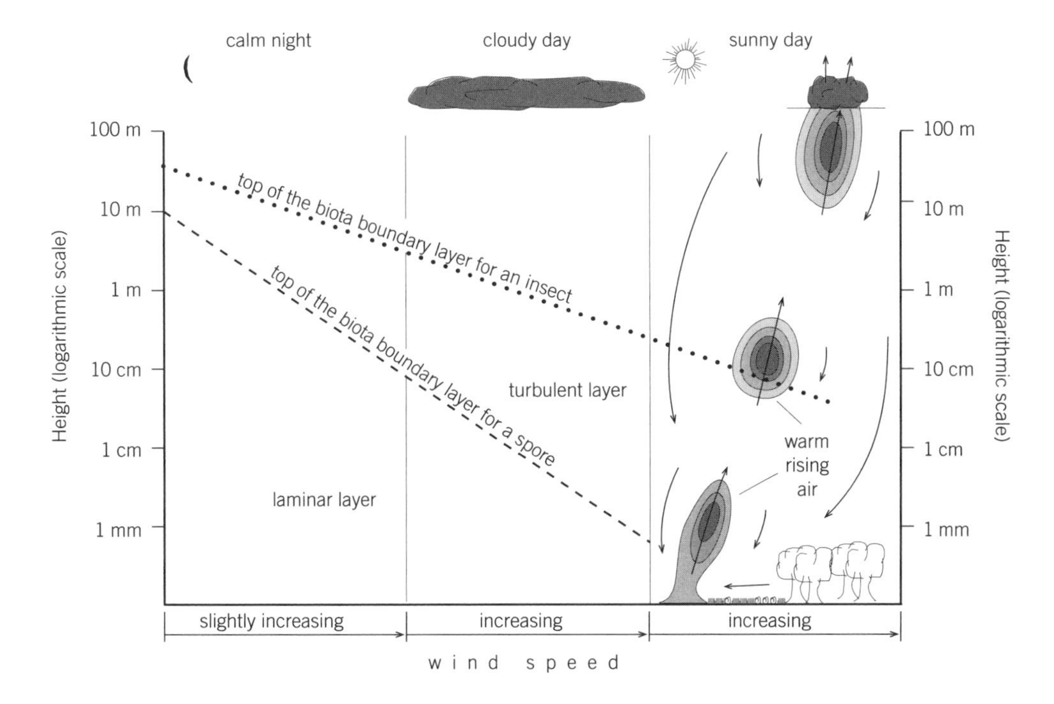

Figure 4.6. Variations in biota boundary layer thickness for spores and insects during different atmospheric conditions.

Approximate height of BBL for a hypothetical spore and insect during calm night, cloudy day, and sunny day conditions. Short dashes represent the top of the laminar boundary layer and thus the BBL for spores. It is unlikely that many spores released during calm nights or calm, cloudy days can ascend into the turbulent PBL. In contrast, those released during warm sunny days need only cross a very thin BBL near the leaf to be entrained in turbulent eddies. Dots represent the BBL for a hypothetical insect. Many relatively strong insects migrate at night because level flight is more efficient in the absence of convection. Weak-flying insects often fly during warm sunny days when they can be lifted high in the atmosphere by thermals. Figure adapted from Gregory 1973.

to the lowest 1 to 2 km of the atmosphere, which roughly corresponds to the depth of the PBL or mixed layer. When winds are relatively strong in the PBL, transport in this stage is primarily horizontal, and can result in long-distance aerial movement of biota. Because winds often vary substantially with height in the PBL, the direction and distance of movement depends on the altitude at which biota flow (Figure 4.7). Layered or stratified structures in the PBL, such as waves, jets, and fronts, can help to concentrate actively flying organisms in the atmosphere and are effective in transporting biota long distances. However, larger scale movements of air masses can be equally important for aerial flow of biota. In contrast, during periods of high pressure dominance and calm winds, movements of biota during the horizontal transport stage can be primarily vertical. As discussed above, organisms can be transferred by convection currents high into the atmosphere and returned by subsiding airflows to the same landscape.

Figure 4.7. Back trajectories for aerial collections of aphids over central Illinois.

The direction and length of 24-hour back trajectories calculated for aerial collections of aphids over Illinois change with altitude throughout the PBL due to variations in wind speed and direction. The trajectory endpoints shift westward with increasing height from the surface to the 800 mb (1900 m) level. The surface backtrack ends in east central Missouri, the 500 m trajectory in southwestern Missouri, the 1100 m in northeast Oklahoma and the highest trajectory ends in southeast Kansas. A low-level jet was present at about 1000 m during the night. Figure adapted from Scott and Achtemeier 1987.

Descent and landing.

The final stage of biota flow involves descent through the BBL and landing. The large number of processes that cause deposition of passively transported organisms (e.g., sedimentation, impaction, turbulent deposition, electrostatic deposition, and washing out by precipitation) have been extensively studied in wind tunnels. The results of these studies are typically generalized for biota (particles) of a given size and shape, irrespective of their taxa. Vegetation is generally an effective filter for pollen, spores, seeds, and the other small biota carried by airflows through the canopy. Although there are biological adaptations that influence the deposition of passively transported biota, these adaptations are less numerous than those that influence takeoff of these organisms. In contrast, very little is known about the descent and landing stages of actively transported biota. Certainly landing must occur at a time when and location where their biota boundary layer exists, so as to avoid physical injury upon impact. Once these organisms descend into their BBL, they can again exert a large degree of control over their destination. Actively flying biota may arrive in unsuitable habitats, and they may or may not take off again.

Symbolic representation of the aerobiology process model.

In the aerobiology process model shown earlier (Figure 4.2), we grouped the diverse array of factors which impact organisms that move within the atmosphere into four components: (1) the life history of organisms, (2) the current vitality of organisms, (3) the environment and ecology of their terrestrial ecosystems, and (4) the airscapes that organisms encounter during movement. The distinction between the first two components primarily relates to the time scales over which they vary. The life history of an organism includes intrinsic factors such as size, weight, morphology, physiology, and the genetic underpinnings that are in large part common to a species. These are the factors that comprise the sequence of stages and processes in the lives of organisms and are relatively invariant over short time scales (<1 life span) in a population. During movement in the atmosphere, these factors determine the capacity of organisms for passive or active transport. Vitality relates to the current condition or health of individuals within the population. It is highly variable both in time and space and potentially can be manipulated by humans for their own benefit. Vitality combines with the life history component to determine an organism's ability to live, grow, reproduce, and move. When the organisms are in their source and destination areas, their vitality is governed by the extrinsic ecology and terrestrial environment, which are highly interrelated. The extrinsic ecology includes resources for nutrition, reproduction, and protection, predators, parasites and pathogens, and mutualists and commensals. The terrestrial environment is composed of the soil, water, biota, and air near the Earth's surface in which an organism lives. Atmospheric state variables (climatic elements) influence both the terrestrial environment and extrinsic ecology. Consequently, they influence the vitality of organisms in the source and destination areas. The vitality of an individual after preconditioning depends on its vitality during the previous aerobiology model stage(s). While the organisms are airborne, their vitality (including survival) also is directly affected by atmospheric state variables. During the takeoff/ascent, horizontal transport, and descent/landing stages, vitality and life history components determine the aptitude (inclination and capacity) of the organism for movement. During preconditioning and impact stages, the vitality and life history components combine to control the growth, development, and reproduction in the life cycles of organisms. Because the time scales associated with takeoff/ascent and descent/landing are short (usually seconds to minutes), the vitality of organisms during these stages is essentially the same as it was at the end of the previous stage in the aerobiology process model, barring physical injury due to collisions.

In addition to influencing the vitality of organisms in the source and destination areas, the terrestrial environment affects the takeoff/ascent and descent/landing stages. Topography and plant architecture interact with atmospheric motion systems to govern the characteristics of the turbulent eddies that help organisms move through air near the ground. The terrestrial environment also influences airscapes at larger scales, often causing local winds and circulations that influence biota during horizontal transport.

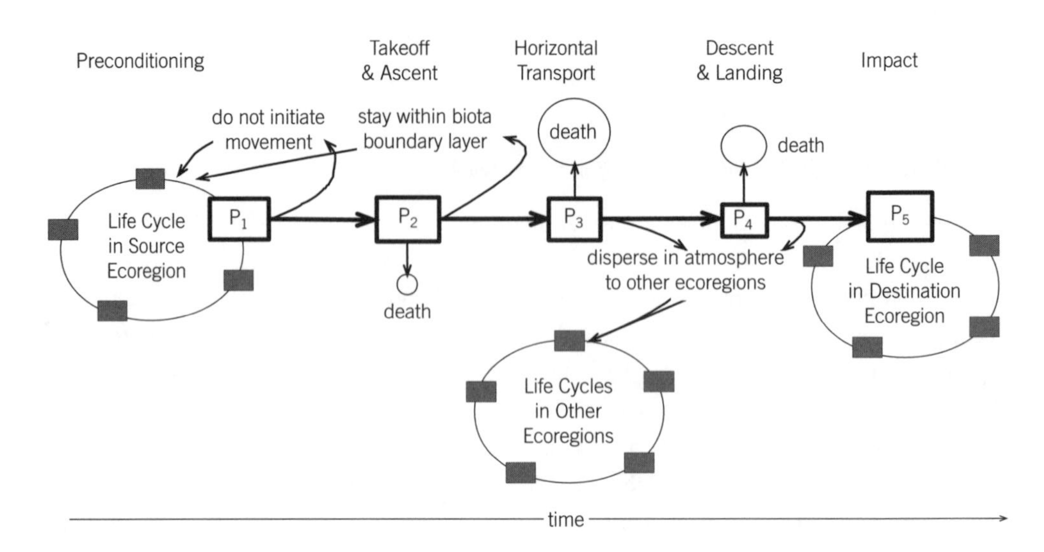

Figure 4.8. Symbolic representation of general aerobiology process model for an age-specific population cohort.

Relationships among components of the general aerobiology process model are shown in Figure 4.2. Here, P_0, P_1, P_2, P_3, P_4, and P_5 represent the age-specific cohort of population at birth, at the end of the preconditioning stage, and during the takeoff/ascent, horizontal transport, descent/landing, and impact stages, respectively. $P_1 = \alpha_1 P_0$, $P_2 = \alpha_2\beta_2 P_1$, $P_3 = \alpha_3\beta_3 P_2$, $P_4 = \alpha_4\beta_4 P_3$, and $P_5 = \alpha_5\beta_5 P_4$. Typically, $P_1 > P_2 > P_3 > P_4$, while P_5 will be highly variable and can range from 0 to a value much greater than P_1. α_1, α_2, α_3, α_4, and α_5 = survivorship factors for age-specific cohort during preconditioning, takeoff/ascent, horizontal transport, descent/landing, and impact, respectively. α_1 and α_5 are governed by the extrinsic ecology and terrestrial environment in the source and destination landscapes. For many aerobiota, $\alpha_2 > \alpha_4 > \alpha_3$. α_2 may be approximately 1.0 when atmospheric conditions are favorable because the time scales associated with takeoff/ascent typically range from seconds to minutes. Ultraviolet radiation and other atmospheric state variables greatly impact survivorship during horizontal translation (α_3). Other factors not shown in Figure 4.2 may also influence the survivorship of specific taxa in the atmosphere. For example, predation by birds can reduce the survival rates of insects and arachnids during takeoff/ascent, horizontal transport, and descent/landing. Aerobiota may crash during ascent/landing reducing α_4 below 1.0, even though the time scales associated with this process may be extremely short.

β_2 = the proportion of the age-specific cohort that take off from the source area. β_3 = the proportion of these aerobiota that ascend into the atmosphere above their BBL. Many organisms stay within their BBL and thus land and continue their life cycle within their source area. β_4 = the proportion of the aerobiota in the planetary boundary layer that travel to the destination area of interest. This is usually a very small proportion of the organisms in the PBL because turbulence acts to diffuse biota within the atmosphere and usually winds do not blow consistently from a source to a specific destination area. β_5 = the proportion of the organisms that remain (do not reinitiate aerial movement) in the destination area after landing. These proportionalities are all highly variable from place to place and through time. β_3 and β_4 may range from 0 to 1 for passively transported and weak-flying organisms, depending entirely on atmospheric conditions. β_2 and β_3 depend on the population dynamics in the source and destination ecoregions as well as the weather.

The airscape is composed of atmospheric state variables and motion systems. The important state variables include air temperature, pressure, moisture, and solar radiation. Extended exposure to ultraviolet radiation and cold temperatures in the atmosphere is lethal to many biota. Thresholds of these state variables also have important impacts on the takeoff of aerobiota. Atmospheric motion systems range in size from tiny eddies to global

Table 4.5. Conceptual frameworks for studying the aerial movement of biota.

FRAMEWORK	PURPOSE
Aerobiology pathway (process) model, Benninghoff and Edmonds 1979	Focus research on the environmental factors that affect biota in each of the stages in the aerobiology process and the coupling of the states in order to predict aerial movement of biota.
Flight boundary layer concept, Taylor 1958, 1974	Provide a context for understanding the interactive control of biological and atmospheric factors over insect flight by dividing the atmosphere into a zone in which an insect can control its direction and distance of flight and a zone above, where it cannot.
Insect migration system model, Drake et al. 1995	Provide a holistic approach to understanding the diverse components that influence the life histories of migratory insects and how they interact to function and evolve within highly variable environments.
Taxonomy of movement behaviors, Dingle 1996	Provide a template for classifying movement behaviors for a wide range of biota to facilitate the understanding of commonalties in the life histories and the evolution of organisms that migrate.
Airscape perspective on the aerobiology process model	Focus research on predicting changes in location of aerobiota among habitats and general principles that govern their movements because they occur within a common medium, the atmosphere.

wind belts. The structure of motion systems, the influence of the landscape on these structures, and their effect on aerobiota during the takeoff/ascent, horizontal transport, and descent/landing stages are explored in the following chapters. Turbulence and wind influence the direction and distance that biota move in the atmosphere. In addition, these motion system characteristics govern the dispersion or dilution within the atmosphere of organisms from a single habitat.

A symbolic representation of the aerobiology process model for an age-specific cohort of a population that moves in the atmosphere from a source to a destination area is shown in Figure 4.8. The model is built on the framework proposed by Aylor (1986), and presented in Figure 2.5, for predicting the long-distance transport of fungal spores. In this model, P represents the number of individuals in the age-specific cohort during each stage. The number of individuals in the cohort during a stage is given as the product of a survivorship factor (α) for that stage and the P from the previous stage. The survivorship factor for each stage is represented as a function of the life history of the species and the current vitality of the individuals that comprise the age-specific cohort. The extrinsic ecology, terrestrial environment, and atmospheric state variables associated with the air- and landscapes in the source area, encountered during aerial movement, and at the destination have important effects on survivorship (see Figure 4.2).

The βs in the model represent the proportion of survivors (αPs) that proceed to the next stage in the process model. Specifically, β_2 represents the proportion of the age-specific cohort that take off from the source area. Frequently some individuals in a cohort continue their life cycles in the source area. This factor is strongly influenced by life history, current vitality, and atmospheric state variables. β_3 represents the proportion of those biota that, after commencing aerial movement, ascend into the atmosphere above their BBL. Many biota disperse locally within their BBL and thus continue their life cycle within the source area. Life history factors, vitality, topography, plant architecture, and atmospheric motion systems usually interact to determine β_3. β_4 represents the proportion of the aerobiota in the PBL that descend into their BBL and land at the destination of interest. Typically only a very small fraction of the large number of organisms (with the exception of birds and some strong-flying insects) that ascend into the PBL land in a particular destination area. This is because turbulence acts to diffuse biota throughout the atmosphere during horizontal transport. Thus the likelihood that biota from one source area encounter winds with the direction and speed to assist them to a specific destination area is quite small. β_5 represents the proportion of the organisms that continue their life cycle in the destination area after landing. Some organisms may resume long-distance movement to other areas.

Summary.

In this chapter we have presented some useful conceptual models for studying the movement of biota of in the atmosphere. Each of these models was created for a distinct purpose. Consequently, each provides a different and important perspective from which to pursue knowledge and understanding in aerobiology. The purposes of these conceptual frameworks and our airscape perspective on the aerobiology process model are summarized in Table 4.5.

In our presentation of the aerobiology process model (Figures 4.2 and 4.8), we have separated the factors that determine aptitude for aerial movement from the direct influences of the air- and landscapes. The factors that determine the aptitude for aerial movement (grouped into the life history and current vitality components) are usually specific to an organism or taxa. Knowledge of how genotypic and phenotypic elements as well as ecological and environmental factors influence the aptitude of biota for aerial movement is critical to a holistic understanding of aerobiology and are explored in many of the references listed in Table 4.3. The distinction between life history and current vitality is based on the time scale over which these factors typically vary. Potentially, many of the factors that determine the vitality of organisms can be manipulated within the context of ecosystem management programs. The atmospheric motion systems that comprise the airscape are described in the next five chapters with an emphasis on the important influence of the landscape on these motion systems. In Chapters 10–13 we provide examples of how researchers might utilize the airscape perspective in the design of programs to measure and study atmospheric movements of organisms among habitats.

CHAPTER 5

Airscapes: Important Factors and Relationships that Influence Movement of Biota in the Atmosphere

Scale in aerobiology.

Knowledge of the atmospheric medium is paramount to understanding the aerial movement of biota. The airscapes that organisms encounter can be separated into two interrelated components: weather and atmospheric motion systems. Weather is the state of the atmosphere as it affects life. Its elements include temperature, humidity, solar radiation, precipitation, and wind. Atmospheric motion systems result from weather. They are the circulations in which organisms move through the atmosphere. Figure 5.1 shows some of the motion systems that commonly assist aerial movements of biota, plotted on time and space axes. The temporal and spatial scales that are relevant for organisms that move in the air among and within ecosystems range from fractions of seconds to weeks and from leaf parts to continents. The concept of sliding or nested scales (Figure 5.2) is useful for connecting atmospheric motion systems and landscape units of different duration and size (Wellington and Trimble 1984).

Traditionally, scientists have had a tendency to compartmentalize events and places, using scale as an artificial divider to group processes and patterns into separate packages. However, in aerobiology research, an indifference to the reciprocal connections between large- and small-scale atmospheric events invites failure. What happens on a leaf is always a direct result of circumstances in the PBL, and to some small degree, will influence subsequent motion systems aloft. Instead of using scale to divide processes and places, aerobiologists must use it to connect them. The nested space and time scale concept emphasizes linkages and the continuity in the aerial movement of organisms between turbulent eddies and extratropical cyclones, between small patches and vast ecoregions, and between landscapes and airscapes.

For example, as a summer convective storm (thunderstorm) traverses a landscape over a period of a few hours it generally produces billions of tiny raindrop-generated motions. Each splash lasts for a fraction of a second and may assist in the liberation of spores and other small aerobiota. The storm's inflows and outflows, which often can be detected well before and after the storm passes, will influence the structure of the turbulent eddies within and above the plant canopy. These eddies in turn may help organisms ascend through their

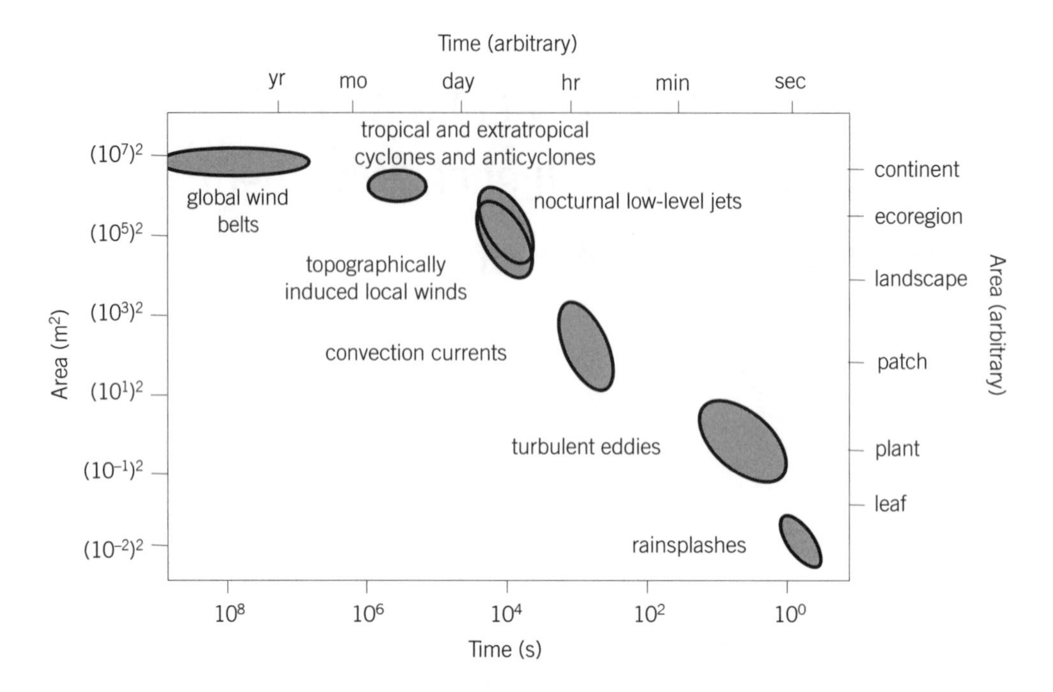

Figure 5.1. Spatial and temporal scales associated with a few of the atmospheric motion systems that commonly influence the aerial movement of biota.

The spatial and temporal scales of atmospheric motion systems are cross-dependent. For example, the strong hurricanes and extratropical cyclones that impact large regions can last for a week or longer while raindrop-generated motions affect the air above parts of leaves for a few seconds at most. In contrast, aerobiological events may be mismatched with respect to temporal and spatial scales. Events of relatively small scales such as the release of pathogenic spores and pollens from a single or group of fields in a landscape over a few hours, may impact large ecoregions downwind over subsequent days and weeks as the aerobiota land. Figure adapted from Levizzani et al. 1998.

biota boundary layers into the PBL above. The eddies may coalesce into strong updrafts and quickly lift biota high into the troposphere where they may be transported far downwind within the storm cell over subsequent hours. Finally, organisms may be quickly deposited in high concentrations on the landscape below by downdrafts and precipitation from the convective storm.

The spatial and temporal scales of atmospheric motion systems are also cross-dependent. Small motions tend to be fleeting while atmospheric systems that influence large areas tend to last from days to weeks. On the other hand, the spatial and temporal scales of aerobiological events are coupled with the rate at which biota move in the atmosphere. Consequently, the release of pathogenic spores, or the takeoff of insect pests from a single field over a period of a few hours, may quickly (1–2 days) impact large landscapes (1000s of km^2) that are downwind. In many respects, understanding and predicting aerobiological events that are mismatched with respect to temporal and spatial scales (events of short duration that impact large areas) are among the most intellectually challenging problems facing society. This is because an extensive amount of forethought, resources, and coordination is

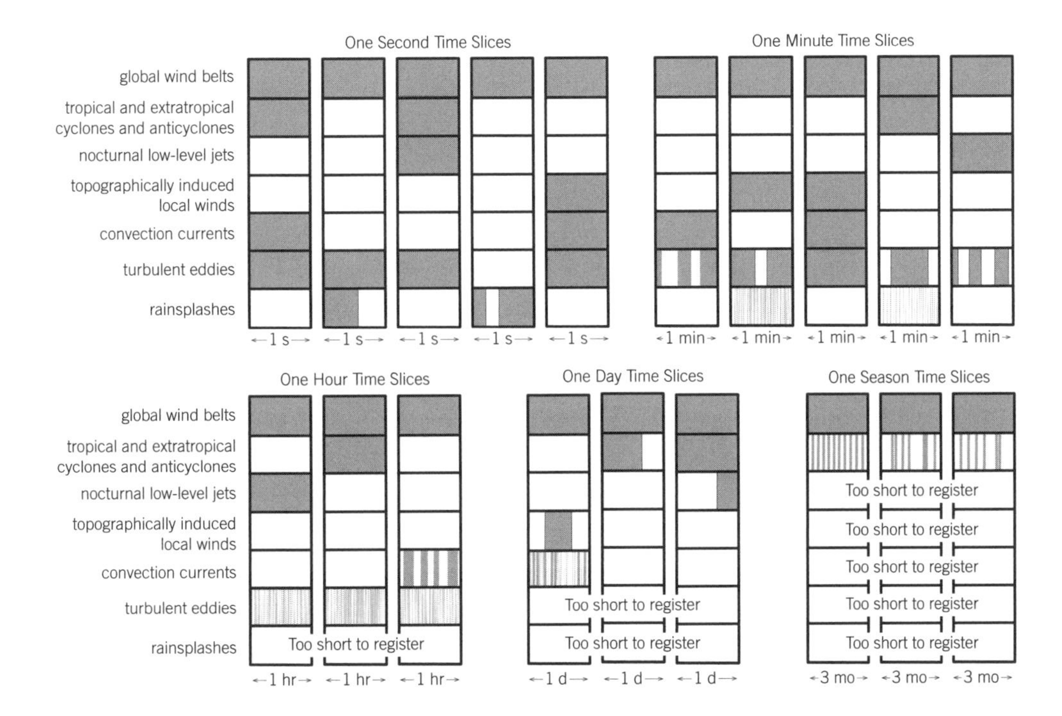

Figure 5.2. Hypothetical "slices of time" ranging from seconds to seasons depicting some combinations of atmospheric motion systems that can occur simultaneously at different scales and interact to influence the aerial movement of biota.

Many atmospheric motion systems are nested within larger scale systems. Over any specified time interval (e.g., seconds, minutes, hours, days, and seasons), motion systems that occur simultaneously but at different scales may interact to influence the movement of biota within the atmosphere. For example, the first 1-s time slice (top row, left) could represent a situation where turbulent eddies combine to form convection currents in a warm airflow associated with an extratropical cyclone that is traveling eastward in midlatitude westerlies. The nested scale concept emphasizes linkages and the continuity in the aerial movement of organisms between arial units and motion systems of different spatial and temporal scales and between landscapes and airscapes. Figure adapted from Levizzani et al. 1998.

required to quickly and simultaneously collect the biological data from many widely dispersed locations that are necessary to achieve an understanding of, and thus the capability to accurately predict, aerobiological events with mismatched scales.

Primer on some basics of atmospheric motion.

Organisms that take off from a single landscape patch tend to be dispersed by atmospheric motion systems because wind and turbulence combine to cause diffusion of biota and other atmospheric constituents throughout the lower troposphere. Variations in the speed and direction of airflow near the Earth's surface generally appears extremely complex. We commonly use wind, or more technically, the mean wind, to denote the horizontal component of air in motion relative to the surface of the Earth that can be responsible for very rapid

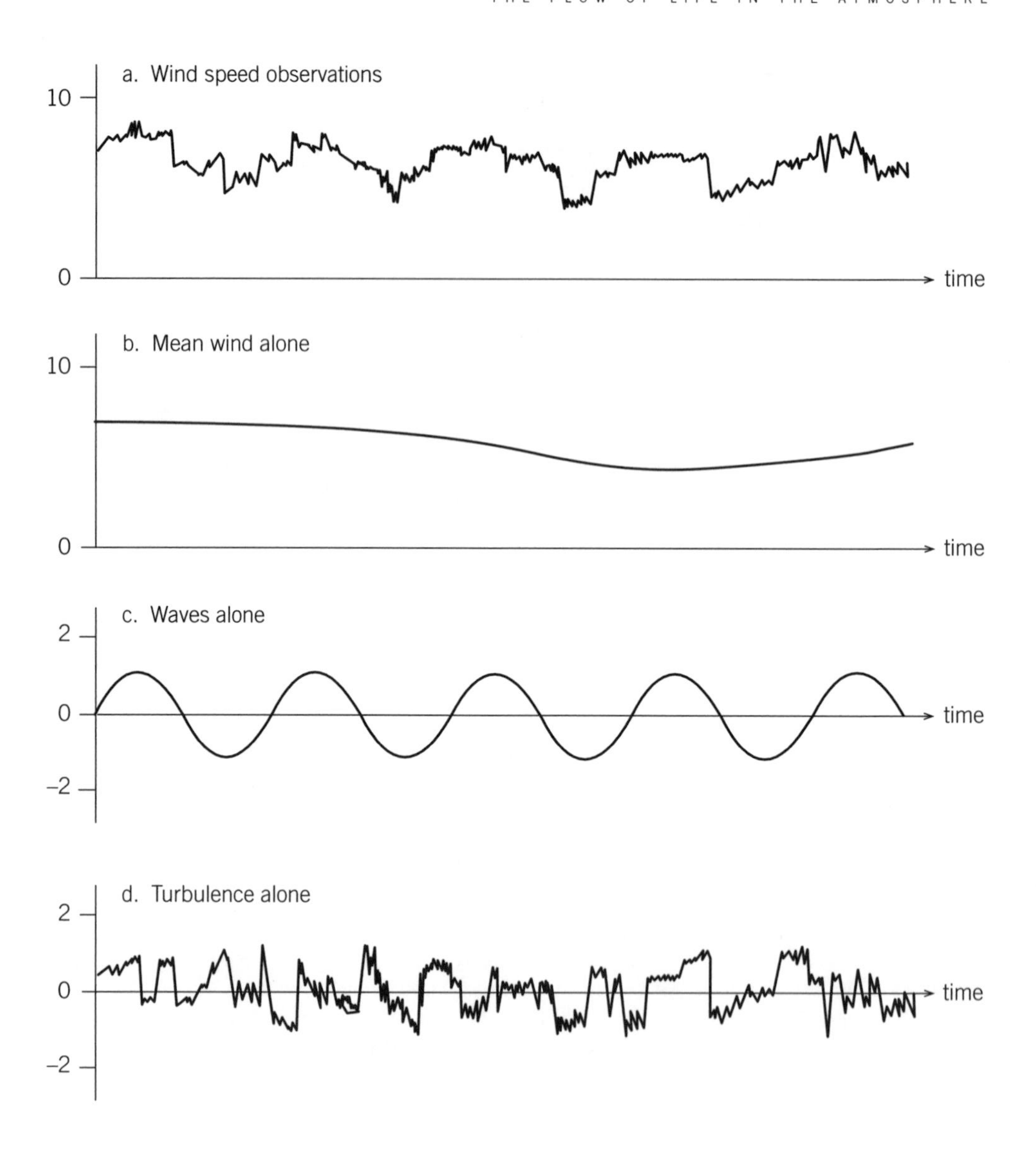

Figure 5.3. Idealization of mean wind, waves, and turbulence in airflow.

A time series of wind speed observations generally appears very complex (a). To better understand its variability and the scales of the motion systems involved, the time series is decomposed into three components: the mean wind, waves, and turbulence. The mean wind is the speed of airflow in the downwind direction (b). It is usually averaged over periods of 10 or more minutes at a location far from obstructions and for a height where the surface roughness elements (e.g., vegetation, rocks, and buildings) have only a moderate effect on the wind. The speed and direction of the airflow associated with waves (c) and turbulence (d) vary tremendously over short temporal and spatial scales. They are generally considered to be superimposed on the mean wind. In the lower atmosphere, waves and turbulence are governed by the size and shape of roughness elements and the speed and direction of the wind in the PBL above. Consequently, measurements of wind speed and direction that are commonly available may not provide sufficient information to understand the influence of the atmosphere on the movement behavior of many organisms. They are not very good indicators of the atmospheric motions that assist the aerial movements of organisms near vegetation and other surface roughness elements within their biota boundary layer. Figure adapted from Stull 1989.

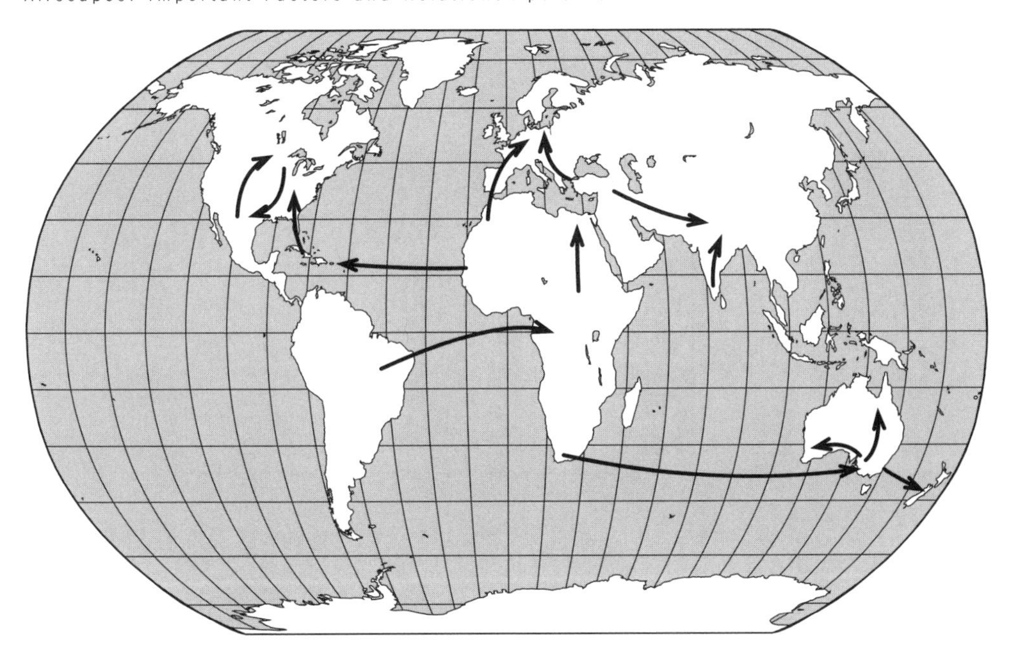

Figure 5.4. Long-distance aerial movement of rust pathogens.

The "global long-distance dispersal highways" for spores (arrows). Plant pathogens produce enormous numbers of spores that are transported passively by winds. Tropical and midlatitude cyclones and anticyclones are usually responsible for long-distance atmospheric movements of these aerobiota. Often they land on nontarget sites, mostly in uncongenial environments and only occasionally on acceptable hosts. Although the seemingly random dispersal of these organisms may appear inefficient, this wastefulness is compensated by prolific production rates. For example, the tobacco blue mold pathogen can produce 60,000 spores per sq cm of infected leaf tissue (Aylor 1986). Nagarajan and Singh (1990) note that each of the global long-distance dispersal highways operate during specific months of the year and that most of the routes are used simultaneously by rust diseases, insects, and birds. They suggest that critical analysis of these "biologically active" aerial pathways is prerequisite to effective management of plant pathogens that are dispersed over long distances. Figure adapted by Nagarajan and Singh 1990.

horizontal transport or advection. Perturbations within the airflow are considered waves if they are periodic and turbulence if they are irregular (Figure 5.3). For most aerobiota, including viruses, fungi, bacteria, soil nematodes, pollen, seeds, and wingless arthropods, their concentration in the atmosphere is primarily controlled by the number of organisms that take off from a source in the landscape and the wind and turbulence within the air layers they enter. Often only a small fraction of the passively transported biota that are liberated from a plant are able to escape the plant canopy into the overlying atmosphere. In addition, the concentration of passively transported biota in the air generally decreases very rapidly over short distances downwind from source areas. Nevertheless, when huge numbers of organisms are released into the air and winds are strong and persistent, even passively transported organisms may be carried 1000s of km in large enough concentrations to be a pest (Figure 5.4).

Pedgley (1990) notes four general ways in which the aerial concentration of biota can increase. It is important to differentiate between increases in the density of insects on the

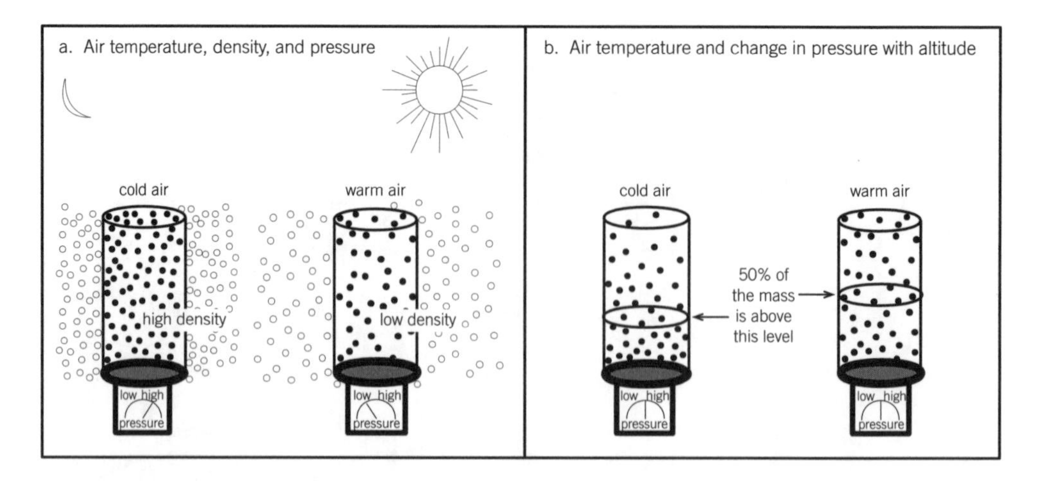

Figure 5.5. Relationships between air temperature and density, and the change in pressure with altitude in the atmosphere.

Temperature is a measure of the average kinetic energy of molecules in a gas. The most important source of kinetic energy in the Earth-atmospheric system is the sun. In (a), as the kinetic energy of unconfined air molecules increases, the volume they occupy expands and their density decreases. Thus the pressure exerted by a column of warm air is less than that exerted by a column of cold air with the same dimensions. If two columns of air have the same amount of mass (b), air density and thus pressure will decrease more rapidly with height in the cold air column than in the warm air column. At the Earth's surface the pressure in the two columns will be equal, but at any altitude above the surface, the pressure (exerted by the mass above the measurement level) in the warmer air column will be greater than the pressure in the cold column at the same height.

ground due to deposition or landing processes and increases in their aerial concentrations. The former often occurs when intense precipitation washes biota out of the atmosphere or where wind speeds are reduced to the lee of obstacles on the ground. In contrast, the numbers of biota in a given volume of air (volume density) can increase: (1) as the air parcel traverses habitats that serve as sources of aerobiota, (2) due to changes in biological and environmental conditions that stimulate takeoff or inhibit landing, (3) if light precipitation washes biota downward in the PBL but biota remain airborne in the surface boundary layer, and 4) due to convergent airflows that help enhance the concentration of organisms in the atmosphere when biota avoid being carried aloft. Atmospheric convergence occurs in the PBL when airflows from different directions merge and where there is a decrease in the wind speed over distance. The "excess" air that "piles up" whenever the winds near the Earth's surface meet and/or slow down is forced to ascend. Some organisms resist being carried aloft, perhaps due to the temperature and pressure decreases they experience when carried upward and/or the reduction of visual stimuli with height. In general, winds in the PBL blow from areas of divergence (where air is usually replaced from above the stratum of interest) to areas of convergence. Thus when convergence zones retain their identity for relatively long periods of time and organisms resist upward movement, the concentration of biota in the converging airstreams can increase substantially (for examples see Rainey 1951, Schaefer 1976, Rainey 1989, Pedgley 1990).

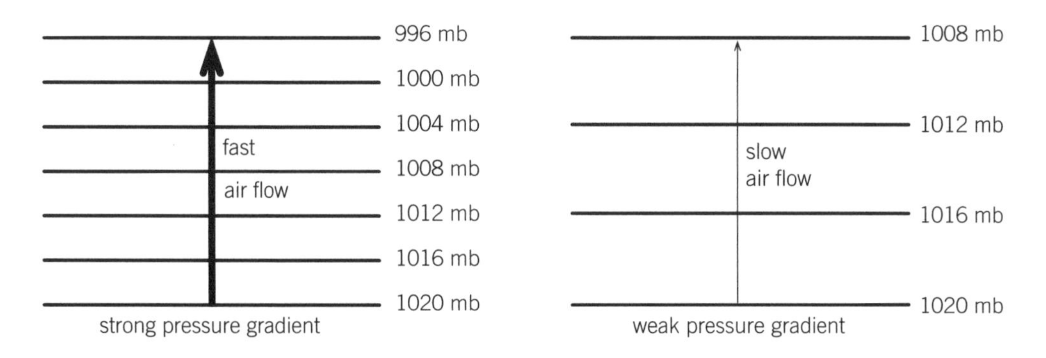

Figure 5.6. Relationship between the air pressure gradient and wind direction and speed.

The pressure gradient is the force that motivates movement. It is the change in pressure over a unit distance. Atmospheric pressure at the surface of the Earth is mapped by constructing isobars which are lines joining stations recording equal pressure. Air will move from areas of high pressure to areas of low pressure down the pressure gradient, and winds will blow at right angles to isobars if no other controlling factors are considered. There is a direct relationship between the strength of the pressure gradient and the speed of airflow.

Atmospheric motions result from the interrelationship among temperature, density, and pressure of air. Although the atmosphere is a combination of gases, it conforms to the ideal gas law. Consequently, any change in one of these properties will result in a change in one or both of the others. The need to add air (increase mass) to a bicycle tire with the onset of cold weather (decrease in temperature) to maintain a rigid tire (constant pressure) illustrates this relationship. Continuous changes in the spatial and temporal distribution of the amount of solar radiation absorbed in the Earth-atmospheric system result in variations of temperature across the Earth's surface and throughout the atmosphere and thus provide the basic inducement for movement. The flows of air in the atmosphere and of water in the oceans provide for the redistribution of this energy from areas where it is highly concentrated (warm) to those with less energy (cold). These motions occur in both horizontal and vertical directions within the atmosphere.

Atmospheric pressure is the downwardly directed force on an area that results from the weight of the overlying air. It is directly related to the mass or density (mass per volume) of the air, and thus changes in mass in the column of air above a site result in changes in atmospheric pressure (Figure 5.5).Variation in pressure over distance is called the pressure gradient. The force that results from horizontal and vertical pressure gradients plays the dominant role in atmospheric motion because it is the factor that initiates the motion of air and to a large extent governs its direction and speed. If the pressure gradient was the only force acting on the flow, air would move down the pressure gradient from high to low pressure and thus perpendicular to isobars, lines drawn on weather maps between points with equal atmospheric pressure. In addition, the speed of movement would be proportional to the magnitude of the pressure gradient (Figure 5.6).

However, once air has been set in motion, other factors influence its direction and speed of flow. Air moving horizontally undergoes an apparent deflection due to the rotation of the

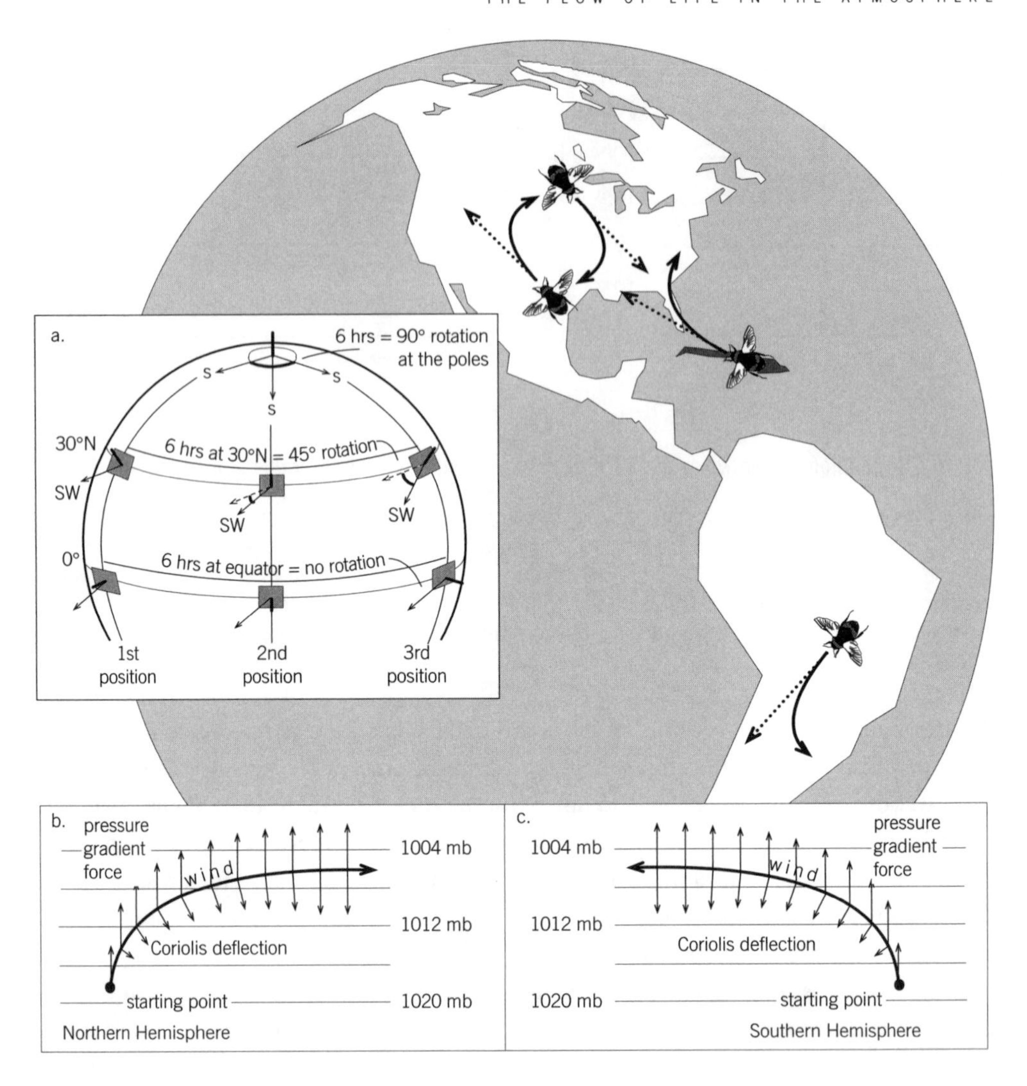

Figure 5.7. Effect of the Earth's rotation on the wind.

(a) A person standing at either pole will spin 360° in 24 hours. A person standing at 30° N or S latitude will experience ½ of a rotation during the same period. At the equator an upright person will experience no rotation during the diel period.

The Coriolis deflection is due to the rotation of the Earth. The magnitude of the deflection increases from the equator toward the poles (a), because it is related to the angular velocity (the rate of change of angular position over time) of a rotating plane normal to the Earth's surface. Note that the dashed arrows at 30° N latitude in the 2nd and 3rd positions on the rotating Earth represent (and thus are parallel to) the solid arrow extending from the 1st position. While the angular positions of the solid arrows at the later time steps have rotated, each still points toward the SW. The Coriolis deflection is always exerted perpendicular to the direction of the airflow (b and c). Thus upper-level (geostrophic) winds that are initially motivated to move down the pressure gradient are continually turned by the Coriolis deflection until they flow perpendicular to the pressure gradient. The turning of the winds is to the right of the pressure gradient force in the Northern Hemisphere (b) and to the left in the Southern Hemisphere (c). Figure adapted from Lutgens and Tarbuck 1989.

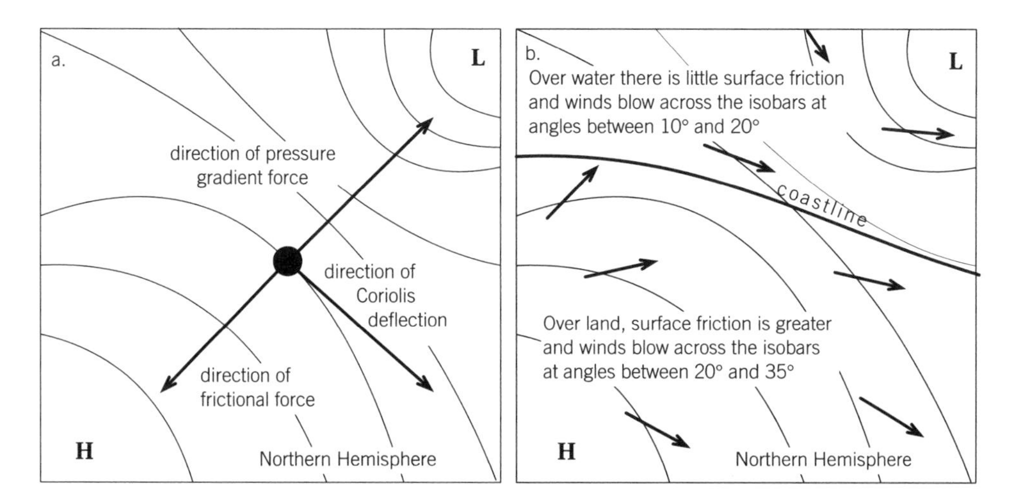

Figure 5.8. Effect of surface friction on wind speed and direction.

Friction between the airflow and the Earth's surface slows the wind, decreasing the Coriolis deflection (a). The relative influence of the pressure gradient force over the direction of flow is thus increased, and winds blow from high to low pressure at a small angle with respect to the isobars (b). This angle is related to the roughness of the underlying Earth surface and is less over water than land. Figure adapted from Hildore and Oliver 1993.

Earth. This Coriolis deflection is always directed perpendicular to the airflow and results in winds bending to the right in the Northern Hemisphere and to the left in the Southern Hemisphere. The magnitude of the deflection increases with increasing latitude from the equator toward the poles and with increasing wind speed. Above the air layers in close proximity to the Earth's surface, a balance between the pressure gradient force and the Coriolis deflection results in geostrophic flow in which the air moves perpendicular to the pressure gradient (Figure 5.7).

Friction results from the contact between the Earth's surface and the air flowing over it. Its direction is opposite to the wind and slows the airflow. Thus friction reduces the deflection due to the Coriolis effect. As a result, air near the Earth's surface generally moves from high pressure toward low pressure areas across the isobars at a small angle. Over land surfaces with high surface roughness this angle ranges between 25° and 35° with respect to the isobars and between 10° and 20° over smoother water surfaces (Figure 5.8). The lowest part of the atmosphere where the friction between physical elements of the Earth's surface (e.g., plants, human structures, and small topographic features) and the air greatly influences the speed and direction of flow is called the surface layer. Warm air rising from the Earth's surface propagates friction upward, retarding the speed of the wind in the lower atmosphere (Figure 5.9). The layer of atmosphere in which the flow is influenced by friction is called the planetary boundary layer (PBL) or friction layer. Consequently, the surface layer comprises the lowest zone within the PBL. The depth of the PBL is highly variable and can be as little as a few 100 m at night or reach 2–3 km during the day. Because the influence of friction decreases with altitude throughout the PBL, the profile of wind speed and direction can vary in a systematic manner from the ground upward to the geostrophic wind. This

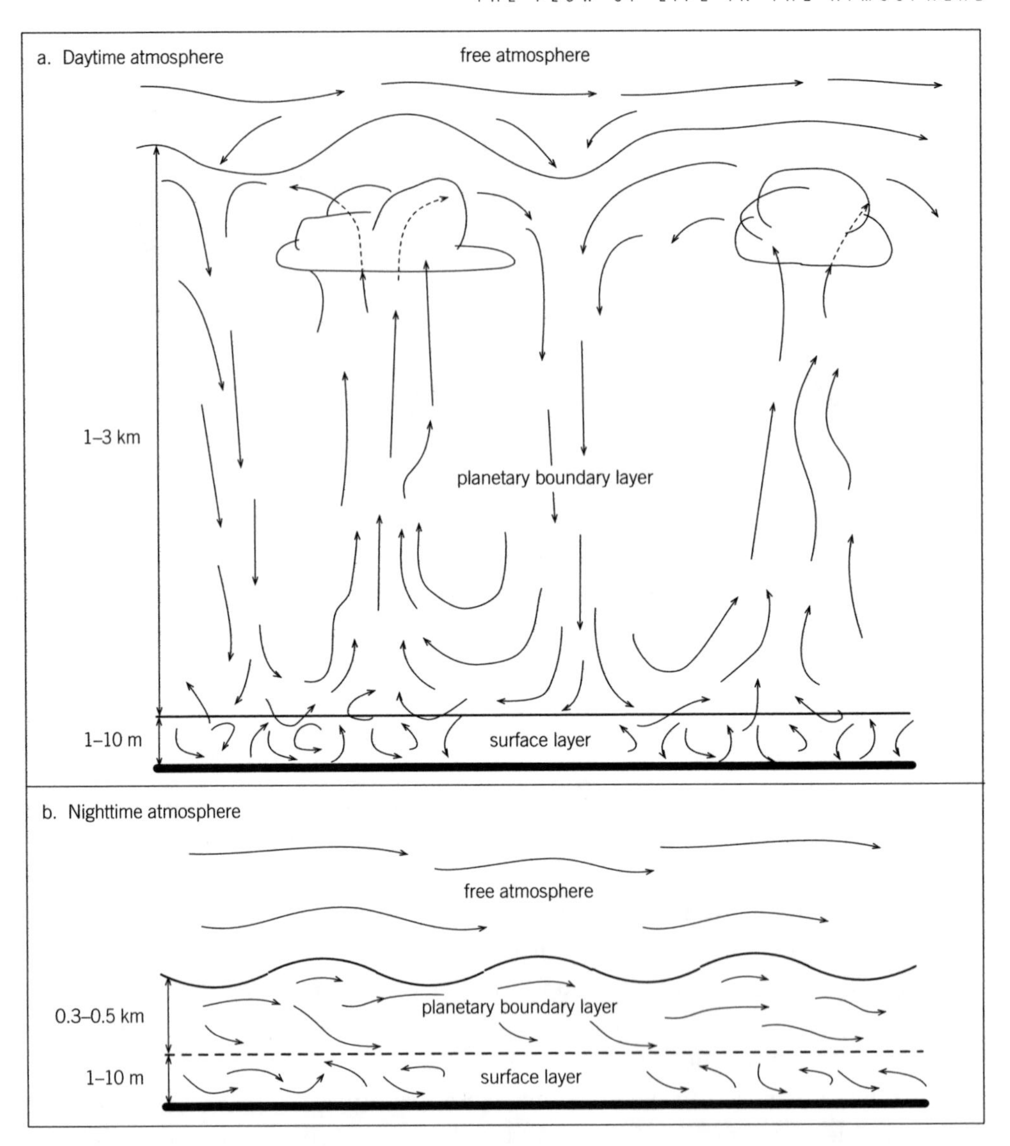

Figure 5.9. Structure of the surface and planetary boundary layers during day and night.

Schematic representation of airflows in the surface layer and planetary boundary layer (PBL) during typical daytime and nighttime conditions. Winds in the free atmosphere are geostrophic. Friction between the airflow and the Earth's surface slows the winds in the PBL. Vertical movement of warm air due to convection during the daytime propagates friction upward and extends the altitude of the PBL to between 1 and 3 km. At night, the height of the PBL decreases to as little as a few 100 meters. The surface layer is the lower portion of the PBL. Adjacent to the ground, vertical gradients of wind speed, temperature, and humidity are often very large and are strongly influenced by the physical properties of the surface. The size, shape, and configuration of the surface roughness elements strongly influence the depth of the surface layer. In an agricultural field, the surface layer generally extends above the tops of plants one or two times their height. Figure adapted from Oke 1990.

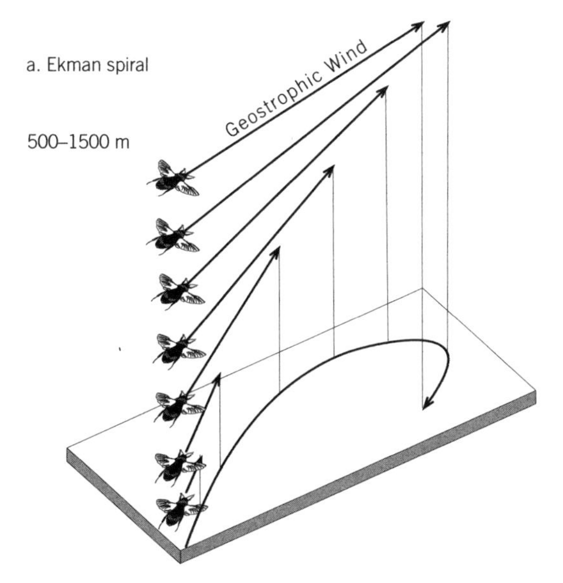

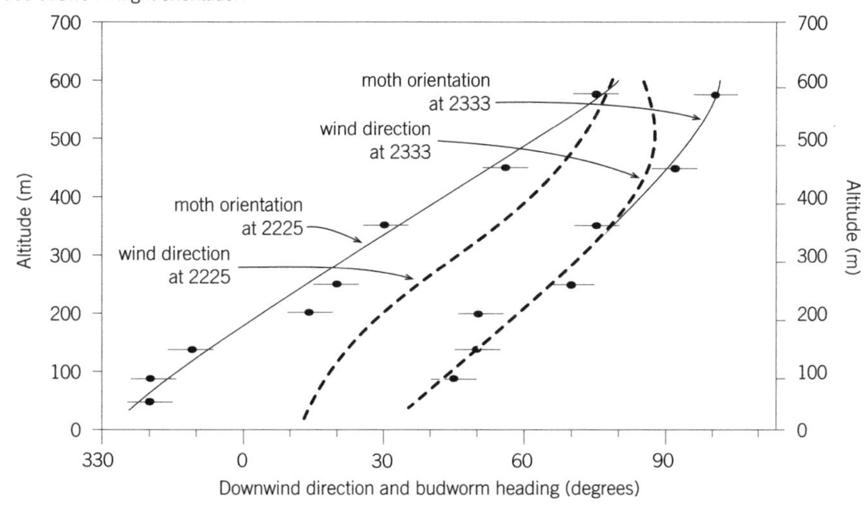

Figure 5.10 The Ekman spiral with an example of its effect on moth orientation in flight.

(a) When convection is minimal, the direction of the wind spirals to the right with altitude in the Northern Hemisphere as the effect of the frictional drag of the Earth's surface is reduced. The direction of the Ekman spiral is to the left in the Southern Hemisphere. Wind speed generally increases with altitude within the PBL. A change in the direction or speed of the wind in either the vertical or horizontal direction is called wind shear. Note how the direction and distance of the insect back trajectories shown for different altitudes in Figure 4.7 correspond to the wind shear associated with the Ekman spiral. Figure adapted from Barry and Chorley 1987.

(b) Variation in spruce budworm moth (*Choristoneura fumiferana*) orientation with altitude and wind direction measured with radar at 2225 and 2333 on 8 July 1973 over New Brunswick, Canada. Moths primarily flew downwind. Dots and horizontal dashes show mean and range of moth orientation. Wind direction veered 50° to 60° to the right over the 600 m of altitude. Wind speed also increased with altitude (not shown). Moth orientation was about 25° to the left of the downwind direction at low altitudes after takeoff with the difference decreasing with altitude. The later flying moths showed direct downwind orientation which followed the Ekman spiral to the right with altitude. Figure adapted from Schaefer 1976.

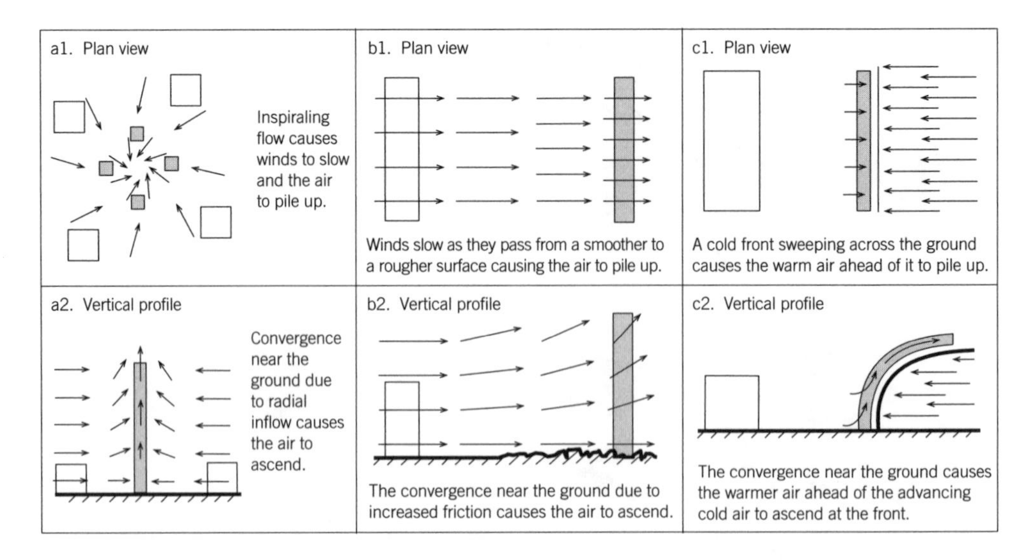

Figure 5.11. Wind convergence in the planetary boundary layer.

Wind arrows for three idealized convergent flow patterns. Plan views show the slowing and contraction in the horizontal plane of air volumes near the ground for: (a1) winds exhibiting a radial inflow pattern, (b1) winds with uniform direction that pass from a relatively smooth surface to a rougher surface, and (c1) air ahead of an advancing cold flow or front. In each case the vertical profiles (a2, b2, and c2) show that horizontal convergence (shaded volumes) near the ground produces ascending air flow aloft. Biota that resist moving upward within these flows are concentrated in the convergence zone. Figure adapted from Pedgley 1990.

turning and acceleration of the wind with height is referred to as the Ekman spiral and is most pronounced during night when thermally induced vertical motions are generally absent (Figure 5.10). The layer of geostrophic flow above the PBL is called the free atmosphere.

Vertical and horizontal motions in the atmosphere are intimately connected. For example, the transfer of heat from the Earth's surface to the overlying air decreases its density, causing it to rise. Because the troposphere is capped by an inversion layer (tropopause) where air temperature begins to increase with altitude due to absorption of solar radiation by ozone, rising air increases pressure aloft and generally results in diverging winds within the upper troposphere. This vertical movement also creates areas of relatively low pressure and thus converging horizontal winds at the Earth's surface. Similarly, air that subsides within the troposphere, because it is cold and dense, produces areas of high pressure and diverging horizontal winds at the Earth's surface, and low pressure and converging winds aloft.

Convergence and divergence of horizontal winds can also generate vertical motions in the atmosphere. Changes in either wind direction and/or speed result in converging or diverging airflows that are manifest across a wide range of spatial and temporal scales. Two examples of atmospheric convergence at large scales are: (1) when airflows in the middle and upper troposphere are turned by the Coriolis effect toward the equator and slow down and (2) where winds near the Earth's surface inspiral around low pressure centers. At smaller

Table 5.1. Recommended books and review articles on atmospheric motion systems.

WEATHER AND ATMOSPHERIC MOTION SYSTEMS (GENERAL):
Barry and Chorley 1987, Stull 1988, Lowry and Lowry 1989, Oke 1990, Geiger, Aron and Todhunter 1995.

WEATHER AND ATMOSPHERIC MOTION SYSTEMS AS THEY RELATE TO THE AERIAL MOVEMENT OF BIOTA:
Gregory 1973, Schaefer 1976, Pedgley 1982, Drake and Farrow 1988, Kerlinger 1995.

scales, convergence occurs where the roughness of the Earth's surface over which the wind is blowing increases downwind thus acting to slow the flow and where winds from different directions meet. In each case, the moving air "piles up" and is forced ascend or subside within the troposphere (Figure 5.11).

In contrast, atmospheric divergence occurs at large spatial scales aloft, when airflows are turned poleward and thus accelerate due to the increasing Coriolis deflection or where air outspirals from surface high pressure cells toward areas of lower pressure. At smaller scales, divergence occurs when the roughness of the Earth's surface, over which wind is blowing, decreases in the downwind direction, as at a coastline. Where divergence occurs aloft, it encourages uplift of air from the Earth's surface. When it occurs near the Earth's surface, subsidence of air from the middle troposphere is enhanced.

Together, the thermally driven motions that are caused by heating and cooling of air and those generated by converging and diverging airflows result in atmospheric circulations that occur over a wide range of spatial and temporal scales. Many of these motion systems are instrumental to the aerial movement of biota. In general, heating of air near the ground and converging winds near the Earth's surface enhance the upward movement of biota within the PBL (e.g., Figure 5.11). When these motions are enhanced by large-scale divergence of flow within the middle troposphere, long-distance movements of biota are likely to occur. In contrast, large-scale converging winds aloft and the associated subsiding airflows tend to be less important to bioflow. Organisms are more often washed out of the atmosphere by precipitation, driven to the Earth by smaller scale downdrafts from storms, or land of their own accord in response to changing environmental conditions (e.g., light, temperature, food, and habitat), than they are carried to the surface by large-scale subsiding airflows. Finally, diverging airflows of all scales near the Earth's surface can provide assistance to the aerial movement of biota. However, these flows are usually slow and tend to disperse organisms throughout the atmosphere.

Summary.

The airscape is composed of complex and interactive atmospheric motion systems that are manifest at a variety of scales. The following chapters provide an overview of some that are important to the aerial transport of biota. The discussion builds on an understanding of the basic factors and relationships presented above. In keeping with our perspective on temporal and spatial scales, we stress their interdependence. The influence of the landscape on the overlying airscape is also emphasized. The discussion of these systems proceeds from the lengthy and large to the fleeting and small. It is not intended to be exhaustive. However, it does include systems that span the range of temporal and spatial scales that influence aerial movements of biota. Those who are interested in expanding their understanding of atmospheric motion systems should consider reading one or more of the books listed in Table 5.1.

CHAPTER 6

Airscapes: Global Scale Motion Systems

Atmospheric circulation systems at the global scale are composed of highly complex and interrelated airflows that transport energy, momentum, and mass (including biota) in both vertical and horizontal directions. The location and strength of the components of these large-scale flows vary at many temporal scales, including diurnal and seasonal periods. They are also greatly influenced by the configuration of the continents and oceans, the physical features of the land surfaces, and ocean circulation patterns. For these reasons, any brief overview of their characteristics must be simplistic.

Tropical circulation.

Atmospheric movement between approximately 30° north and south of the equator is dominated by large 3-dimensional circulations commonly known as Hadley cells (Figure 6.1). These cells are driven by surface heating in low-latitude areas and the release of latent heat in the atmosphere above. Under clear skies in the tropics, incoming solar radiation is intense and heats the Earth's surface rapidly. Much of this energy is transferred to the air immediately above the ground, decreasing its density and thus causing it to rise. The upward motion, composed of numerous small-scale eddies and convection currents, is widespread and results in a large equatorial belt of generally low pressure at the Earth's surface called the equatorial trough. The ascending air provides an extremely important mechanism for assisting the upward movement of biota in the PBL. Condensation in the rising air releases energy, causing the air to accelerate upward in the troposphere. Numerous convective storms with abundant precipitation result, creating some of the wettest environments on Earth and providing an important mechanism for washing biota out of the atmosphere. The tropopause marks the top of the troposphere where vertical motions are inhibited because air temperature begins to increase with height due to absorption of ultraviolet radiation by stratospheric ozone.

The ascending flow increases air pressure in the upper troposphere above equatorial latitudes, and thus creates horizontal pressure gradients between the equator and the poleward latitudes aloft in both hemispheres. As air flows down this pressure gradient, it is bent

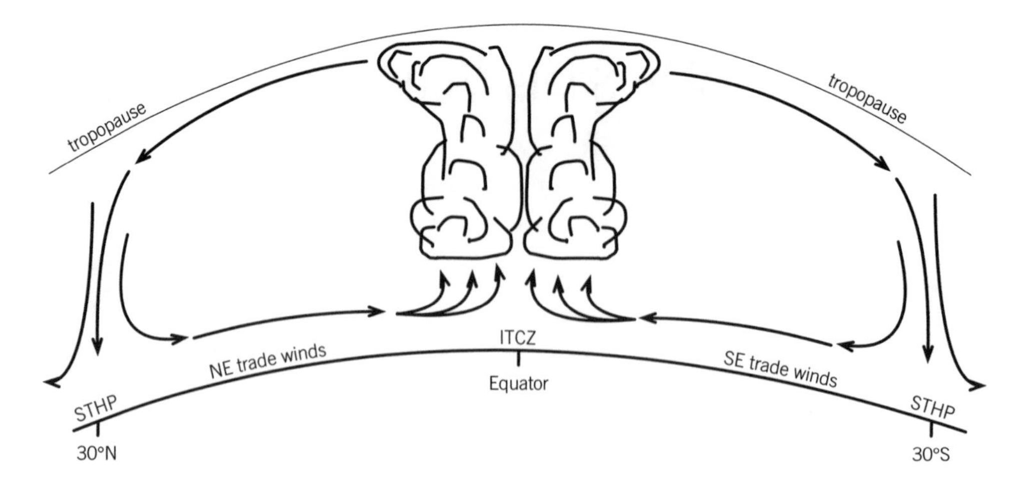

Figure 6.1. Hadley cell circulation.

Schematic representation of the structure and position of Hadley cell circulations for the equinoxes. The vertical scale is greatly exaggerated. The position of the circulation cells shift northward between April and August and southward between October and February. The intertropical convergence zone (ITCZ) extends around the Earth, although it is often discontinuous. The subtropical high pressure cells (STHP) are most pronounced over the subtropical oceans. Consequently we commonly acknowledge five subtropical high pressure cells. The Hadley cell circulation features are strongest in the hemisphere that receives maximum incoming solar radiation. Figure adapted from Hidore and Oliver 1993.

into westerly winds (turned toward the east) by the Coriolis deflection. At about 30°N and S latitudes, a number of factors induce this flow to descend toward the Earth's surface. These factors include: radiative cooling which causes the air aloft to become more dense as it moves poleward; convergence as air streams poleward and is increasingly bent eastward by the Coriolis deflection; convergence of longitude with increasing latitude which squeezes the air as it moves poleward; and the effect of midlatitude circulations.

As the air that comprises these flows subsides toward the Earth's surface, it experiences increased pressure due to the increase in the mass of the atmosphere above. In response to the increasing pressure, the volume containing a given mass of air decreases (density increases), and to compensate, its kinetic energy or temperature increases. This commonly results in clear skies and warm, high pressure cells (anticyclones) at the surface near 30°N and S latitude. These subtropical high pressure (STHP) cells are most pronounced over the oceans and are the dominant circulation features at these latitudes. Over land surfaces, the clear skies associated with descending air aloft result in pronounced radiative heating of the ground and thus warming of the air near the surface. For this reason, many of the our planet's hottest and driest terrestrial environments are located under the poleward wings of Hadley cells (i.e., near the Tropics of Cancer and Capricorn). Rapid upward movements of hot surface air often occur within the PBL in these hot, dry ecoregions to create low pressure at the ground. However, the zone of upward motions and low pressure is limited by the descending flow of air from above and is generally too shallow for condensation and cloud formation.

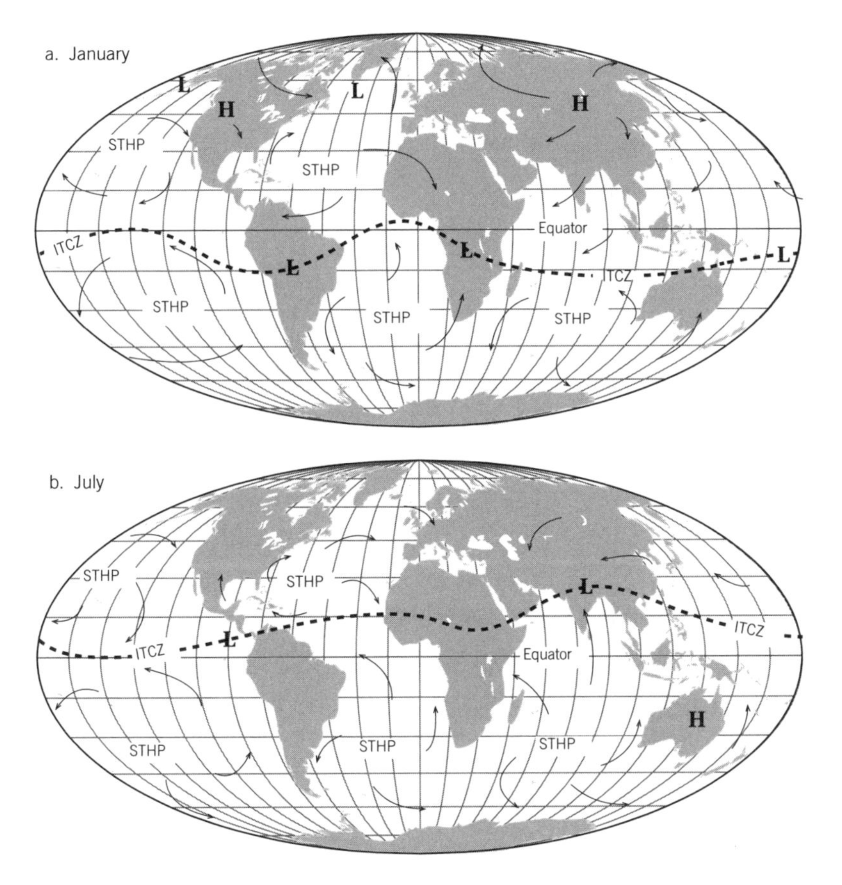

Figure 6.2. Schematic representation of mean air pressure and surface winds showing the position of the STHP cells and the ITCZ.

Air pressure and winds at sea level for January (a) and July (b). The arrows represent the general direction of airflow. The dashed line represents the position of the intertropical convergence zone (ITCZ) which is governed by the distribution of incoming solar radiation. The ITCZ is primarily south of the equator in January and north of the equator in July. Its position is also influenced by the configuration of the continents and the location of mountain ranges within the low latitudes. The subtropical high pressure (STHP) cells are pronounced in the high sun (or summer) hemisphere. The trade winds diverge from the STHP cells and converge into the ITCZ.

The Hadley cell circulations are completed by return flows of air near the Earth's surface from the subtropical high pressure cells into the equatorial trough. As air subsides on the equatorward margins of the STHP cells, it begins to converge toward the equator from both hemispheres and is bent into easterly surface winds (turned toward the west) by the Coriolis deflection. These easterly surface winds, called the trade winds over the oceans, are the most consistent and extensive of the global wind belts. They are often characterized by a low-level temperature inversion (trade wind inversion) because the air warms as it sinks. The belt of convergence near the equator can extend around most of the Earth and is commonly referred to as the intertropical convergence zone (ITCZ). Over land surfaces, the front that forms where the air from both hemispheres meet is called the intertropical front (ITF). The extent and position of this front generally fluctuates on both daily and seasonal time scales.

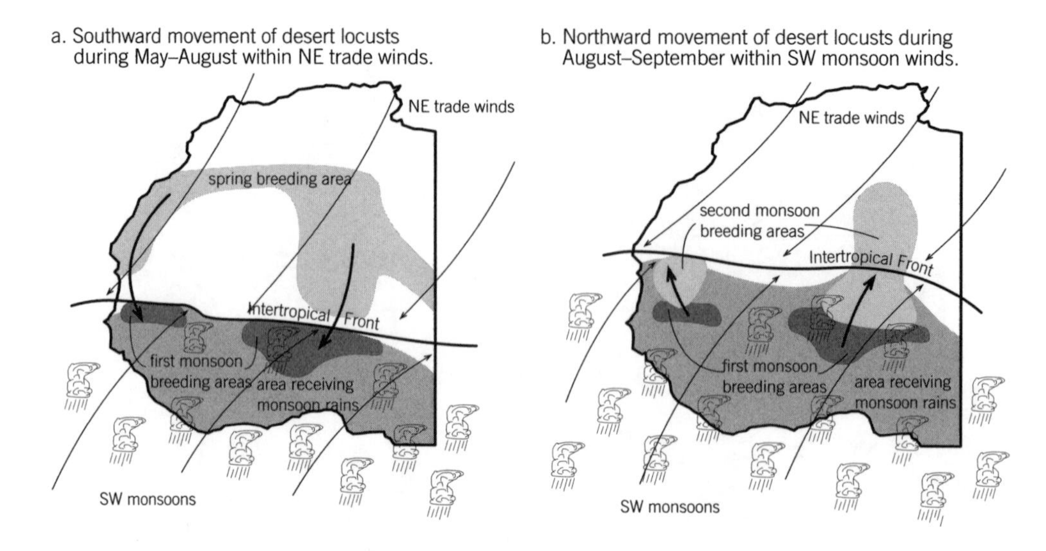

Figure 6.3. Summer movements of desert locusts within the northeast trade and the southwest monsoon winds.

With the onset of dry conditions during May and June in West Africa (a), the desert locust, *Schistocerca gregaria,* leaves its spring breeding area and flies equatorward (heavy arrows) within the northeast trade winds. The locusts cross the ITF and breed where the southwest monsoon winds bring rain. This moisture-laden air that originates in the southeast trade wind belt over the South Atlantic Ocean is turned eastward by the Coriolis deflection once it flows north of the equator. During August and September (b), the locusts move northward (heavy arrows) to a second breeding area on the SW monsoon winds. Figure adapted from Farrow 1990.

Because the Hadley cells are driven by surface heating, the seasonal variation in the spatial distribution of incoming solar radiation impacts their strength and location. Consequently, during April through August, the ITCZ is primarily in the Northern Hemisphere, and the high pressure cells in the North Atlantic and North Pacific oceans are pronounced and centered poleward of 30°N. During October through February, the ITCZ, for most of its extent, shifts south of the equator and the STHP cells in the Indian, South Atlantic, and South Pacific Oceans move poleward and strengthen. A comparison of maps of sea level air pressure and winds for January (Figure 6.2a) and July (Figure 6.2b) reveals the seasonal variation in the surface features of the Hadley cell circulations.

The seasonal changes in the position of the ITF is of great importance to many biota in the tropics because an ecoregion generally experiences a dramatic decrease in precipitation after the ITF moves away. The drying occurs because rising air motions that cause condensation, and subsequent precipitation at the front, are replaced by subsiding airflows and clear skies associated with the poleward wing of the Hadley cells. Because winds at the surface within the ITCZ generally blow toward the intertropical front, many biota use these airflows to move from ecoregions that are experiencing resource shortages due to drying to environments with more abundant precipitation (Figure 6.3).

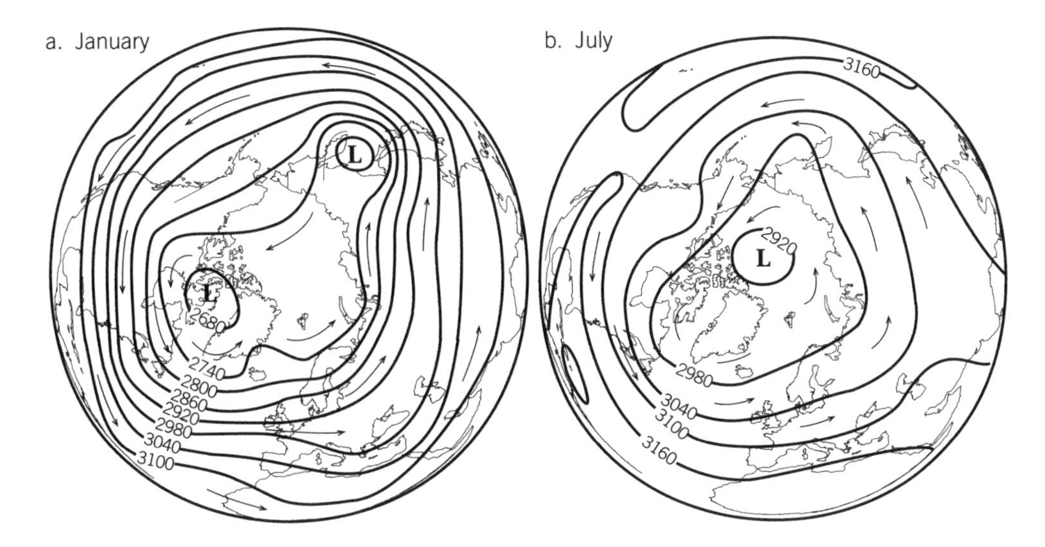

Figure 6.4. Schematic representation of the height contours of the 700 mb surface for the Northern Hemisphere during January and July.

The pressure gradient directed from the subtropics to the poles in the free atmosphere creates a circumpolar vortex of westerly geostrophic winds above the mid- and polar latitudes. Because the latitudinal temperature gradient at the Earth's surface, which is responsible for the pressure gradient aloft, is most pronounced during winter, the circumpolar vortex of strong winds in the middle troposphere is expanded over the midlatitudes (a). In summer (b), the horizontal temperature gradient is much less, and the winds associated with the circumpolar vortex are weaker and restricted to polar latitudes. Height contours are in meters.

Midlatitude circulation.

Poleward of the descending wing of the Hadley cell circulations, transport of energy from subtropical to high latitudes is accomplished by horizontal wave-like motions called Rossby waves. These waves are embedded in a vortex of winds that blow from west to east aloft around the poles. The latitudinal temperature gradient at the Earth's surface, between the subtropic and the polar areas in each hemisphere, creates a horizontal pressure gradient in the mid- to upper troposphere with high pressure above the warm subtropical air and low pressure over cold polar areas (see Figure 5.5). Air driven poleward by the pressure gradient is turned to the east by the Coriolis deflection and creates a circumpolar vortex of winds in the geostrophic flow above polar and middle latitudes (Figure 6.4). Waves often develop in the circumpolar vortex during periods when the temperature gradient between the subtropics and polar areas is pronounced. These waves usually extend vertically throughout most of the troposphere and transport cold air equatorward and warm air poleward (Figure 6.5). The zone of rapid temperature transition in the lower troposphere is called the polar front because it is the boundary between the equatorial and polar air masses. It is often an area of strong winds and precipitation. Wind speed reaches a maximum above the front but below the top of the troposphere. This core of fast moving air, called the polar jet stream, may be a few hundred meters thick, a few thousand meters wide, and a few thousand

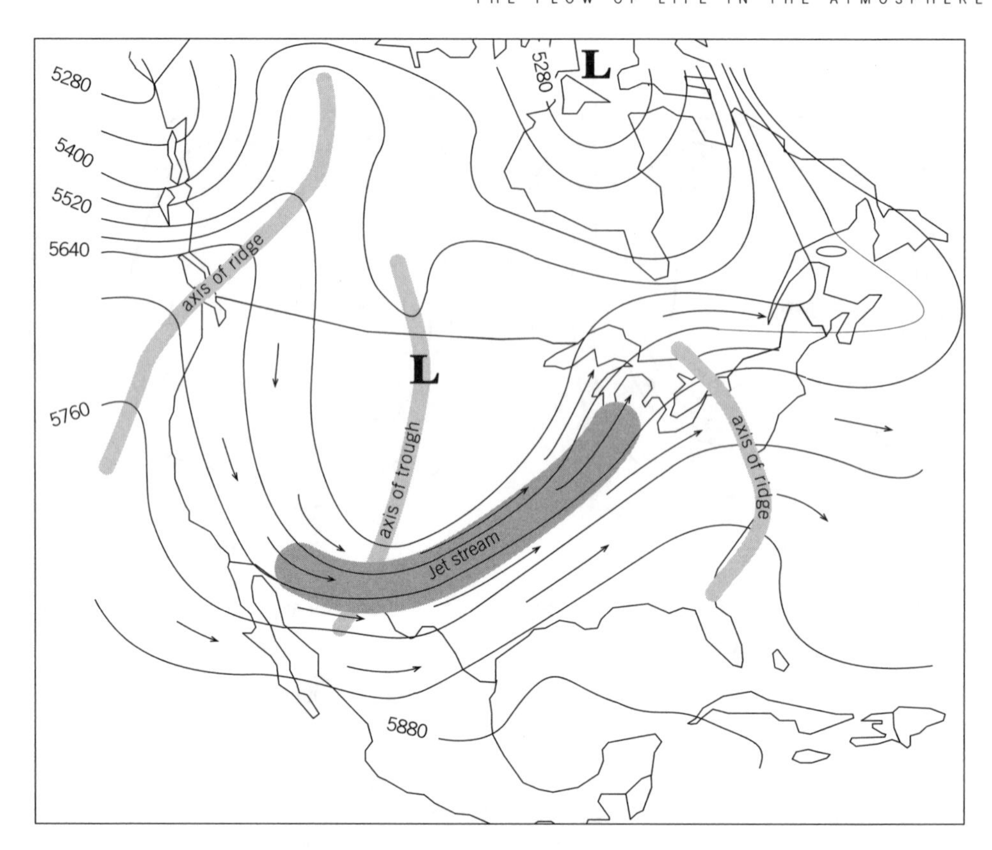

Figure 6.5. Height contours for a 500 mb pressure level in the middle troposphere over North America.

The map shows a typical winter flow pattern when a Rossby wave in the circumpolar vortex is anchored by the Rocky Mountains. A trough of low pressure in the middle troposphere extends over the central United States while ridges of high pressure are above western North America and the southeastern United States. Arrows represent wind vectors. Winds are strongest in the jet stream on the downwind side of the trough (dark shading) over the central United States. Midlatitude cyclones tend to form near the bottom of the trough and track northeastward beneath the jet stream. Anticyclones generally form in the cold surface air beneath the low and the downwind side of the ridge in the 500 mb level airflow.

kilometers long (Figure 6.6). The polar jet stream is strongest on the downwind side of troughs in the circumpolar vortex. The air in this sector of Rossby waves accelerates as it flows poleward causing divergence in the middle troposphere. The resulting upward movement of air from near the Earth's surface is instrumental to the formation of extratropical cyclones.

The Rossby waves are always in a state of flux. As many as four to six Rossby waves can occur in the circumpolar flow above the midlatitudes. Often these long waves are anchored by terrain features such as the Rocky Mountains (see Figure 6.5) and the Tibetan Plateau. At other times, they move with the westerly flow. Because the strength of the flow and the position of the polar front are governed by the horizontal temperature gradients between subtropical and polar latitudes, they vary with the seasons. During winter, when the temperature gradient from the subtropics to the pole is large, winds in the upper troposphere

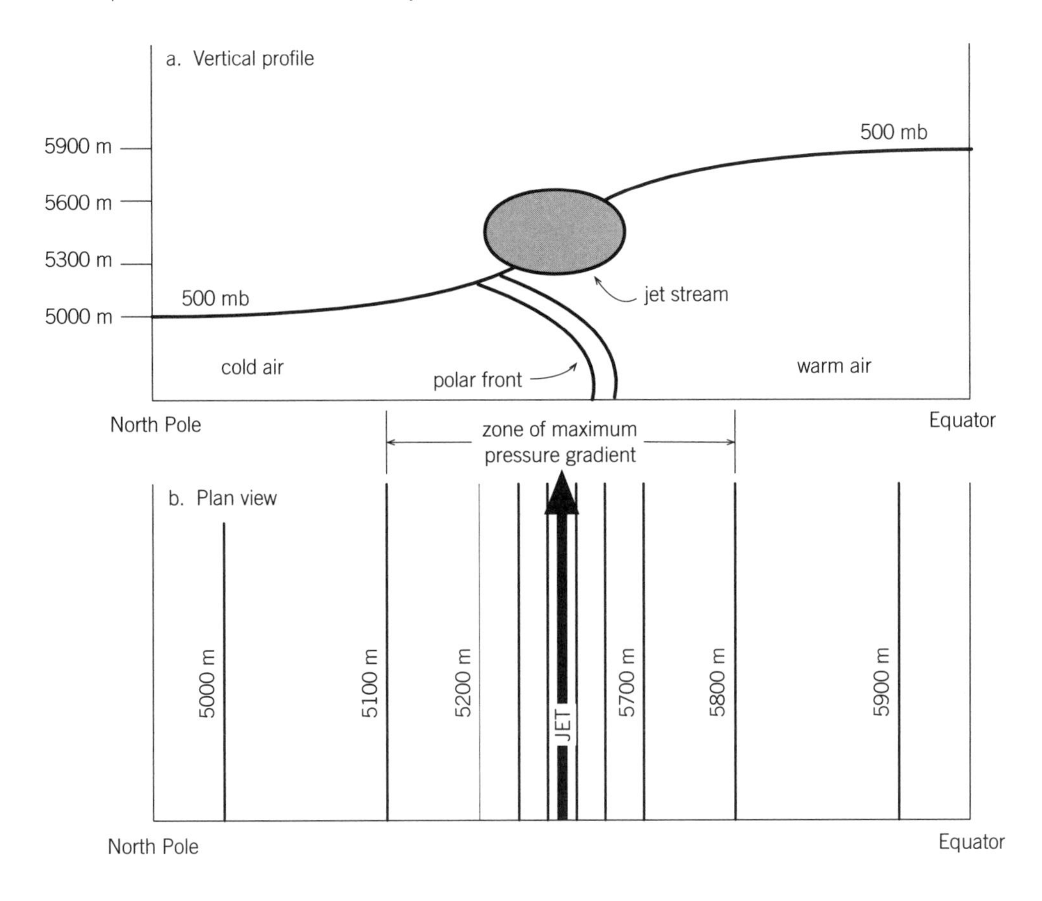

Figure 6.6. Profile and plan views of the polar jet stream in the middle troposphere.

Vertical profile (a) and plan view (b) of the 500 mb surface between the North Pole and the equator. The polar front is the boundary between cold polar and warm tropical air. The polar jet stream is located within the circumpolar vortex at about the 500 mb pressure level above the polar front where the horizontal pressure gradient is steepest (Figures 5.6 and 6.5).

are very strong, and the Rossby circulation dominates the midlatitudes (see Figure 6.4a). In the summer hemisphere, where the subtropics to pole temperature gradient is much smaller, the winds aloft are weaker, and the circumpolar pattern of airflow contracts as the Hadley cells expand poleward (see Figure 6.4b). Sometimes the flow in the midlatitudes is primarily zonal (west to east), and little mixing of tropical and polar air masses occurs. During these times, the temperature gradient from the subtropics to poles generally increases. At other times, the westerly flow assumes a meridional pattern, with pronounced north-south flow which redistributes energy from surplus to deficit areas (see Figure 6.5), thus reducing the temperature gradient from the subtropics to the polar latitudes. The airflow aloft in the midlatitudes shifts back and forth between zonal and meridional patterns at irregular intervals. In a general sense, the Rossby waves and the global-scale atmospheric and oceanic circulations represent flows that occur within the Earth-atmospheric system as

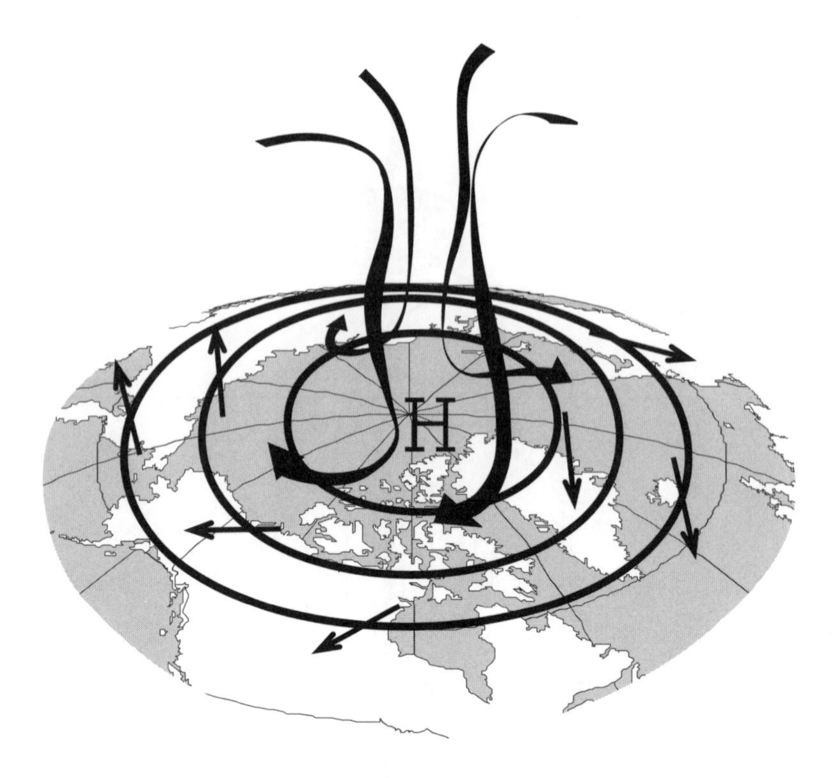

Figure 6.7. Subsidence of cold air above the polar latitudes in the Northern Hemisphere.

The cold dense air subsides to the Earth's surface in polar regions, creating high pressure and diverging northeasterly winds above the Arctic Ocean and southeasterly winds over Antarctica. These flows probably assist the long distance transport of only a few aerobiota (e.g., birds). Not only are these winds usually very cold, but also the land and water surfaces over which they blow are frozen for most of the year and thus do not serve as source ecoregions for very many aerobiota.

it continually tends toward a state of thermal equilibrium. Obviously, equilibrium across the Earth is never realized because of temporal and spatial variations in solar energy receipts; however, thermal equilibrium at large spatial and annual temporal scales is achieved by the meridional flows insofar as no one area consistently gains heat at the expense of another over a year.

Because the winds in the middle and upper troposphere influence the airflow in the PBL, the general direction of surface winds in the midlatitudes is westerly (midlatitude westerlies). The strength and direction of the midlatitude westerlies are far more variable than those of the trade winds because the wind at any given time and place in the midlatitudes is affected by eastward moving high (anticyclones) and low pressure (cyclones) cells. These anticyclonic and cyclonic eddies, or short waves, travel in the circumpolar westerly flow along the Rossby waves. These large-scale eddies embedded in the Rossby waves are associated with extensive advection (horizontal transport) of air within the PBL from the tropics into the midlatitude and polar latitudes, and vice versa, and thus provide an important mechanism for long-distance transport of biota.

Polar circulation.

Subsidence of cold dense air is the dominant atmospheric motion in polar areas. Radiative cooling of the air above polar areas increases its density and results in widespread subsidence and high pressure at the Earth's surface (Figure 6.7). The downward flow of air diverges equatorward at the ground and is turned toward the west by the strong Coriolis deflection, resulting in a polar belt of northeasterly (Northern Hemisphere) and southeasterly (Southern Hemisphere) winds. These airflows are generally too cold, and move in the wrong direction (away from the poles), to be conducive to long-distance atmospheric transport of biota.

Summary.

To a great extent, the global scale motion systems govern the distribution of ecosystems on Earth and thus, in a general sense, the distribution of sources and destinations with suitable hosts for many aerobiota. The air mass concept provides a useful approach for linking global scale circulations with the distribution of the major plant and animal assemblages (herein called biotic divisions or biomes). Air masses are vast pools of air (1000s of km in diameter) with relatively homogeneous temperature and moisture conditions in the horizontal plane in their lowest layers (i.e., PBL). Sometimes air masses are stationary, and then they acquire heat and moisture characteristics through exchanges of energy and water with the underlying landscapes. For example, an air mass that sits over the interior of a continent in the polar latitudes during winter, will become cold and dry, while an air mass that lingers over an ocean in the low latitudes will acquire substantial heat and moisture from the warm water. At other times, the circulation of the atmosphere is more active and air masses flow from their source areas over other zones, where they influence temperature and moisture regimes. In turn, as air masses move away from their source areas, their characteristics are gradually modified by exchanges of heat and moisture with the surfaces over which they traverse.

The strength, latitudinal position, and configuration of the circumpolar vortex and the Hadley cell circulations dictate when and where air masses are stationary, and when, and over which areas of the Earth, they will flow. Regional climates are often delineated by global circulation features and air mass dominance (i.e., the types and frequencies of air masses within a region and how these air masses and their frequencies change with the seasons). Together, latitude, continentality and oceanality, elevation, soil, and climate largely govern the spatial distribution of the major plant and animal assemblages. However, within the midlatitudes during any given year, it is the sequence of air masses (i.e., weather) advected from tropical and polar latitudes that primarily dictates the timing of the cyclical changes in the biotic components of an ecosystem. The seasonal progression of temperature and moisture conditions within a biotic division (e.g., prairie, coniferous forest, and tropical savanna) induces cohort synchrony in both plant and animal populations. For example, an early contraction of the circumpolar vortex and associated polar jet stream during spring will allow the penetration of warm tropical air masses into the interior of midlatitude continents. The associated warm weather will hasten the phenological development of many plants and animals throughout large biotic divisions, setting the stage for synchronous long-range movement by a large number of individuals and species once they enter a "dispersal-ready" state. The importance of the seasonal progression of temperature and moisture conditions to the synchronization of migration and dispersal of many plants and animals is a focus of Chapter 12, a plan to characterize and quantify the flow of biota between the subtropical and midlatitude continental interior ecoregions of North America.

CHAPTER 7

Airscapes: Large-Scale Motion Systems

Most day to day changes in weather are caused by large-scale atmospheric motion systems. These systems are eddies, with either cyclonic or anticyclonic winds embedded in the global scale circulations. In the middle and polar latitudes, these large-scale weather systems usually appear as well-defined low- and high-pressure areas on surface weather maps, while in the tropical latitudes they may be weaker and less pronounced. These migratory weather systems are typically between 300 and 3500 km across and have lifetimes that range from 2 to 12 days. Extratropical cyclones and anticyclones tend to move eastward embedded in the circumpolar westerly flow within the middle and upper troposphere. In contrast, tropical systems tend to be carried westward by the trade winds. Weather systems that are characterized by large-scale upward motions, and thus low pressure and converging winds at the Earth's surface, are generally called disturbances or cyclones. Together, tropical and extratropical cyclones are responsible for the redistribution of a tremendous amount of energy from tropical to middle and polar latitudes, and they also cause the majority of precipitation on Earth as well.

Tropical cyclones.

The low pressure systems that form in the belt of the trade winds over the oceans are generically called tropical cyclones. They develop from bands of convective cells (tropical disturbances), which under favorable conditions can grow into tropical depressions (with surface wind speeds between 10 and 17.5 m s^{-1}), then tropical storms (17.5 to 33 m s^{-1}), and finally hurricanes or typhoons (> 33 m s^{-1}). The energy source for these cyclones is the tropical ocean. The energy is transferred to the air through sensible and latent heat fluxes. Sensible heat is released from the warm ocean water and causes the overlying air to warm and rise. Latent heat is transferred to the air above the ocean surface through evaporation, and it is released to the atmosphere when the water vapor in the rising air condenses and forms clouds. Therefore, tropical cyclones generally develop between the latitudes of 5° and 20° over warm oceans where surface temperatures exceed 27°C (Figure 7.1). They are generally not present closer to the equator, even though ocean temperatures are sufficiently high,

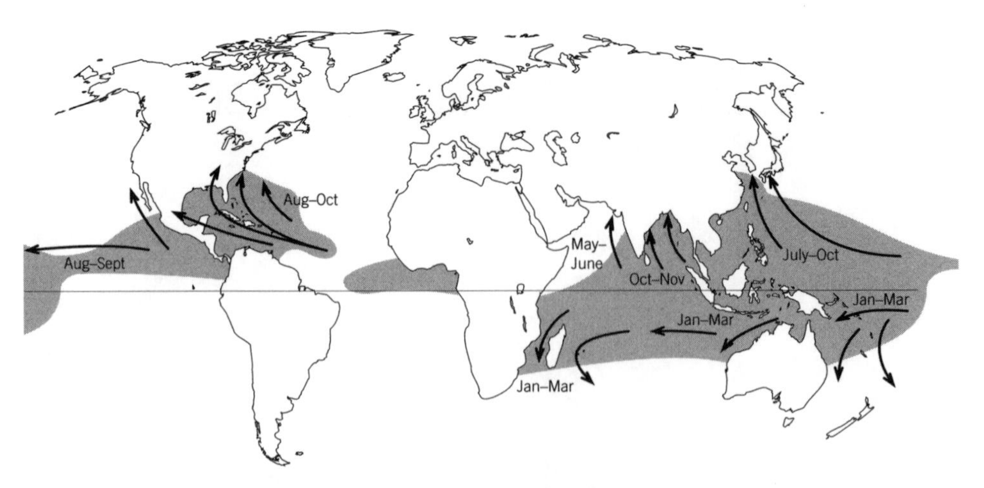

Figure 7.1. Common hurricane tracks and principle months of occurrence.

Shading denotes areas where sea surface temperatures exceed 27°C during the warmest months. Arrows represent common hurricane tracks. Tropical cyclones seldom form within 5° of the equator because the Coriolis deflection is too weak (e.g., North Atlantic off the west coast of Africa). Ocean temperatures are typically too cool for hurricane development in the Eastern South Pacific and South Atlantic Oceans where the ITCZ remains near the equator during the high sun season in the Southern Hemisphere. Figure adapted from Gray 1979.

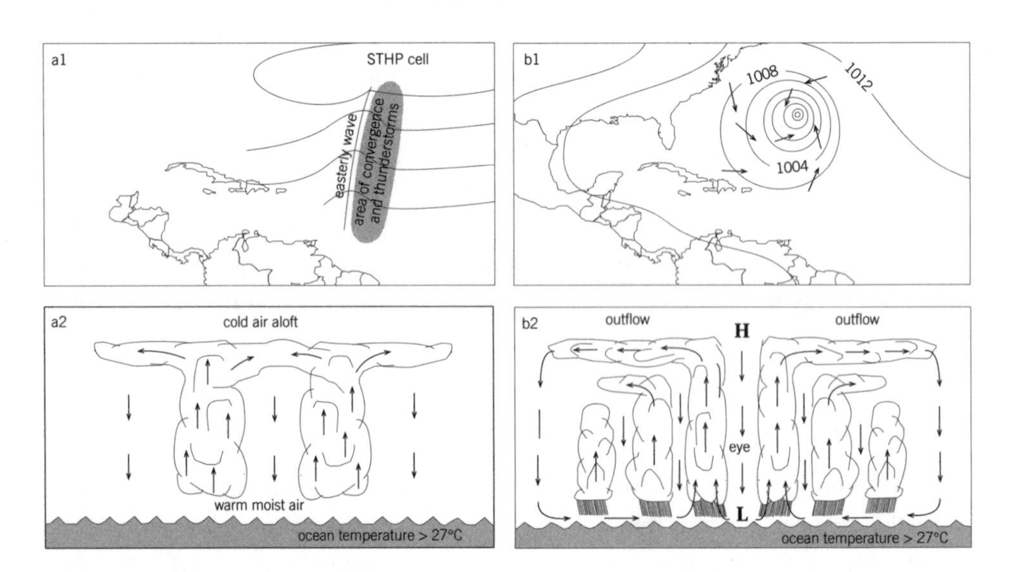

Figure 7.2. Development of a hurricane.

Thunderstorms form in the area of convergence on the upwind side of an easterly wave within the belt of trade winds over the North Atlantic ocean (a1). Cold air associated with an upper-level trough from the midlatitudes enhances vertical motions, helping to organize the circulation within the group of tropical thunderstorms (a2). The rapidly rising warm air causes the surface pressure to drop, creating a low pressure center and strong converging winds immediately above the warm water (b1). The air gains latent heat from the ocean through evaporation while it converges into the low pressure center. As the air rises within the convective cells this latent heat is released and warms the upper troposphere, creating an area of high pressure and diverging air flow aloft (b2). As long as the upper level outflow exceeds the surface inflow the storm will intensify. The outwardly flowing air subsides to the ocean surface on the margins of the storm system. As it nears the ocean surface it converges toward the area of low pressure at the center of the hurricane.

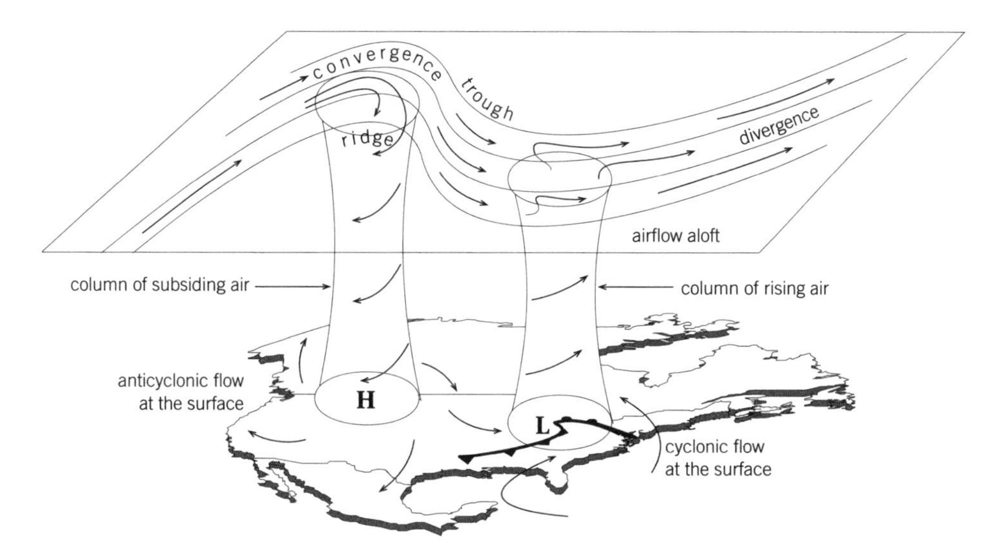

Figure 7.3. Relationship between airflow in a Rossby wave aloft and anticyclones and cyclones at the Earth's surface.

High and low pressure centers at the Earth's surface must be supported by upper-level convergence and divergence of air. The convergence of air in a ridge aloft causes subsidence which enhances high pressure, clear skies, and diverging winds at the Earth's surface. In contrast, divergence of air on the downwind side of a trough in the circumpolar vortex enhances upward movement of air from below. This results in strong converging winds around low pressure centers at the Earth's surface. Figure adapted from Lutgens and Tarbuck 1998.

because the Coriolis deflection is too weak to promote cyclonic winds (see Figure 5.7a). Tropical cyclones can develop from waves that form in the trade winds and move from east to west across the tropical oceans. The convective storms within these easterly waves can become organized if the subsidence associated with the trade wind inversion is weak. In this situation, a circulation can develop that traps much of the latent heat within the system (Figure 7.2). Intensification from tropical storm to hurricane status is often stimulated by upper tropospheric low pressure systems from the middle latitudes. Cold air above the convective storms can induce the vertical uplift, and subsequent outflow of air, away from the center of the cyclone. This air cools as it flows outward and subsides to the ocean surface on the outer margins of the cyclone. The outflow of air aloft allows subsequent vertical motion within the convective cells and convergence of warm moist air immediately above the ocean surface into the tropical cyclone center, thus completing the circulation. There is a strong connection between the seasonal movements of the equatorial trough/ITCZ and tropical cyclone genesis, which is revealed by the high frequency of hurricanes and typhoons between January and March in the Southern Hemisphere, and during August through October in the Northern Hemisphere. Initially tropical cyclones usually drift slowly from east to west across the warm oceans between the STHP and the ITF. As they intensify, their speed of movement increases and their path curves gradually poleward around the western margins of the STHP cells. Hurricanes decay rapidly once they track over land because they lose their source of latent heat and because the increase of surface friction slows their winds and promotes vertical wind

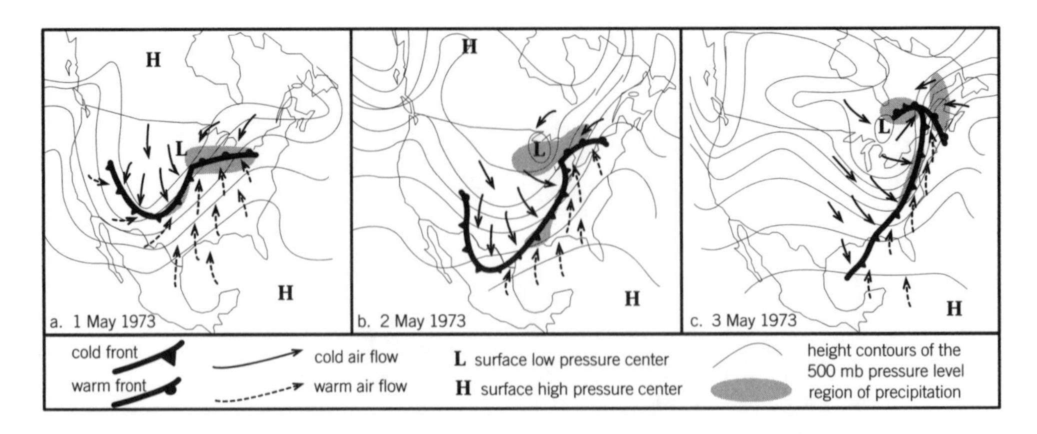

Figure 7.4. Weather fronts, surface winds, and height contours of the 500 mb pressure level of an extratropical cyclone over North America.

The sequence of maps above depicts the primary synoptic weather pattern that draws warm tropical air from the Gulf of Mexico into the continental interior of North America. It consists of a high pressure area (STHP cell) over the southeast coast and western North Atlantic and a depression with a trailing cold front traversing across the High Plains, then the Midwest, to the east coast. Warm air diverging from the western flank of the STHP cell flows poleward into the continental interior within the warm sector of the extratropical cyclone, and the air converges into the low pressure center. The speed of the airflow in the PBL is relatively high in the warm sector of the depression. Low-level jet streams of super-geostrophic winds 50 to 100 km wide at the top of the PBL often parallel the cold front within the warm air flow. Spores, pollen, insects, birds, and many other biota move from the subtropics into the midlatitudes each spring on these warm winds. The travel time from Texas to the Great Lakes for passively transported biota can be as little as 36 hours if movement is continuous or about 3 days when transport occurs only at night. Precipitation associated with frontal systems is frequently present in the interior of the continent and provides a mechanism to wash biota out of the atmosphere. The same synoptic weather systems also provide opportunity for transport from the North American interior to the Gulf of Mexico in autumn. Southward movement of biota primarily occurs in the strong near-surface winds behind the cold front. Downwind displacement may exceed 300 km in a single night (Johnson 1995).

shear. However, hurricanes and typhoons can travel far poleward when they remain over the warm ocean currents that parallel the east coast of continents in the midlatitudes.

Extratropical cyclones.

Low pressure systems or large eddies that form on the polar front, the boundary between subtropical and polar air masses, are often called midlatitude, or more technically correct, extratropical cyclones. These large-scale systems act to redistribute energy from subtropical latitudes poleward. Generally, extratropical cyclones develop as wave disturbances on the polar front, where strong divergent airflows, aloft on the downwind side of a trough in the Rossby waves, enhance the convection from below that commonly occurs along the boundary between the cold polar and warm subtropical air masses (Figure 7.3). The convection forms a depression or area of relatively low pressure at the surface and initiates convergence of the two air masses. In the Northern Hemisphere, the air filling the surface low pressure center has a counterclockwise trajectory, while in the Southern Hemisphere the Coriolis deflection

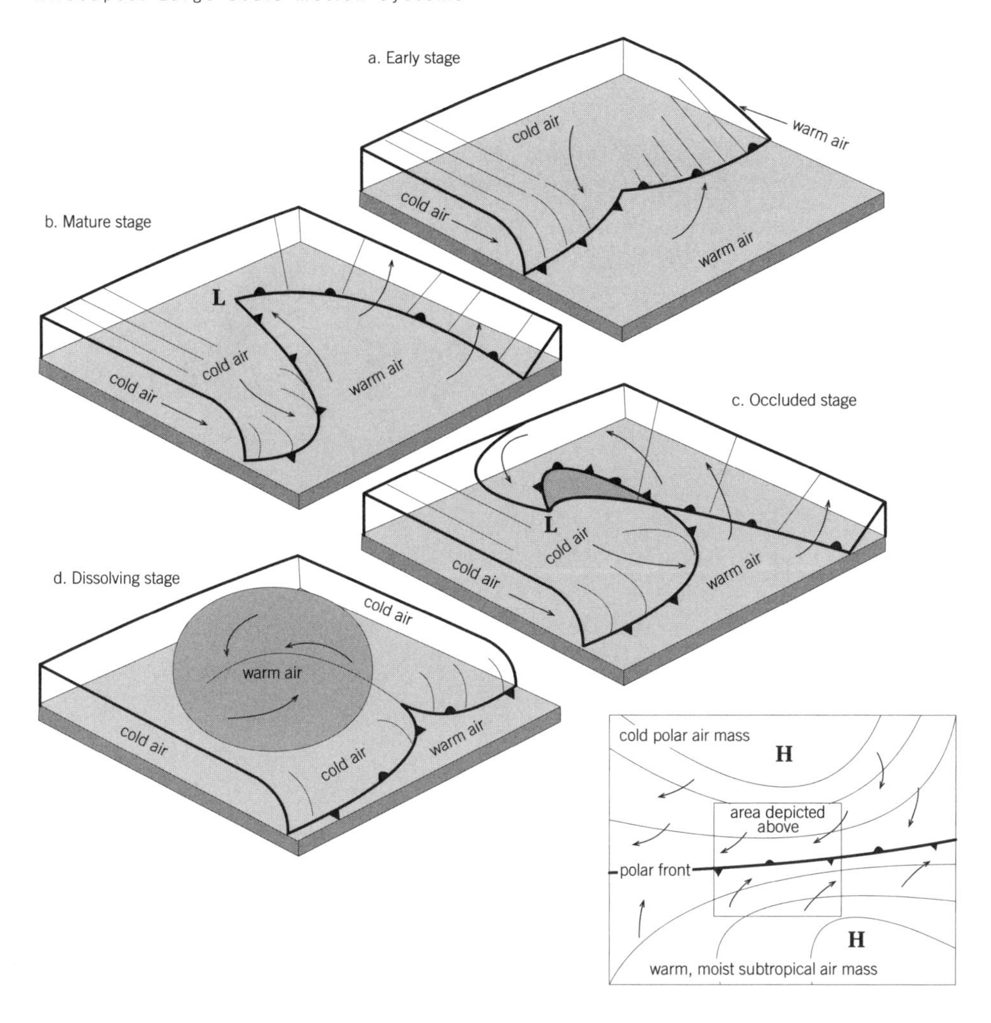

Figure 7.5. Stages in the development of an extratropical cyclone.

Cyclones develop as wave disturbances on the boundary between the cold polar and warm subtropical air masses (lower right). Cold polar air advances rapidly equatorward to the west of the apex of the wave while tropical air advances poleward on the eastern flank of the wave (a). During the mature stage of development (b), an extratropical cyclone has an extensive warm sector. However, the polar air advances more rapidly than the subtropical air, and gradually the warm sector becomes narrower as the subtropical air is lifted upward at both the cold and warm fronts. Eventually, the cold front overtakes the warm front lifting the warm air mass entirely off the Earth's surface (c). This occlusion (occluded front) starts in the center of the depression and spreads outward. In the middle troposphere above the occluded front, there is generally an extensive area characterized by thick clouds and precipitation where the subtropical and polar air masses mix. During the dissolving stage of an extratropical cyclone's life (d), the warm and cold fronts merge completely, the cyclone's travel comes to a halt, and the depression at the surface fills. Figure adapted from Strahler 1966.

results in a clockwise inspiral. The interface of the air masses develops a wave-like shape with the apex of the wave located in the center of the depression where the uplift is most intense. The resulting extratropical cyclones are typically 1500–3000 km in diameter and usually last between 4 and 7 days. During its life, the surface low pressure center typically tracks to the east and poleward beneath the jet stream on the downwind side of the Rossby wave (Figure 7.4).

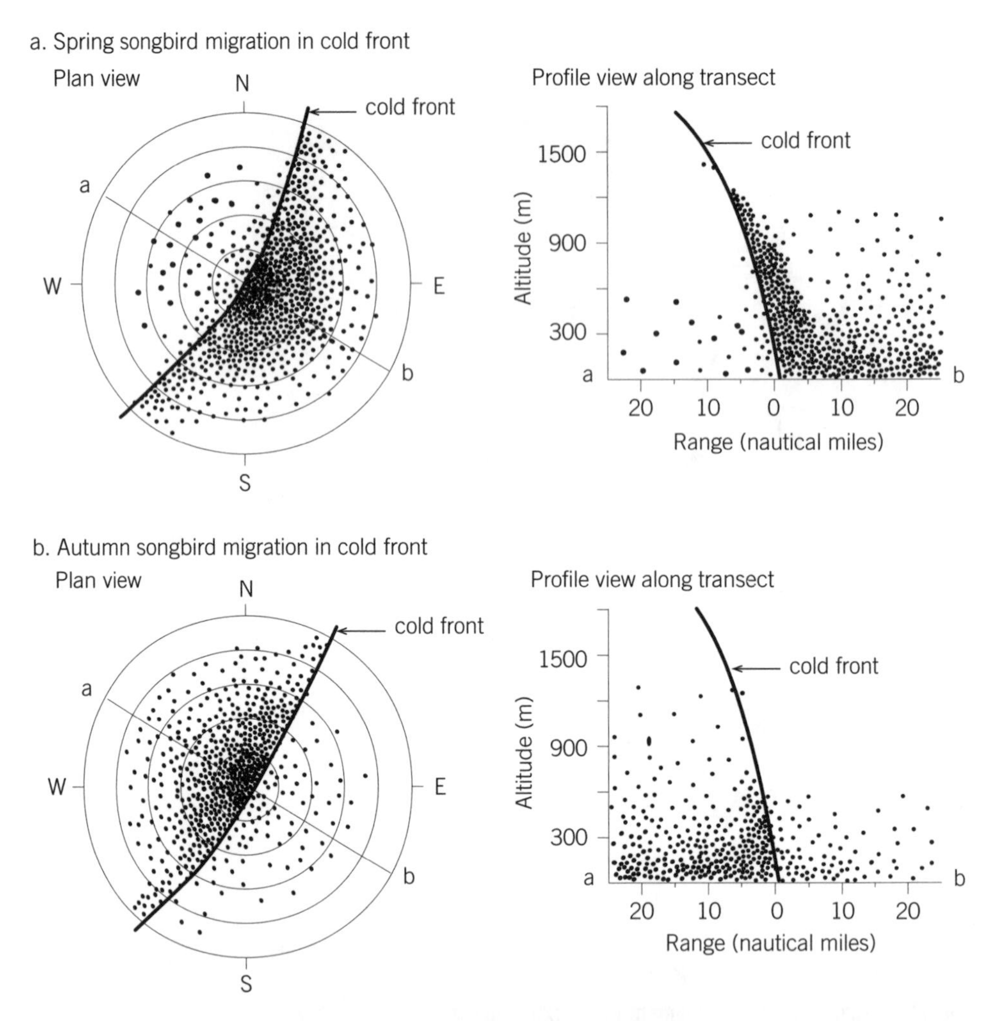

Figure 7.6. Schematic representation of migrant songbirds in airflows associated with cold fronts showing differences in their flight tactics between spring and autumn.

(a) Cold front traversing the southeastern United States during spring. Radar observations during spring often reveal high concentrations of songbirds flying toward the northeast assisted by southwesterly winds ahead of a strong cold front traversing the southeastern United States. The flight direction of birds in the immediate vicinity of the front is more often easterly; perhaps they have been overtaken by the cold air mass and are trying to reenter the warm sector of the cyclone. Many of these birds also appear to be lifted by the rising airflows at the front. Behind the cold front, only a few strong echoes from birds moving toward the northeast are observed. Figure adapted from Gauthreaux 1991.

(b) Cold front traversing the southeastern United States during autumn. Radar observations during autumn show that migrating songbirds are frequently concentrated in the airflows behind cold fronts sweeping through the southeastern United States. Prior to the passage of a front, relatively few echoes from birds are displayed on the radar, and most of these are moving to the northeast within the warm sector of the extratropical cyclone. As a front passes, the winds become northwesterly, and the density of songbirds migrating south often increases dramatically. Figure adapted from Gauthreaux 1991.

The cyclonic wave that forms at the Earth's surface traps warm air of subtropical origin between the cold, rapidly advancing air mass to the west, and less mobile cold air to the east and north. A warm front is formed as the subtropical air mass advances and slides up and over the more dense, cold air to the east of the low pressure center. The slope of the warm front is very gentle so that the resulting cloud system and band of precipitation is usually broad (500–700 km). To the west of the warm air sector, the polar air advancing from the northwest forms a cold front where it undercuts the subtropical air, forcing the warm air upward (Figure 7.5). Many strong-flying moths and birds take advantage of these airflows, flying in the southwesterly winds within the warm air sector during spring and in the northwesterly flow behind the cold front in the autumn (Figure 7.6). The southwesterly winds in the warm sector of the cyclone, immediately ahead of the cold front, are generally quite strong, and low-level jets with speeds in excess of 20 ms^{-1} frequently form at the top of the PBL, especially at night. The slope of the cold front is steep in the lower troposphere and the rapid uplift of the warm moist air can result in severe weather. Generally the horizontal extent of the band of precipitation at the cold front is on the order of a few 100 km.

Extratropical anticyclones.

High pressure cells in middle and polar latitudes may be induced by thermal and dynamical forces, alone and in combination (refer to chapter 5), that cause air to subside to the Earth's surface. Cold air over continental interiors in the upper middle and high latitudes subsides because it is dense. This creates the large areas of high pressure that form the centers of continental polar air masses. Sometimes these cold air masses are relatively stationary on the poleward side of the polar front, and at other times they stream equatorward beneath troughs in the Rossby waves. During winter, continental polar air masses can be very cold and dry, because much of the land surface over which they lay is frozen, while polar air masses that form over the oceans are more moderate. In contrast, polar air masses of both continental and polar origin are cool and more moist in summer. These air masses dominate weather in much of the midlatitudes during winter and are present in the midlatitudes, albeit less frequently, throughout the rest of the year. During summer, anticyclones that are created in the subtropics by the Hadley cell circulation often extend poleward into the midlatitudes to form ridges in the Rossby circulation aloft (see Figure 6.5).

High pressure at the Earth's surface can also be caused by dynamical forces associated with the convergent flow of air aloft on the upwind sides of troughs in the Rossby waves, where winds slow as they flow equatorward (see Figure 7.3). As the air piles up, subsidence occurs, and areas of high pressure form at the surface, essentially between low pressure centers of extratropical cyclones. The thermal and dynamical mechanisms that cause anticyclones can be interdependent. The flows of large, cold air masses equatorward and warm air masses poleward create the Rossby waves in the circumpolar vortex. The subsidence that occurs due to convergence of air in the middle troposphere acts to enhance the high surface pressure within the western flanks of the cold air masses streaming equatorward, and within the eastern flanks of the warm air masses diverging poleward from the STHP cells.

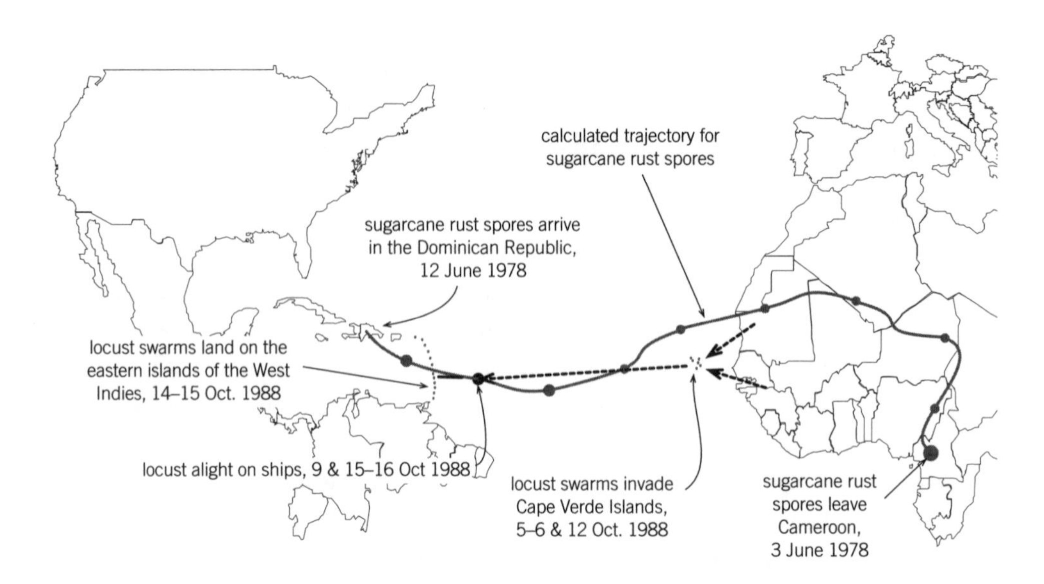

Figure 7.7. The crossing of the Atlantic Ocean by desert locusts and sugarcane rust spores on the trade winds.

In July 1977, sugarcane rust first appeared in the Americas. The most likely scenario for its introduction is that urediospores of *Puccinia melanocephala* were transported on winds from Cameroon in western Africa to the Dominican Republic, where they spread rapidly throughout the island. Wind trajectory analysis suggests that the spores left Cameroon on 3 June 1978 and arrived in the Dominican Republic nine days later. The rust subsequently spread throughout the equatorial latitudes of the Americas from Florida to Venezuela by wind transport of urediospores (Purdy et al. 1985).

In early October, 1988, desert locusts (*Schistocerca gregaria*) arrived on the western African coast between Mauritania and Guinea-Bissau. They invaded the Cape Verde Islands beginning on 5 October and were subsequently sighted by sailors from ships on the Atlantic Ocean. Swarms of locusts reached the eastern islands of the West Indies on the 14th along a 1,100 km front stretching from St. Croix to Guyana and Surinam. The locusts were assisted near the end of their journey by tropical storm Joan (Ritchie and Pedgley 1989, Rainey 1989). Figure adapted from Purdy et al. 1985 and Rainey 1989.

Aerial movements of biota in large-scale motion systems.

There are numerous records of biota that have moved long distances assisted by airflows associated with tropical cyclones and extratropical cyclones and anticyclones (refer to works listed in Table 4.3). One of the most spectacular of these journeys is the crossing of the Atlantic Ocean by bands of desert locusts (*Schistocerca gregaria*) in October, 1988 (Ritchie and Pedgley 1989, Rainey 1989). The locusts left the west coast of Africa, and flying within the trade winds, traversed approximately 4500 km of ocean. They arrived in huge numbers in the West Indies 4 to 6 days later, embedded within a tropical cyclone. Other biota, including sugar cane rust (Purdy et al. 1985) and perhaps spores causing the sigatoka disease of bananas, (Stover 1966) also were spread from Africa to the West Indies by the Northeast Trades (Figure 7.7).

The movement of cereal rusts in North America provides a well-documented example of the importance of extratropical cyclones and anticyclones to the long distance transport of

a. 1923 — gradual aerial dissemination of wheat stem rust spores by anticyclone

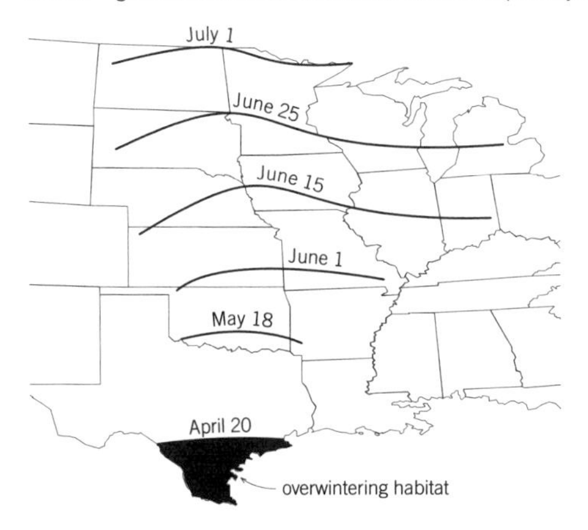

b. 1925 — rapid aerial dissemination of wheat stem rust spores by extratropical cyclone

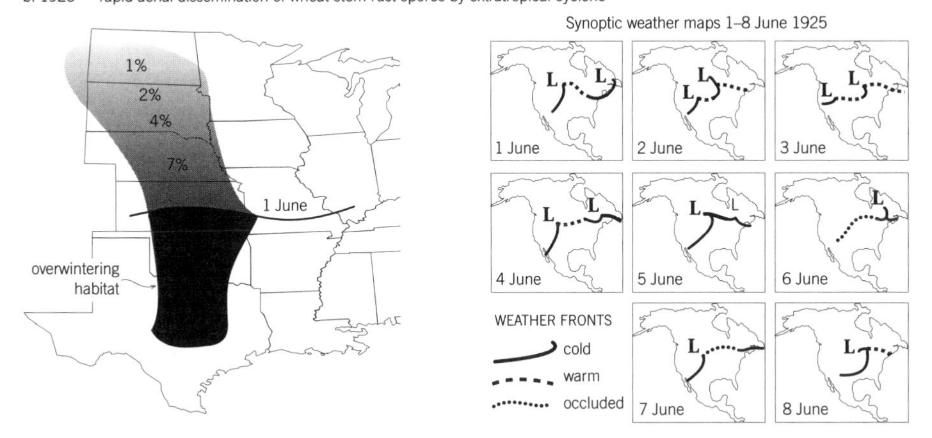

Figure 7.8. Northward movements of wheat stem rust spores within the United States.

(a) In 1923, wheat stem rust spread 3,000 km from northern Mexico and Southern Texas to Canada over a period of two months. The rust was abundant in its overwintering habitat on 20 April. Spores were trapped at an altitude of 500 m between Laredo and San Antonio, Texas, on 4 May. By 10 May, the rust had moved into southern Oklahoma, and on 31 May, spores were caught in St. Joseph, Missouri. The spores reached Kansas, southern Nebraska, and southern Illinois by 4 June. Rust was found on wheat in southern South Dakota and southern Minnesota on 22 June, and by 1 July, the wheat stem rust spores had reached the Canadian border. Figure adapted from Stakman and Harrar 1957.

(b) During the first few days of June, 1925, wheat stem rust suddenly spread poleward from its overwintering habitat in southern Kansas, Oklahoma, and Texas to North Dakota. On 1 June, a frontal system with strong southerly flow developed (surface winds > 10 m/s) throughout the High Plains and continued until 7 June. A huge spore cloud moved 1,000 km northward over a front stretching more than 600 km from east to west. During this period, numerous spores were caught in traps throughout the previously rust-free area. Rain fell in Nebraska, the Dakotas, and western Minnesota each day of the week. On 8 June, the cold front moved eastward over the Great Lakes region. Field observations revealed that infection throughout the region was almost simultaneous. The prevalence of rust decreased from south to north; about 7%, 4%, 2%, and 1% of the culms of susceptible wheat varieties were infected in Nebraska, southern South Dakota, northern South Dakota, and North Dakota respectively. Figure adapted from Stakman and Harrar 1957.

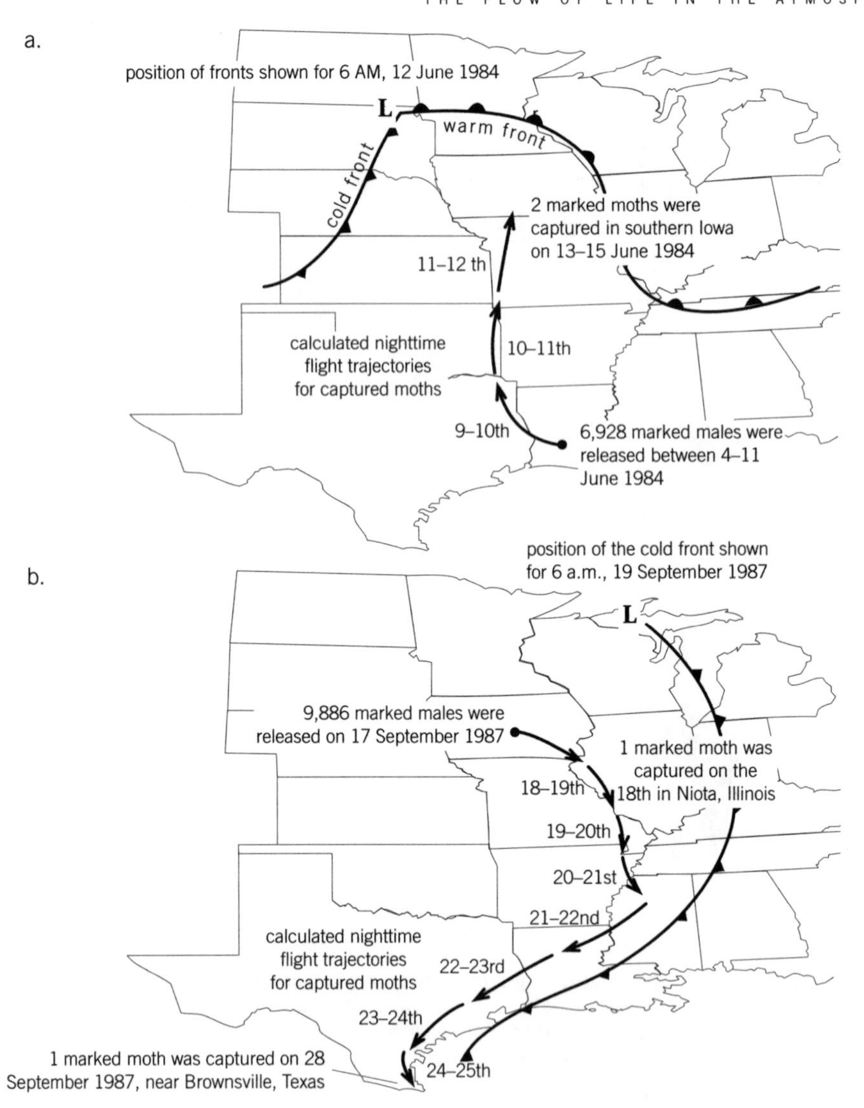

Figure 7.9. Round-trip migration of black cutworm moths in North America.

(a) Likely route traveled by the black cutworm moths during their northward aerial migration. They flew within the warm southerly winds of a midlatitude cyclone that was traversing the High Plains.

(b) Likely route traveled by the black cutworm moths during their equatorward aerial migration. At first they flew within the relatively warm, northwesterly winds behind the cold front of a midlatitude cyclone that was traversing the High Plains. Then they traveled on the warm northeasterly winds diverging from a strong anticyclone that followed the cyclone across the continental interior of North America.

Each spring, many noctuid moth species that are indigenous to the Gulf Coast states and Mexico expand their range northward into the Cornbelt where they do extensive economic damage. However, some are unable to maintain permanent populations in the North American continental interior region due to cold winter temperatures. Each autumn, some of these noctuids return equatorward to overwinter in the subtropics. Mark-release-recapture studies with laboratory-reared black cutworm moths (*Agrotis ipsilon*) have proven that these noctuids can migrate from Louisiana to Iowa in the spring and from Iowa to Texas during autumn. The long-distance migration of these moths occurs at night in the PBL often in the downwind direction on strong airflows associated with midlatitude cyclones and anticyclones within North America. Figures adapted from Showers et al. 1989, 1993.

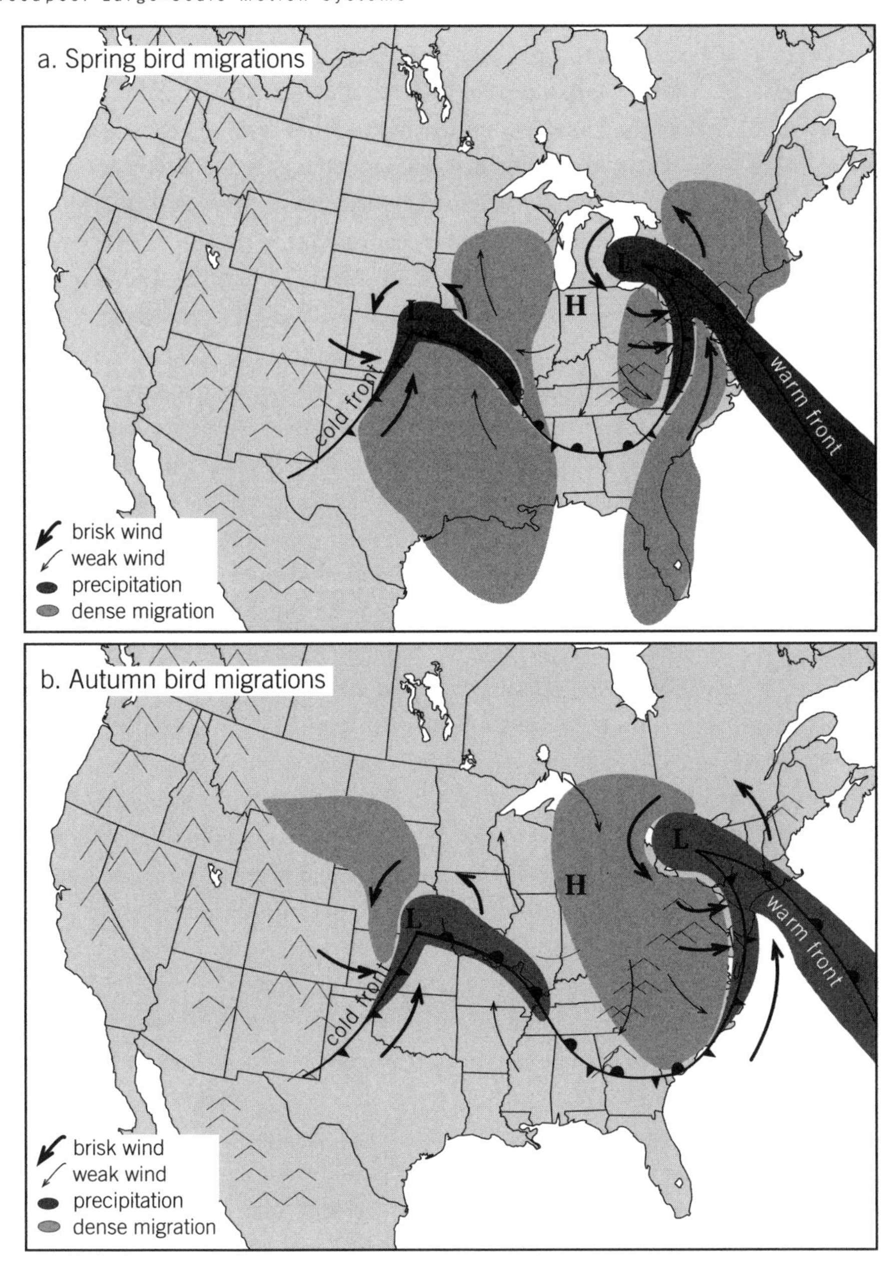

Figure 7.10. Synoptic weather patterns conductive to bird migrations within North America.

(a) Spring migration routes of birds within winds associated with extratropical cyclones and anticyclones in North America. During spring, poleward migrations most often occur in the warm sector of an extratropical cyclone. Birds also tend to fly in the southeasterly winds ahead of warm fronts and in the westerly flow immediately behind cold fronts.

(b) Autumn migration routes of birds within winds associated with extratropical cyclones and anticyclones in North America. During autumn, equatorward migrations frequently occur in the northerly winds on the back side of low pressure centers (cyclones). Birds also tend to fly in the northwesterly flow behind cold fronts and in the northerly winds within the east and central portions of high pressure centers (anticyclones) that follow the passage of a cold front. Figures adapted from Richardson 1978.

biota (Stakman and Harrar 1957). The wheat rusts *Puccinia graminis* and *P. Recondita* (= *P. Triticina*) are unable to survive either the cold winters in the northern portions of the Great Plains in North America or the hot, dry summers in the southern United States and Mexico. Each year, spring-sown wheat in the northcentral United States and southcentral Canada receives spores from rusted autumn-sown wheat in Mexico and Texas. In turn, winter wheat in the south becomes infected during autumn by spores from the north. In some years, the rust is spread gradually by anticyclones taking a succession of short jumps, with intervening stops where the inoculum multiplies locally. In other years, it is transported by extratropical cyclones over a distance of a 1000 km or more in 1 to 2 days (Figure 7.8). The same airflows that carry passively transported biota back and forth between the midlatitudes and the subtropics also provide invaluable assistance to strong-flying insects, such as black cutworm moths (Figure 7.9), and birds (Figure 7.10) that make the same journeys each year.

Although tropical cyclones have provided valuable assistance to the long distance aerial movement of biota, especially among islands in the Pacific Ocean (e.g., Visher 1925) and to crops that were introduced into areas in the tropics and subtropics where they had not previously been grown, we speculate that their role in transporting biota to suitable habitats is less significant than that of extratropical cyclones. There are a number of reasons why this should be the case. First, tropical cyclone genesis typically occurs over warm ocean waters far from land, and thus source habitats of most aerobiota. Although these systems frequently track over islands and near coasts of continents, their winds diminish rapidly once they pass over a large land mass. In contrast, the genesis of extratropical cyclones in the Northern Hemisphere generally occurs over or near large land masses where cold continental polar and warm subtropical air masses meet. Second, tropical cyclones are typically smaller in area (diameter ≈650 km) than their midlatitude counterparts (diameter ≈1500–3000 km). Third, although tropical cyclones have converging winds above the ocean surface and diverging winds in the upper troposphere, much of the airflow within the system is vertical and often at speeds that are extremely high. Unlike extratropical cyclones, they do not involve the large scale horizontal convergence of air masses from subtropical and polar latitudinal belts. Finally, tropical cyclones generally form during late summer or early autumn when surface temperatures in the tropical oceans are highest. Consequently, when they assist the movement of biota from low to middle latitudes, opportunities for colonization and rapid population growth are limited, because many of their potential hosts are beginning to senesce with the approach of winter. In contrast, extratropical cyclones and the associated large-scale advection of both polar and tropical air masses are frequent in the spring and autumn seasons when many biota need to move among habitats due to seasonal changes in the availability of resources.

CHAPTER 8

Airscapes: Landscape-Induced Motion Systems

Local winds that blow across landscapes can be either mechanically or thermally driven. The dimensions of these circulations are often influenced by surface characteristics and/or the extent of the geographic assemblage of ecosystems that comprise the landscape. Thus these circulations typically range from less than 1 to 100s of km across, and generally have lifetimes that last from a few hours to a day. Some of these local winds are circulations with horizontal dimensions that exceed their vertical dimensions while others are more cellular-shaped, stretching upward within the PBL. Relief in a landscape can induce mechanical winds and circulations that are pronounced when the large-scale flows in the PBL are relatively strong. In contrast, thermal circulations typically occur during calm atmospheric conditions when skies are clear and where there is spatial variation in the rate of surface heating. Both passively and actively transported biota are known to move considerable distances within terrain-induced airflows (see references in Table 5.1, especially Schaefer 1976, Pedgley 1982, and Drake and Farrow 1988). Many of these local circulations provide organisms the opportunity to move back and forth between habitats within a single diel period.

Mountain winds.

Hills and mountains can act as obstacles to large-scale atmospheric flows, creating a large variety of atmospheric structures including orographic waves and downslope winds. Their mechanically induced effect on the wind depends on both the characteristics of the topography (height, length, width, spacing, and arrangement of hills and mountains into ridges and valleys) and those of the airflow (direction relative to the topographic barrier, vertical profiles of wind, and atmospheric stability). In the simplest case when the large-scale flow is relatively weak, topographic barriers deflect winds up and over or around them. In stronger flows, the air to the lee of the barrier can overturn and/or oscillate in the vertical plane (Figure 8.1). Strong katabatic (downslope) winds can occur on leeward slopes of mountains when large-scale atmospheric flows are favorable. These winds can be either cold (e.g., called a bora, fall wind, or gravity wind) or warm and dry (e.g., called a foehn or

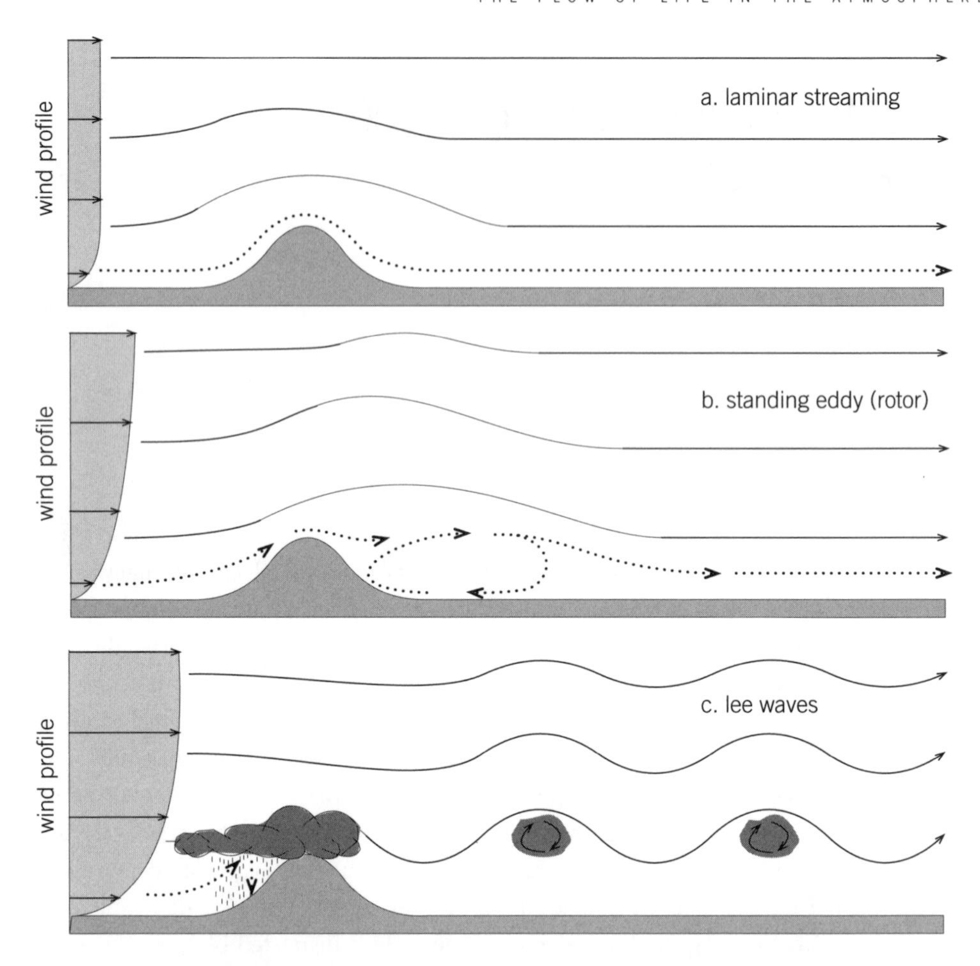

Figure 8.1. Common types of orographic waves.

The effect of orographic barriers (mountain ranges) on the flow of air and biota near the Earth's surface depends on the vertical profile of wind speed in the PBL. When wind speeds are low, air flows over (and around) a mountain in laminar-like streamlines (a). The dashed line represents the path of biota moving within the airflow near the ground. The topographic obstruction affects the flow for only a short distance downwind. When winds in the PBL are stronger, rotors can form in the flow to the lee of mountains (b). The direction of the winds close to the ground in the lee of the obstruction is opposite to the flow above and generally much weaker. Where these stationary eddies are long-lasting, snow and biota can accumulate on the lee side of orographic barriers. Waves form in the PBL to the lee of topographic barriers when winds speeds are very high and can propagate for 100s of km downwind (c). Clouds and precipitation are common near the crest of the barrier. Under these conditions, biota transported in the atmosphere near the Earth's surface may be trapped on the windward side of a mountain and washed out of the air by precipitation. Figure adapted from Atkinson 1989.

chinook), depending in large part on the temperature of the air mass at the bottom of the slope prior to the wind.

In clear, calm weather, the temperature difference between air over slopes and that at the same altitude in the free air over valleys can cause mountain/valley wind circulations. In the daytime, radiative heating of mountain slopes results in warming of the air near the slopes to a greater extent than the air at the same height above sea level over adjacent valleys. This

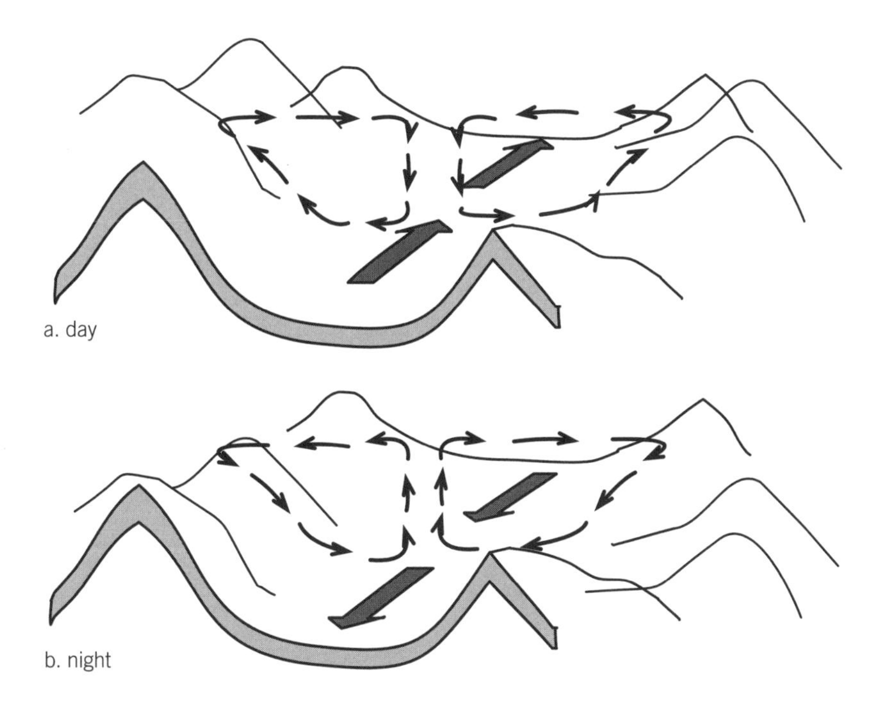

Figure 8.2. Mountain-valley circulations during day and night.

During clear, calm mornings, solar radiation heats high mountain slopes (a). Some of this energy is transferred to the air immediately above these surfaces, creating temperature and thus pressure gradients between this air and that at the same altitude over the center of the valley. As a result, air moves horizontally away from the slope and then downward, creating high pressure at the surface along the valley axis. The circulation is completed by a flow of air directed up the mountain slopes. Altitudinal temperature gradients that also develop along the main axis of long valleys during calm, clear days make mountain-valley circulations 3-dimensional (only the surface winds directed along the main axis of the valley are shown above). During warm summer days, many biota travel high into mountain regions on the upslope surface winds. The ladybugs that are frequently present in spectacular numbers on high mountain snowfields in the midlatitudes are the most noted example (Mani 1962). The fallout of insects and other biota to snowfields on mountaintops are influenced by: (1) enhanced turbulence above mountain crests where the upslope winds from the different sides of the mountain converge with the geostrophic wind, (2) cold air temperatures in the midtroposphere, and (3) attraction to the bright snow surface. Radiative cooling of high altitude slopes during clear, calm nights causes a reversal of the circulation in mountain valleys (b). Cold dense air from high elevations forms a density current that flows down the mountain slopes (downslope wind). Air rising above the valley axis and then flowing toward the slopes aloft can complete this nocturnal circulation. Because air temperatures are generally too cool at night in many mountainous areas for takeoff, downslope winds likely provide less assistance to aerobiota than do upslope winds. Figure adapted from Oke 1990.

produces a horizontal flow of air at high elevations away from the slope and toward the center of the valley, then a subsiding flow above the valley's axis, and finally air movement upslope, creating a valley wind above the sloping surface (Figure 8.2a). On warm, sunny days, upslope winds transport a wide variety of aerobiota high into mountainous areas, often to alpine ecosystems in which they are unable to survive. The number of insects that can accumulate on the snowfields near the peaks of mountains is amazing. Fallout rates between

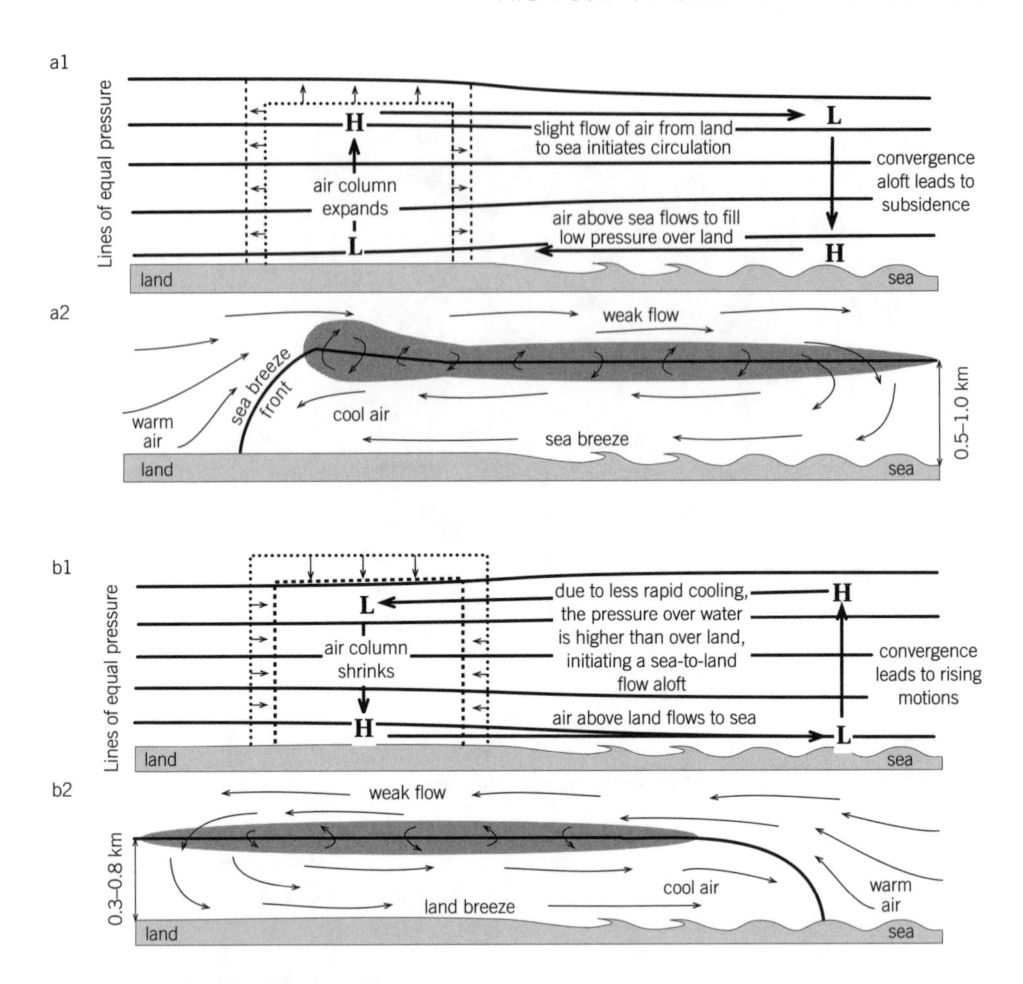

Figure 8.3. Sea-land breeze circulations.

As the land heats up during clear days, convective currents distribute heat upward to heights of 1 to 3 km in the PBL. This heating of the air column over the land results in an expansion of the column vertically and to a lesser extent horizontally, producing a zone of high pressure aloft over the land relative to the zone at the same altitude over the water (a1). This difference in pressure aloft initiates a weak flow offshore resulting in convergence and subsidence over the water. Consequently, a zone of relatively high pressure develops at low levels over the water. As a result, the horizontal pressure gradient at the surface is directed from sea to land at low altitudes causing a cool sea breeze and rising motions over the land to complete the sea breeze circulation. The inland penetration of the sea breeze often results in a sharp zone of convergence known as the sea breeze front (a2). Vertical motions of 1 to 2 m/s have been observed along sea breeze fronts and often, especially when the prevailing wind is from an opposing direction, a head may develop at the top of the sea breeze front. The hatched areas are zones where entrainment and mixing between ambient and sea breeze air occur.

During clear nights, radiative cooling of the land surface tends to create a more pronounced vertical pressure gradient in the cooler air above the land than in the relatively warmer air above the water (b1). The resulting horizontal pressure gradient aloft initiates a weak onshore flow and a corresponding counter airflow, or land breeze, near the Earth's surface (b2). Land breezes are notorious for assisting the aerial movement of mosquitoes and biting flies from wetlands and livestock enterprises in coastal regions to shorelines, where they can be a nuisance to humans and have detrimental economic impacts on the tourist industry.

a. Spruce budworm moths in sea breeze

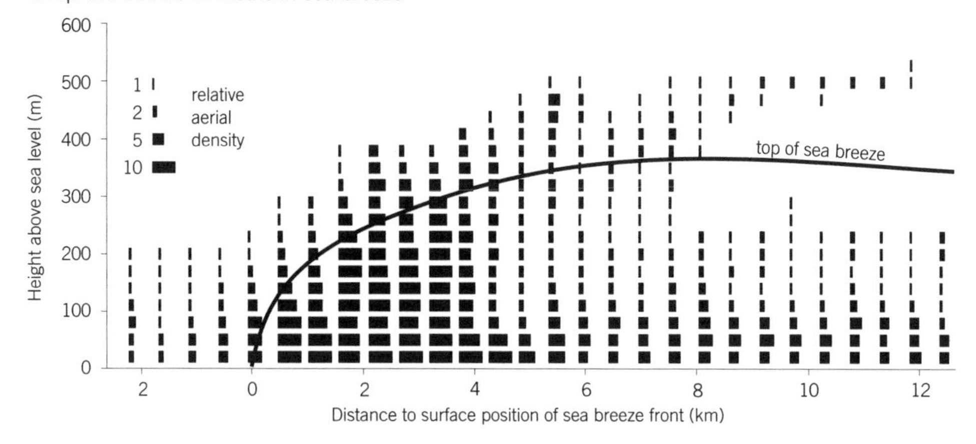

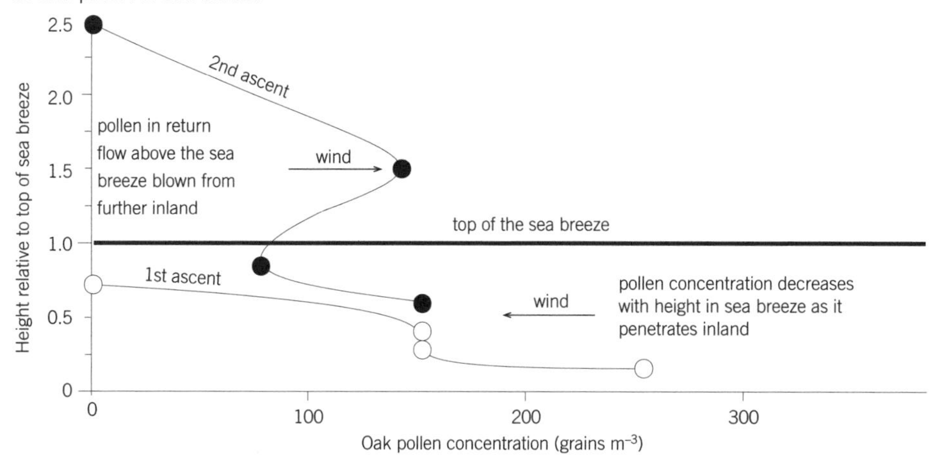

Figure 8.4. Biota concentrations in sea breeze circulations.

(a) Density of spruce budworm moths (*Choristoneura fumiferana*) in a sea breeze in New Brunswick, Canada, on 10 July 1976 measured with airborne radar (Schaefer 1979). Moths were heavily concentrated at the sea breeze front, defining the nose that sloped backward with altitude. The orientation of the moths above the top of the sea breeze was essentially in the opposite direction to those below. There was a dense shallow layer of moths extending 10–20 km backward from the front within the sea breeze. While gliding over Hampshire, U.K., Simpson (1967) encountered numerous swifts (*Apus apus*) between 150 and 250 m altitude on 26 June 1966. The birds were soaring in the upwardly deflected return airflow ahead of the front, presumably eating insects. Figure adapted from Schaefer 1979.

(b) Vertical profiles of oak pollen concentration within a sea breeze over Long Island, NY, on 18 May 1972 (Raynor et al. 1974). Like the moth density shown above, the concentration of pollen behind the front decreased with height in the sea breeze flow. The pollen above was likely liberated by turbulence and strong winds associated with the passage of the sea breeze front farther inland and lifted aloft within the return flow. Figure adapted from Raynor et al. 1974.

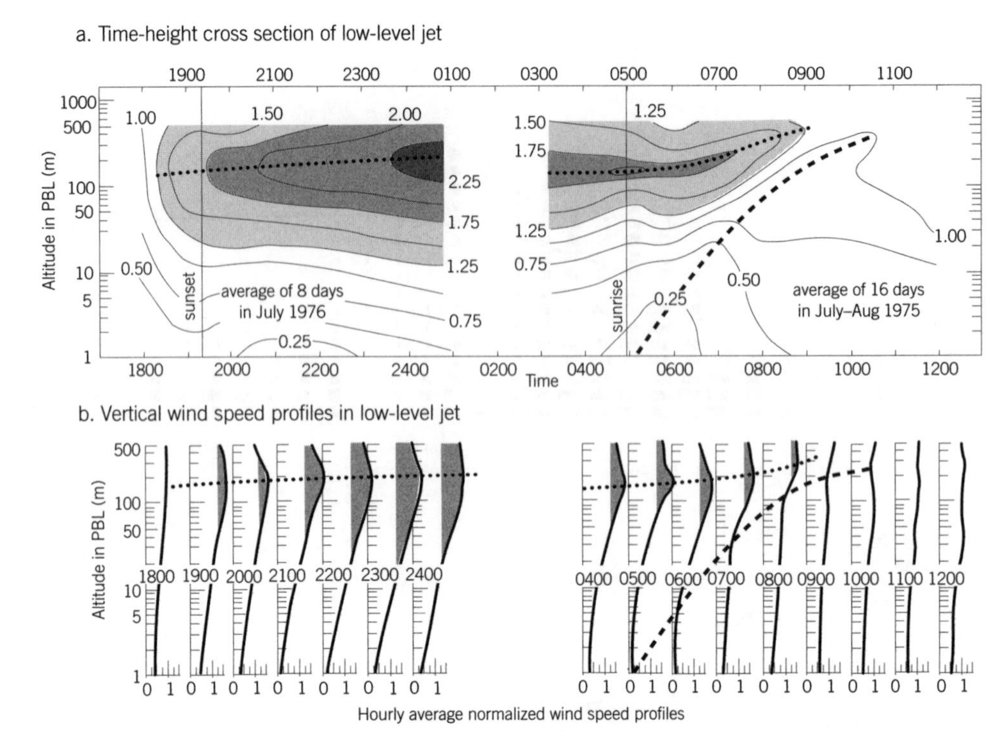

Figure. 8.5. Time-height cross section and vertical profiles of averaged normalized wind speed showing a low-level nocturnal jet over Illinois.

Wind speeds are expressed as approximate multiples of the winds observed at 500 m in well-mixed, daytime conditions. Measurements are averaged for 24 clear, calm nights in Sangamon County, Illinois, during summer. Note the growth and decay of the low-level nocturnal jet stream indicated by the dotted line. The dashed line shows the growth of the convectively mixed layer after sunrise. Figure (a) from Sisterson and Frenzen 1978.

88 and 1125 insects per $m^2 \cdot day^{-1}$ have been measured in the Pyrenees (Antor 1994) while Mani (1962) recorded 40 insects per m^2 in a 20-minute period at 3200 m in the Himalayas. During clear, calm nights, radiative cooling of high altitude sloping surfaces causes a reversal of this thermally induced circulation and results in a downslope (mountain) wind above the sloping surface (Figure 8.2b).

Coastal winds.

Differential radiative heating of land and water surfaces often results in sea (or lake)/land breeze circulations, especially when skies are clear and the PBL is relatively calm. During the morning hours the air over the land heats up to a greater extent than the air over the water body, creating horizontal pressure gradients perpendicular to the coastline (Figure 8.3a). The contrast in pressure aloft initiates a weak airflow directed from land to sea. The concurrence of subsiding air above the water surface and rising air above the land results in a pronounced horizontal pressure gradient in the lower PBL, producing a relatively strong, low-level airflow from water to land (sea breeze). When the offshore wind is relatively strong, a sea breeze front can be generated near the coast. It may remain stationary, or move slowly

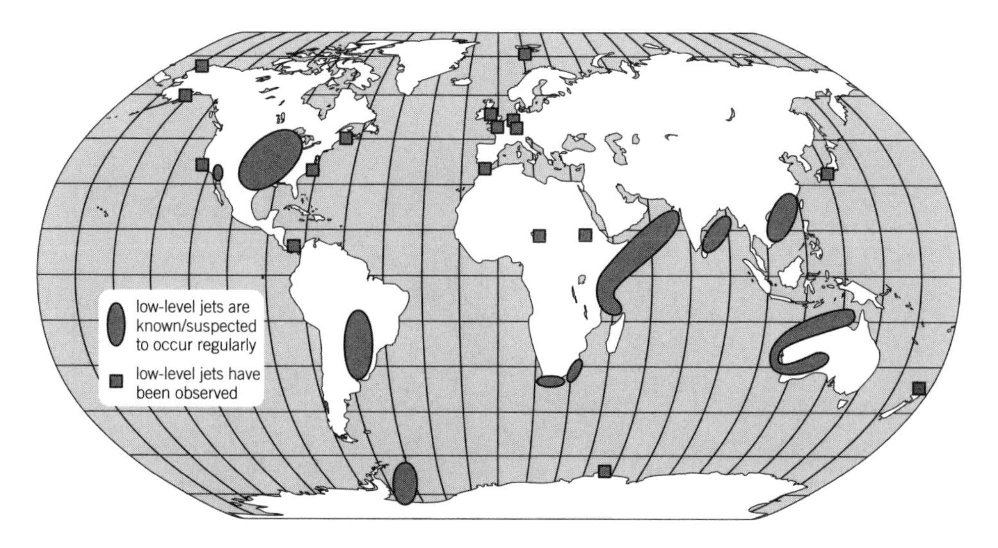

Figure 8.6. World distribution of low-level jets.

Low-level jets have been observed frequently over parts of North America, South America, Africa, Asia, Australia, and Antarctica. The jets occur most frequently to the lee of a large mountain range or at coastlines where pronounced temperature gradients develop. In the middle latitudes, low-level jets are more frequent during the summer months than at other times of the year (Stensrud 1996). Figure adapted from Stensrud 1996.

inland during the day, as long as the sea breeze itself continues to blow behind the front and warm air ascends along the frontal surface. High concentrations of biota can accumulate over time within the convergence zone at the sea breeze front or within the nose of the current (Figure 8.4a). Other biota can be lifted by the enhanced turbulence ahead of the front and travel out over the water body within the counter airflow (Figure 8.4b). There they may subside and travel back to land within the sea breeze. A reversal of the circulation (land breeze, Figure 8.3b) occurs at night due to the more rapid radiative cooling of land than water. A land breeze is generally less intense than a corresponding sea breeze because there is no solar heating to enhance convection at the land breeze front.

Nocturnal low-level jets.

Radiative cooling of warm ground surfaces on clear nights often creates air temperature inversions (increases in air temperature with altitude) in the absence of strong horizontal pressure gradients which advect cold air or enhance turbulent mixing. These inversions originate adjacent to the ground as sensible heat is transferred from the lowest air layer to the surface to replace the energy radiated to the atmosphere. Because of this heat loss, the temperature of the air immediately above the surface decreases below that of the overlying air, and thus the thermally-induced turbulence (convection) near the ground ceases. As radiative cooling continues throughout the night, the inversion layer (the nocturnal PBL) expands upward. Because turbulent mixing among air layers in the lower atmosphere decreases as the inversion deepens, the speed of the airflow in the upper portion of the PBL

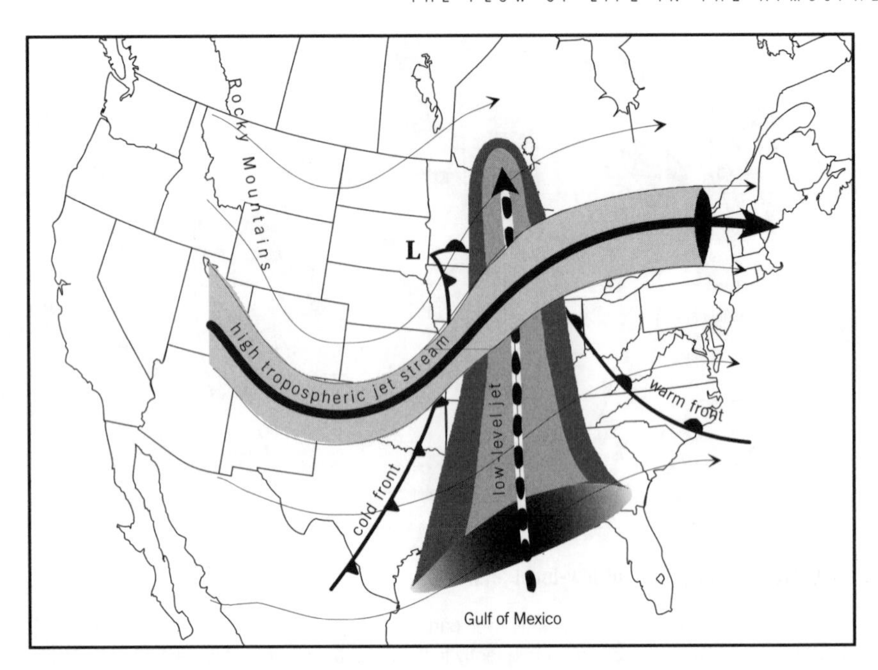

Figure 8.7. Low-level jet within the warm sector of an extratropical cyclone over North America.

An extensive analysis of hourly observations from a network of 31 wind profilers indicates that the formation of strong low-level jets over the central United States is promoted by the coincidence of landscape-scale influences on the nocturnal planetary boundary layer and large-scale airflows associated with extratropical cyclones (Mitchell et al. 1995). Weak low-level jets occur at any time day or night; also, their coincidence with extratropical cyclones is not particularly strong, suggesting that nocturnal boundary layer conditions or extratropical cyclones can cause these jets. In contrast, the strongest low-level jets occur approximate six times more frequently within a few hours of midnight than during the day and are usually located in the warm sector of an extratropical cyclone. The average direction of low-level jet flow over the Great Plains is south to southwest; however, about 25% of the jets have a northerly component. Mean low-level jet duration is about five hours for the weak jets and two hours for the strongest jets. Figure adapted from Newton 1967.

is no longer retarded by upwardly moving thermals from lower, slower moving airstreams. Relieved of this restraint, the wind in the PBL gradually speeds up, resulting in a wind profile that increases with height throughout the temperature inversion (Figure 8.5). The rapid acceleration of the wind after the release of this frictional constraint can result in wind speeds near the top of the temperature inversion that exceed the speed of the geostrophic wind immediately above the PBL (Blackadar 1957). The zone of supergeostrophic wind speed is generally referred to as the nocturnal low-level jet.

Large-scale landscape features have important influences on atmospheric processes that dictate the geographical distribution of nocturnal low-level jets and consequently the pathways of the biota flows they assist. Diurnal oscillations of buoyancy forces and temperature gradients in the air above sloping terrain, horizontal temperature gradients at coastlines, and/or flow blockage by mountain ranges due to increased stratification of air at night can interact with the diurnal cycles of convection to create low-level jets. As a result, low-level jets are common at night in many parts of the world (Figure 8.6). The pronounced pressure gradient ahead of a cold front often enhances landscape influences to induce the

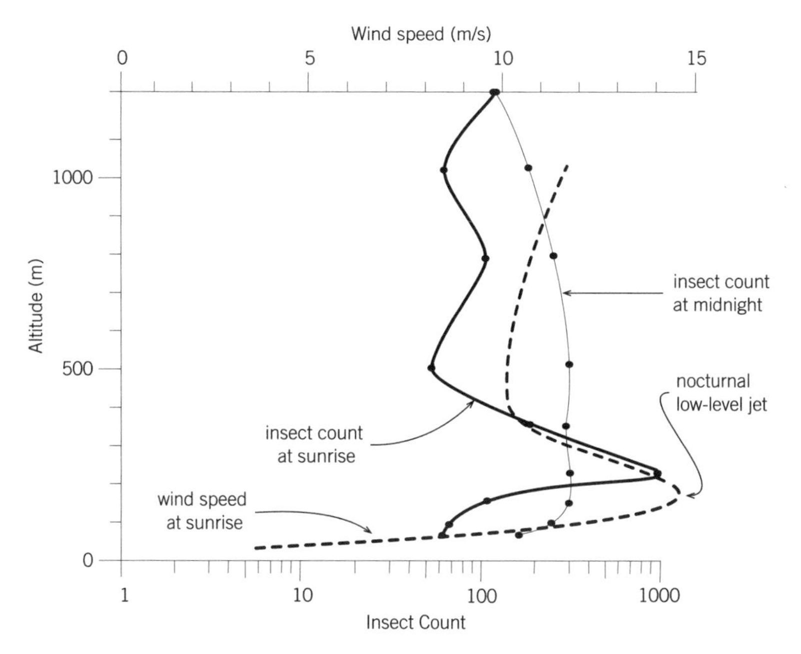

Figure 8.8. Insects concentrated in a nocturnal low-level jet.

Radar observations from midnight show moths distributed throughout the lower 1200 m of the atmosphere on 25 September 1980 over central-western New South Wales, Australia. By 0500, the moths were concentrated in a 100 m deep layer centered about 250 m above the ground. A low-level jet was also present in the PBL ahead of a weak front approaching from the west. Wind speeds were maximum between 100 and 300 m. There was strong shear in the wind direction below 300 m. A pronounced temperature inversion was present during the previous evening with air temperatures at the top of the inversion 10°C warmer than those at the ground. It had likely deepened throughout the night and by dawn the top of the inversion was probably near the altitude of the low-level jet (i.e., the top of the PBL). Drake (1984, 1985) suggests that the insect layer concentration was formed in response to the surface temperature inversion rather than to the low-level wind speed maximum. Figure adapted from Drake 1985.

formation of a low-level jet within the warm sector of extratropical cyclones. For example, Bonner (1968) found that low-level jets were present in about 30% of the soundings of PBL winds that were made within the central United States over a two year period. Strong low-level jets were primarily a nighttime phenomena that occurred in large-scale southerly flow and were most frequent in spring and late summer, although they occurred in all seasons of the year. Bonner found that a strong pressure gradient across the Great Plains with advection of air northward from the Gulf of Mexico into the central United States to the east of a cold front was the weather pattern most conducive to large-scale low-level jet formation (Figure 8.7).

Nocturnal temperature inversions, such as those that initiate the formation of low-level jets, inhibit upward movement of passively transported biota out of their flight boundary layer (see Figure 5.9). Microorganisms liberated within inversion layers tend to remain concentrated near the ground or settle out of the atmosphere due to gravity. In contrast, strong-flying insects have frequently been observed to form layer concentrations in nocturnal low-level jets at the top of deep inversion layers. Drake (1985) suggests that these insects

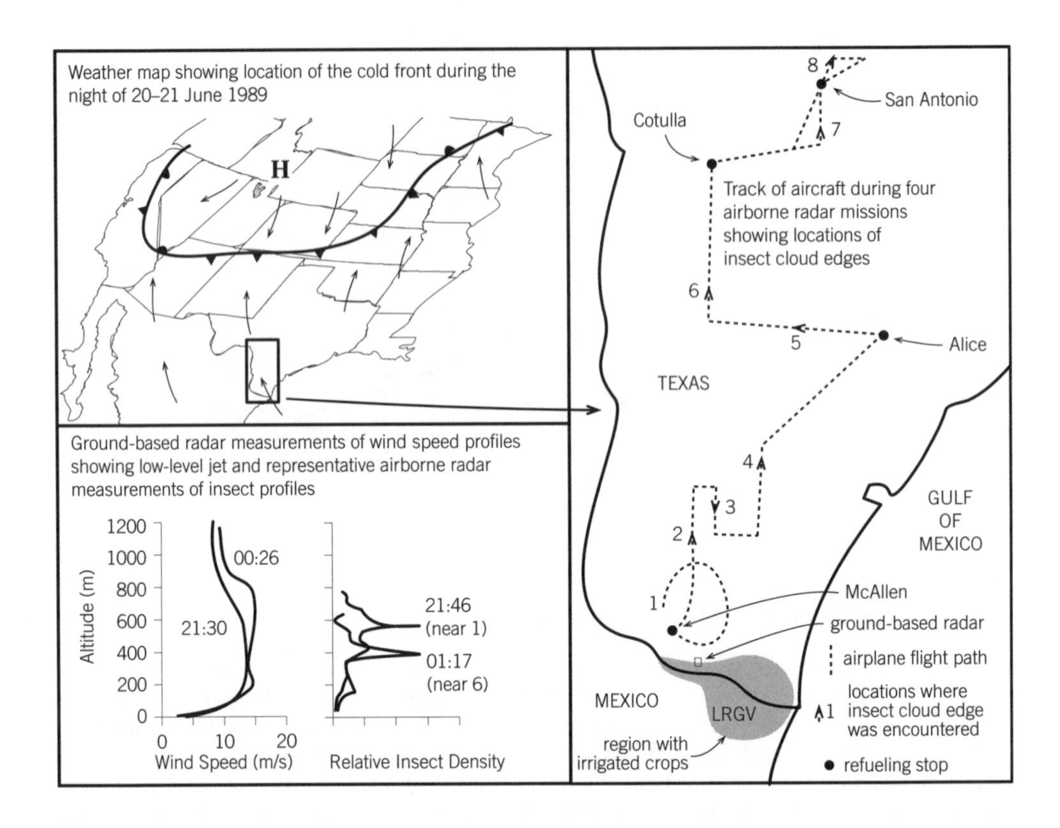

Figure 8.9. Airborne radar observations of moths migrating poleward in a low-level jet.

An irrigated maize-growing area in the Lower Rio Grand Valley (LRGV) of northern Mexico and southern Texas can act as a large insectary. Tactics designed to control populations of the fall armyworm (FAW) and corn earworm (CEW) moths throughout Texas and the central U.S. can be ineffective due to long-distance immigration of these pests from the LRGV. In 1989, the LRGV produced an estimated 7.9 billion FAW and 6.9 billion CEW moths (Pair et al. 1991). On the night of 20–21 June 1989, during the period of peak emergence, a cloud of FAW was tracked with ground-based and airborne radars from the LRGV to San Antonio, a distance of about 400 km (Wolf et al. 1990). Large numbers of flying insects were detected 200 to 700 m above the ground in a low-level jet. An edge of the insect cloud, marked by a rapid change in moth density, was clearly detected by the airborne radar on 8 separate occasions during the night. Citrus pollen on CEW captured in Oklahoma has also provided direct evidence of poleward moth migrations from the LRGV and elsewhere in the North American subtropics (Lingren et al. 1994).

During the past decade, USDA Agricultural Research Service scientists have harnessed a vast array of technologies and methodologies to improve knowledge about and the capability to predict aerial movements of these moths. In addition to employing the above-mentioned ground-based scanning radar, airborne radar, and pollen analysis, they have used extensive soil sampling to estimate larval populations in source areas, pheromone trap lines to measure adult populations in source and destination areas, and mylar tetrahedral balloon (tetroon) trajectories, air parcel trajectories, vertically looking radar, and WSR-88D (NEXRAD) radar to study the movements of these important pests (e.g., Westbrook et al. 1995a, 1995b). Figure adapted from Wolf et al. 1990.

exercise a high degree of selection and control over their flight altitude. They avoid cold air layers that limit their flight activity. In addition, whenever the temperature at night in the lower atmosphere is within the range that is conducive to flight activity, strong-flying insects respond to vertical temperature gradients to find the warmest air layers, which, when a low-level jet is present, are also zones of strong airflows (Figure 8.8). Migrating nocturnal moths have been tracked 100s of km in these winds (Figure 8.9).

Weak-flying insects, such as aphids, have also been found in relatively high concentrations in nocturnal low-level jets (Berry and Taylor 1968). Johnson (1969) suggests that weak-flying aerobiota carried high in the PBL by convection currents during the day may descend after nightfall (as the atmosphere cools) into the warm jet stream at the top of the temperature inversion and continue to fly throughout the night.

Convection currents.

Free convection results from the unequal absorption of radiant energy by landscape units. On sunny days, transfers of energy from surfaces that heat efficiently (generally characterized by high absorptivity for solar radiation, e.g., dark soil, and/or low thermal capacity, e.g., dry sand), warm the overlying air more rapidly than the air overlying cooler surfaces (generally characterized by high solar reflectivity, e.g., snow, high thermal capacity, e.g., wet soil, and/or mixing, e.g., water body). As a result of heating, the warm air expands, becoming less dense than the cooler air both beside and above it. Consequently, the resulting "thermal bubble" starts moving upward through the surrounding cooler, denser air. The continuous production of these buoyant eddies over a landscape unit constitutes a convection current. Eventually, the rising air loses its buoyancy as it mixes with the cooler air aloft. Near the ground surface, cooler, surrounding air flows into the space previously occupied by the warm rising air, where it too may be rapidly heated by the landscape unit (Figure 8.10).

When large-scale horizontal pressure gradients are small and consequently winds are calm, these convection currents can originate near the bottom of the biota boundary layer and extend through the PBL into the upper troposphere. Various types of convection currents may develop, depending on the rate of surface heating of the landscape units, the diversity and arrangement of the units on the landscape, the relief and topography (i.e., roughness) of the units, as well as the mechanical turbulence and environmental lapse rate in the lower atmosphere. They include thermals (small masses of warm rising air detached from the surface), plumes (columns of warm rising air), whirls (narrow twisting, quickly rising columns of warm air), and cells (columns of rising warm air interspaced with columns of subsiding cooler air).

Convection currents are generally a daytime phenomena in the atmosphere, most active during late morning through mid-afternoon. Under fair-weather conditions, the upper limit of convection is the top of the PBL, which generally does not exceed 1 to 3 km. Convection currents generally act to disperse aerobiota vertically throughout the lower atmosphere. They create a local circulation pattern of updrafts scattered among downdrafts. Consequently, fair-weather convection usually produces a highly heterogeneous (patchy) horizontal distribution of aerobiota and generally is considered an inefficient mechanism for

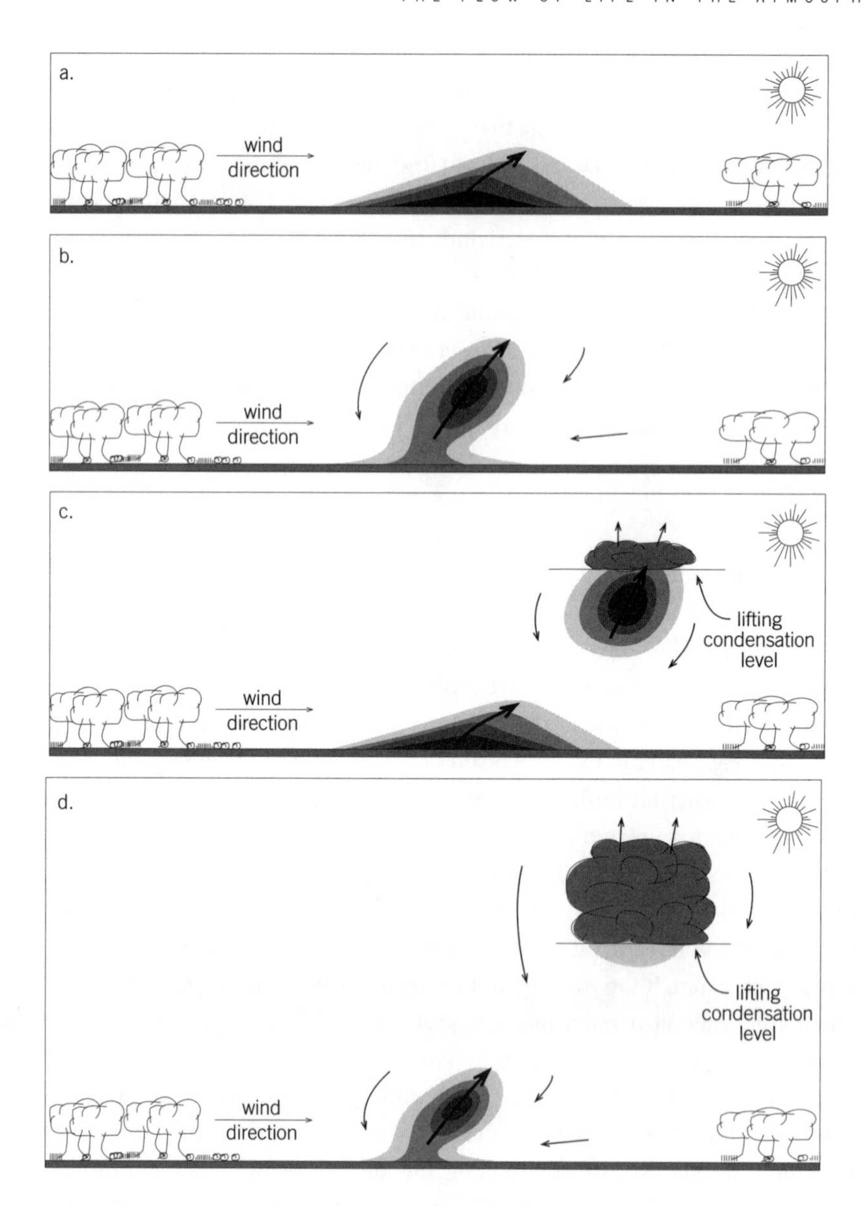

Figure 8.10. Development of thermals and convection currents.

(a) On clear days, local variations in the Earth's surface produce pockets of relatively warm air above materials that are good absorbers of solar radiation. (b) As these thermals grow upward within the surface layer of the atmosphere, they are distorted in the downwind direction. The rising warm surface air is replaced by colder, denser air from above. (c) If the decrease in temperature with altitude in the surrounding air is large enough, the "thermal" will detach from the surface and ascend within the PBL. As it rises, it will cool adiabatically, and it may become saturated and form into a cumulus cloud when it reaches the lifting condensation level. (d) The rate of convective uplift depends on the moisture content of the thermal and the stability of the surrounding atmosphere. A convection current is formed when the production of thermals above a patch or landscape unit is continuous. Many biota ascend high into the PBL assisted by thermals while others avoid aerial movement when the atmosphere is turbulent. For example, soaring birds and insects generally migrate during midday when thermals are prevalent while those that have the capacity for strong powered flight migrate at night or early in the morning when the atmosphere is stable.

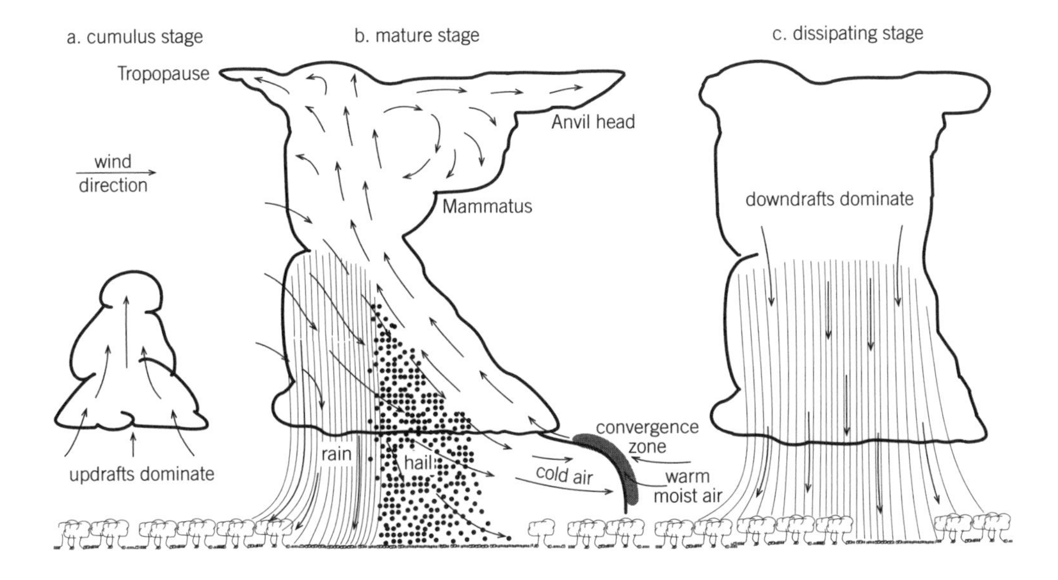

Figure 8.11. Stages of convective storm (thunderstorm) development.

Air-mass thunderstorms tend to develop when the lower atmosphere is highly unstable. Therefore, when warm, moist air rises from the surface, condensation quickly occurs, releasing energy and enabling vertical acceleration of the air within the cumulus cloud (a). The formation of circulation patterns within the convective cell marks the mature stage (b). Downdrafts of cold air originating in the upper center of the cell produce a cool current that spreads outward above the ground surface and lifts the warm, moist air in front of the cell. A convergence zone develops at the outflow front where biota that resist upward movement can become concentrated. Movement of the cell over warm ground helps to propagate continued activity. The dissipation stage occurs when the updrafts weaken, reducing the release of latent energy within the cell, and downdrafts prevail.

transporting organisms over wide areas. Occasionally, when convection is organized into polygonal or linear cells and rolls, horizontal convergence of air into the cell walls can act to concentrate strong-flying insects that resist being carried aloft.

Convection currents generated by differential solar heating of landscape units often produce precipitation. Showers can develop in moist air masses which are sufficiently unstable so that convection currents cause condensation and the development of thick cumulus clouds. Convective storms (often called thunderstorms) develop within an air mass when atmospheric instability is extreme and air layers "overturn" in order to achieve a more stable density stratification. Condensation in strong convective updrafts characterize the early stages of convective storms, while strong downdrafts in columns of precipitation mark their dissipating stages (Figure 8.11). Generally, individual air-mass thunderstorms are of short duration, lasting less than 2 hrs and traversing less than 100 km. A squall line forms as a line of convective storms. In the midlatitudes, squall lines often develop along cold fronts. However, they are also common within the warm, moist air as far as a few hundred kilometers ahead of a cold front. A squall line of prefrontal thunderstorms can extend over 1000 km in length and cause extremely severe weather including tornadoes.

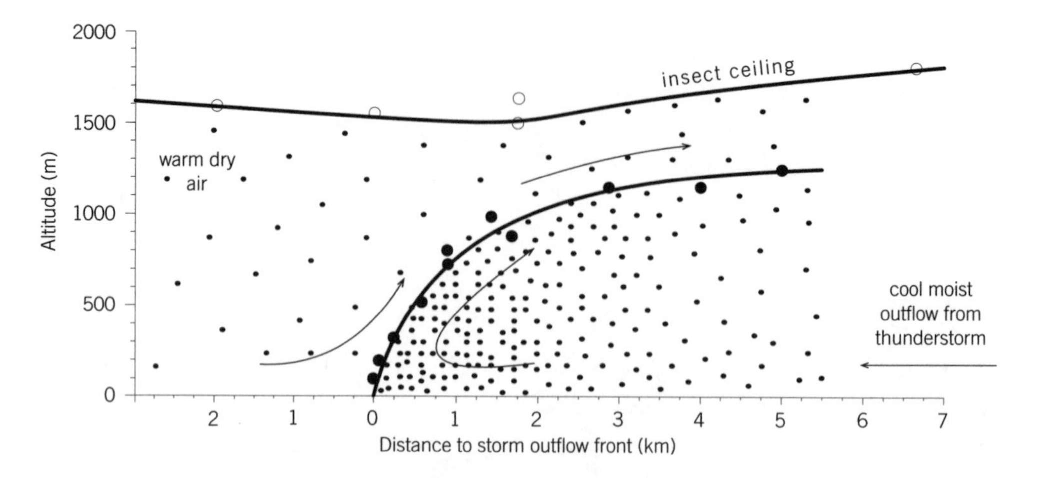

Figure 8.12. Insects concentrated by the outflow from a convective storm.

The concentration of insects increased 60-fold with the passage of a cool outflow front from a rainstorm in the Sudan (Shaefer 1976). There was a dramatic increase in turbulence and wind speed and a shift in wind direction as the cool current approached. Prior to the passage of the front, the orientation of the insects in the air was generally uniform. This uniformity diminished at low levels immediately ahead of the front and insect orientation was random within the cool current. The storm outflow had been collecting insects for about an hour prior to the radar observations. After the passage of the front, the insect concentration gradually decreased, but nonetheless remained about 13 times the concentration in the air before the passage of the cool outflow current. Figure adapted from Schaefer 1976.

Aerobiota that are lifted gradually over time into a storm cell by convective currents and storm updrafts may be deposited in high concentrations on a landscape by sudden strong downdrafts, or washed out of the atmosphere by the accompanying precipitation. For example, very high and localized concentrations of moths have been noted numerous times in association with convective storm cells (see references in Table 5.1). Much of the inflowing air to a convective storm is derived from the surface boundary layer, and convergence in these updrafts could likely produce a 10-fold increase in concentration, over an hour, where insects successfully resist being carried aloft (Pedgley 1990). Strong downdrafts within a storm cell can produce cold outflows that spread laterally across the landscape as density currents. The outflows are usually strongest on the forward side of a moving storm where they can form rapidly moving fronts. As the cold, dense air advances, it lifts the warmer air in its path, creating a thin convergence zone at the leading edge of the outflow. Consequently, storm outflows are effective mechanisms for lifting weak-flying and passively transported organisms from a vegetation canopy, through their biota boundary layer, and into the PBL above. It seems likely that many of these biota are swept into the parent storm cell and subsequently returned to the landscape by downdrafts or washed out by precipitation. However, aerial concentration of actively flying biota can occur in storm outflow convergence zones when organisms resist being carried aloft (Figure 8.12).

CHAPTER 9

Airscapes: The Atmosphere's Surface Layer and Microscale Motion Systems

In the atmosphere's surface layer, the vertical transfers of momentum, heat, moisture, and other constituents, including biota, are extremely complicated. The depth of this layer is highly variable because it depends on the wind speed in the PBL above, the roughness of the Earth's surface below, and the temperature gradient between the surface and air. Typically, the surface layer constitutes the lowest few meters of the PBL (see Chapter 5). As a general rule, it becomes shallower as the speed of the wind in the PBL increases, and it grows thicker where roughness elements on the Earth's surface become taller and more closely spaced or when the sensible heat flux is directed upward from the ground to the air. Within the surface layer, atmospheric constituents are transported along their concentration gradients by both molecular agitation (diffusion) and the mass movement of molecules (turbulence). To understand the types of airflows that biota encounter as they move within the lower atmosphere, it is often useful to subdivide the surface layer into three thin zones or boundary layers: the laminar, the turbulent surface, and the roughness boundary layers.

Laminar boundary layer.

The streamlines of airflow in very close proximity (i.e., millimeters) to the objects, or roughness elements, that compose the Earth's surface are primarily straight and horizontal (see Figure 4.4). Vertical transport of atmospheric constituents through this laminar layer occurs along their concentration gradients and is due to random movements of molecules. Thus, in effect, diffusion tends to equalize the vertical distribution of a constituent by reducing its gradient. The molecular diffusion process is generally much less efficient than the turbulent transfer in the atmosphere above. For this reason, the large fluxes of heat and moisture from the ground to the atmosphere that are common during warm, sunny midday hours, require extremely large vertical temperature and vapor gradients within the laminar boundary layer. Because even small microorganisms (e.g., bacteria, viruses, and spores) floating in the laminar layer will fall to Earth under the influence of gravity, their transport through this layer must depend on energy from sources other than the airflow. As noted in Chapter 4, many small aerobiota are launched into the atmosphere by their parent plants while the particu-

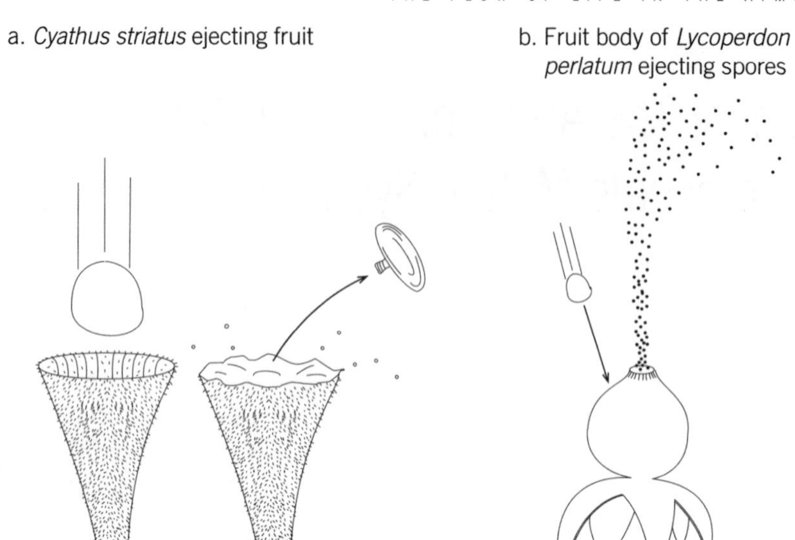

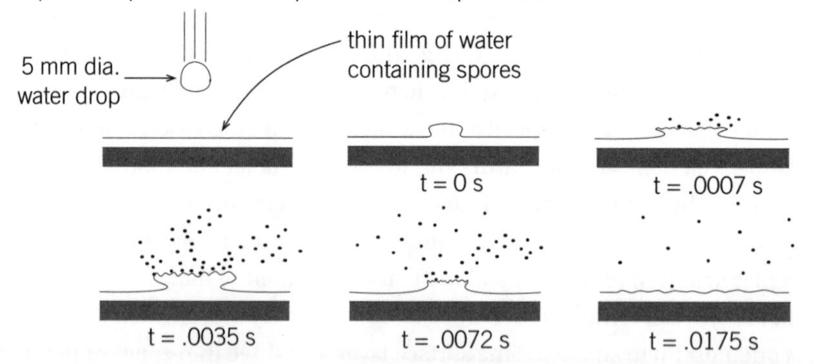

Figure 9.1. Takeoff of microorganisms assisted by raindrops.

Raindrops can liberate fungal spores and other microorganisms through a variety of mechanisms. Some organisms, such as *Cyathus striatus* (a) have elaborate splash cups to facilitate rain-assisted takeoff. Raindrops hitting bellows-shaped fruit bodies provide a mechanism to launch spores (b). The velocity of the puff emerging from a *Lycoperdon perlatum* ostiole has been measured at 10 m/s (Gregory 1949). The same mechanism is likely operated many thousands of times each growing season. Raindrops also can cause takeoff of dry spores through other means: by tapping leaves and other plant parts, causing them to shed their spores and by creating miniature radially moving gust fronts that propel spores at speeds that may exceed 50 m/s for very brief periods and distances (Hirst and Stedman 1963). Droplets containing spores may spray outward from a raindrop incident on a plant part or water film (c). Some of the droplets may be carried downwind and evaporate, leaving the spores they contained to be transported further by eddies. Figure adapted from Brodie (1951), Ingold (1971), and Gregory (1973).

lar presentation of others on exposed plant parts allow them to rely on gravity to fall through the laminar boundary layer.

Although tiny in both temporal and spatial scales, motions induced by raindrops are surprisingly effective at propelling organisms through the laminar boundary layer. In large part, this is because roughly one billion raindrops large enough to produce a splash are incident per square meter in areas that receive approximately 100 cm of rain annually (e.g., many humid tropical and midlatitude areas; Gregory 1973). The physical characteristics of the Earth's surface (e.g., ground slope and vegetation type and structure) and drop size have important influences on the effectiveness of rainsplash as a liberation mechanism.

The impact of raindrops can liberate spores and other microorganisms through a number of mechanisms, such as rain-driven air bellows, shaking and tapping leaves and other plant parts, and generating small air puffs which blow microorganisms from dry surfaces (Figure 9.1). In addition, takeoff and ascent of microorganisms within rainsplash is very common, not only due to the abundance of raindrops but also because a single drop can generate between 100 and 5000 splash droplets. In still air, these splash droplets typically travel 8 to 20 cm outward from the center of the splash, with a few droplets landing as far as 1 m away (Gregory 1973). All but the smallest of droplets are able to carry microorganisms, and thus the splashes of 10s of 1000s of raindrops that fall on a field during a single storm provide an efficient mechanism for the rapid and extensive aerial transport of biota among leaves, plants, and nearby fields. Typically most microorganisms transported in rainsplash are deposited with the droplets on the ground or on neighboring plant parts; however, where it is windy, microorganisms may be entrained in the atmosphere above the laminar layer when their host droplets evaporate. Consequently, they are in a position to be carried upward out of the plant canopy and afar, first by turbulent eddies and then by the many and varied larger-scale atmospheric motion systems that they encounter.

Turbulent surface boundary layer.

Above the laminar boundary layer, transfer of atmospheric constituents is primarily accomplished by turbulence, which while much more efficient than molecular diffusion, is also chaotic. Turbulence is the gustiness, or variation from the mean wind speed, that we constantly feel in the atmosphere. It consists of eddies, which are irregular swirls of motion or microscale circulations (Figure 9.2a). Often eddies of different scales are superimposed on each other and they coalesce as they move upward in the turbulent surface boundary layer. Their structure is strongly influenced by the size and configuration of the surface roughness elements. As the eddies move, they carry atmospheric constituents, including small aerobiota. Although the airflow in the turbulent surface boundary layer has a mean "downstream" direction when averaged over periods greater than a few seconds, the instantaneous fluctuations in the direction of movement of the eddies are random (see Figure 5.3).

Turbulent eddies are generated by mechanical and buoyant forces in the lower atmosphere. Whenever the wind blows, its speed decreases downward toward the Earth's surface (lowest air layer adheres to the surface), creating a vertical wind speed gradient (Figure 9.2b). Mechanical forces are caused by friction in the air as the faster airstreams move over the slower air layers and result in turbulent mixing of the air. Buoyant forces are caused by air density gradients. Near the Earth's surface these forces primarily result from radiative heating and cooling of the ground. In the absence of heating and cooling, air is compressed (its density increases) toward the Earth's surface due to gravity. Whenever the heating of air near the ground produces a vertical temperature gradient that is large enough to cause an

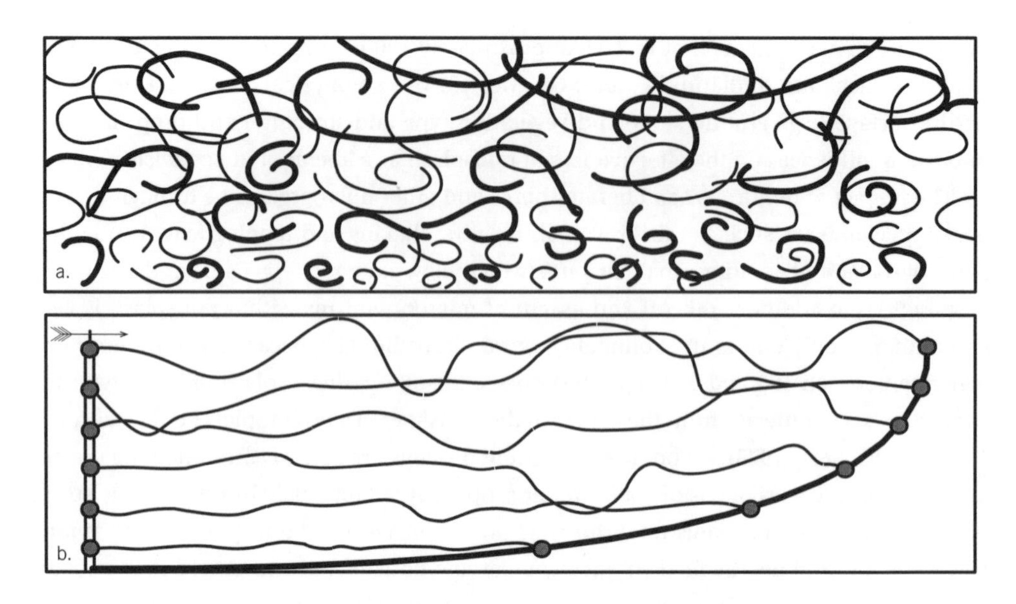

Figure 9.2. Structure of turbulent eddies in the surface layer of the atmosphere.

Instantaneous streamlines (a) show the quasi-circular structure of eddies in the turbulent surface boundary layer of the atmosphere. Three generalizations can be made about the variation of eddies with height above the ground. First, the average size of eddies increases with altitude. Second, the horizontal distance traveled by eddies increases with height. Third, the relative variability of eddy size decreases with altitude within the turbulent surface boundary layer. Downwind trajectories of visual tracers (b) released simultaneously from a mast show the variation in the paths of wind streamlines. The heavy line shows the wind speed profile that generally increases logarithmically with height above the Earth's surface. Figure adapted from Lowry and Lowry 1989.

increase in air density with height (greater than approximately –9.8°C per km—the dry adiabatic lapse rate [DALR]), buoyant forces are directed upward (free convection) and the atmosphere is considered unstable. Conversely, the atmosphere is considered stable when the decrease in air temperature with altitude is less than the DALR and the vertical air density gradient acts to repress vertical motions. At night, the atmosphere is usually stable and temperature often increases with height (inverted lapse rate), causing mechanical turbulence (forced convection) to dominate unless winds are calm. During daytime, when the atmosphere is unstable, free convection can be the main source of turbulence if wind speeds are very low. However, because the vertical gradient of wind speed is pronounced near the Earth's surface even during light winds, both forced and free convection are important in the daytime within the turbulent surface boundary layer.

Momentum from turbulent eddies is used to move microorganisms entrained within them from one position to another in the atmosphere. For passively transported organisms, individual movements are random in duration and direction and are a result of the isotropic velocity fluctuations of eddies in the lower atmosphere. Because the atmosphere possesses physically meaningful averages over space and time (e.g., mean wind direction and speed), the aerial movement patterns that result from the individual random movements form coherent trajectories (Figure 9.2b). Mechanical turbulence tends to disperse organisms in the downwind direction while buoyant forces tend to encourage organized vertical motion of air and biota.

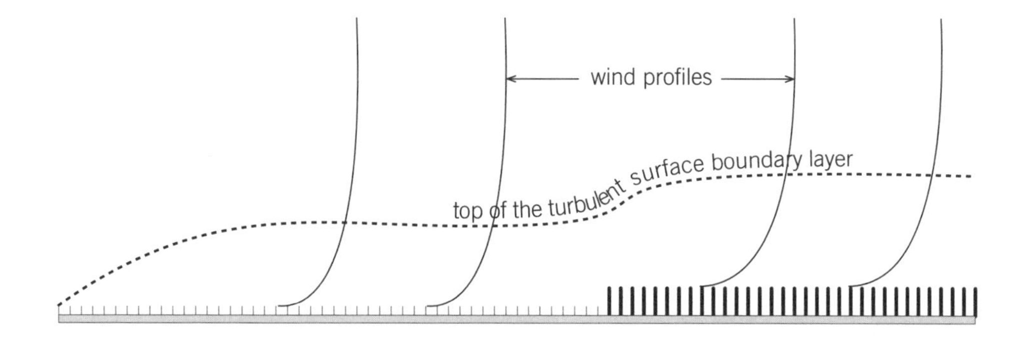

Figure 9.3. Adjustment of the wind profile to changes in surface roughness.

Surface roughness, surface heating, and the wind in the PBL above generate intense microscale turbulence in the surface layer. Despite the variability of turbulent eddies over time scales of less than a few seconds, the turbulent surface layer can appear relatively homogeneous when viewed over periods of many minutes. When the roughness elements on the Earth are relatively uniform, the depth of the turbulent surface boundary layer and the wind speed profile reach an accommodation with the ground cover shortly downwind of the leading edge of the surface unit. Once established, this structure is typically maintained within the flow until a surface unit with a different type or configuration of elements is encountered. The structure of flow within the turbulent surface layer will be maintained through time until surface heating/cooling or changes in wind speed/direction in the PBL above cause it to change. Figure adapted from from Monteith and Unsworth 1990.

As mentioned above, mechanical and buoyant forces, and consequently the structure of turbulent eddies in the lower atmosphere, are strongly influenced by spatial variations in characteristics of the landscape that affect the exchanges of energy (heating of the air) and momentum (frictional drag on the wind) between the Earth's surface and the atmosphere flowing over it. In the simplest situation, where the PBL winds are moderate and the Earth's surface below is relatively uniform and flat (e.g., mowed grass or calm water body), the atmosphere's surface layer is composed of a thin laminar boundary layer beneath a thicker zone of turbulent flow. Within a short distance downwind from the leading edge of such a surface, the structure of the eddies within the turbulent surface layer, as characterized by the wind speed profile, adjusts to the size and configuration of the roughness elements. Once established, this accommodation by the wind speed profile to the surface roughness is maintained, until another surface with different types or configurations of roughness elements, is encountered (Figure 9.3). Because the physical mechanisms that maintain turbulence and the vertical transfers of atmospheric constituents in the surface layer are the same as in the PBL above, the upper limit of the turbulent surface boundary layer is somewhat arbitrary. In practice, however, it is usually defined as the height below which 90% of the total vertical change in the mean horizontal wind speed within the PBL occurs.

Roughness boundary layer.

Under most circumstances, the types and configuration of surface roughness elements on the ground are heterogeneous, and it is convenient to consider a roughness boundary layer composed of turbulent eddies between the laminar and turbulent surface boundary layers.

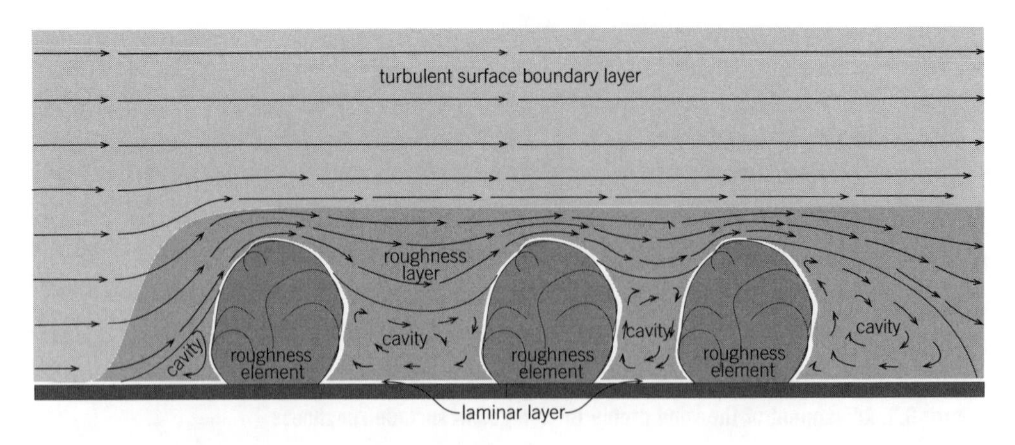

Figure 9.4. Airflow within the roughness boundary layer.

The roughness boundary layer, a zone of deformed airflow, develops within the atmosphere's surface layer when the size, shape, and/or configuration of roughness elements on the Earth's surface is heterogeneous. The speed of the wind within the roughness boundary layer is usually much slower than that in the turbulent surface boundary layer above. Turbulent eddies move among and over the roughness elements along constantly changing pathways. When the wind speed is low and the roughness elements are rounded and short, the flow is not greatly deformed. More complexly shaped objects with sharp edges and large height to base ratios (e.g., buildings) deform the airflow. Small cavities develop upwind, between, and downwind of plants and other roughness elements where the speed and direction of the airflow is somewhat independent of the wind above. The low wind speeds within the roughness layer of a plant canopy, and especially within the cavities near roughness elements, allow many weak-flying biota to take off and land successfully and control the direction of their aerial movements. Figure adapted from Levizzani et al. 1998.

For example, the roughness boundary layer concept is often applied to the air spaces within rows of crops or below the tops of trees in a forest. In these landscapes, a laminar boundary layer is usually present immediately above the soil and around individual plant parts, while the airflow above the canopy constitutes the turbulent surface boundary layer. Within the roughness layer, eddies move along constantly changing pathways through the spaces among the elements in a manner that is analogous to the flow of water among and across rocks in a shallow stream. Each roughness element represents an obstacle that the eddies must avoid. As a result of the increased friction between the airflow and the Earth, the speed of the wind within a plant canopy is generally much slower than that above, and the streamlines of airflow are deformed. Small cavities generally develop immediately upstream and downstream of objects, and the movements of the air within these pockets are much slower, and somewhat independent of the flow in the surrounding airstream (Figure 9.4).

Atmospheric and biota boundary layers.

In most situations there is no clear correspondence between the biota boundary layer concept and various meteorological boundary layers in the atmosphere's surface layer. Although they are related, the biological and meteorological approaches represent different perspectives for viewing the same motion systems. A good understanding of both schemes,

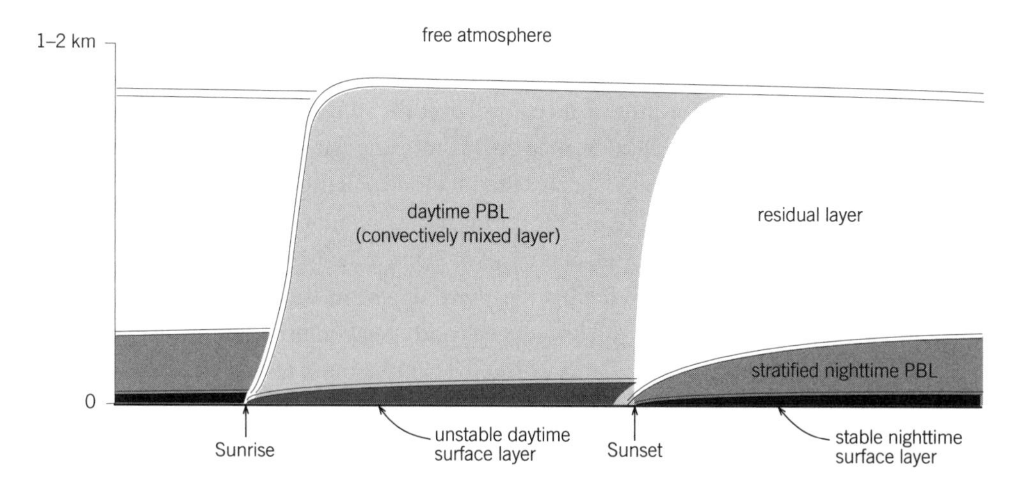

Figure 9.5. Evolution of the planetary and surface boundary layers throughout a diurnal period characterized by clear skies and moderate winds.

On clear mornings, solar radiation rapidly heats the Earth's surface, and vertical air temperature gradients within the surface layer begin to increase in the absence of strong winds. By mid-morning, the buoyant forces exerted on eddies within the surface layer are usually large enough to overcome gravity. During these unstable atmospheric conditions, the warm eddies that form within the surface layer and are displaced vertically by turbulence tend to coalesce into thermals and continue moving upward. This air in turn is replaced in the surface layer by cooler eddies that subside from above. The vertically moving airflows tend to distribute atmospheric constituents, including biota, throughout the PBL, and consequently, the daytime PBL is often called the "mixed" layer. This convectively mixed layer usually extends 1 to 3 km above the ground and is maintained on clear days until about sunset. Nightfall brings radiative cooling of the ground and a decrease in the vertical temperature gradient in the surface layer. During clear sky conditions, a rapid transition occurs between the daytime and nighttime PBL. Often an inversion develops at night within the surface layer, and it can extend upward for a few 100 m, creating a stratified PBL. It is at the top of this inversion layer that the wind speed maxima called low-level jets, which are instrumental to long-distance travel of many biota, develop. Figure adapted from Stull 1988 and Garratt 1992.

however, provides insights that permit a wider understanding of the interactions between biological and meteorological processes that govern the aerial movement of biota.

The correspondence between the two perspectives is better for organisms at either end of the continuum, from passively transported to actively flying aerobiota than for those in between (refer to Table 4.4). As mentioned above, many microorganisms rely on energy from their parent plants or gravity to traverse the laminar boundary layer. For these organisms, the zone of laminar flow represents their biota boundary layer in which they can exert some control over their movement. Outside the thin laminar layer, they must rely entirely on energy from the turbulent airflow for transport. In contrast, the biota boundary layer for strong-flying birds may encompass the entire PBL during most atmospheric conditions, with the exception of weather systems characterized by extreme air temperatures, strong winds, and/or heavy precipitation.

For the majority of the aerobiota taxa in between, the wind speed and structure of the turbulence within the turbulent surface and roughness boundary layers influence the thickness of their biota boundary layer. For example, some weak-flying insects (such as aphids and leafhoppers) tend to avoid takeoff at wind speeds greater than about 0.5–1.0 m s^{-1}, probably because they have little or no control over their flight direction at greater wind speeds. Many of these insects launch themselves into the wind to gain lift. When they encounter an airstream with a speed greater than their own flight speed, their flight trajectory is quickly turned downwind. Frequently during daylight hours, wind speeds above plant canopies exceed the upper limit for takeoff of weak-flying organisms. During these circumstances, the roughness layer within the plant canopy represents their biota boundary layer. Not only does the roughness layer with its reduced wind speeds allow weak-flying organisms to take off and navigate among the plants within the canopy, but it also enables weak-flying biota that leave the canopy to land without incurring damage due to crashing.

A diurnal periodicity in the structure of turbulence is common in the atmosphere's surface layer under many weather conditions (Figure 9.5). Consequently, the depth of the biota boundary layer for many passively transported and weak-flying organisms may change throughout the diurnal period. During clear, calm nights, radiative cooling of the Earth's surface usually creates stable atmospheric conditions. In the absence of strong large-scale atmospheric motion systems, a temperature inversion may develop, inhibiting free convection (thermals) and extending the biota boundary layer of strong-flying organisms through the surface layer to the top of the PBL. Thus it is not surprising that many strong-flying insects, and the majority of birds that use powered flight to migrate long distances travel at night, when vertical motions in the PBL are minimal.

During the unstable atmospheric conditions that predominate during daytime, the biota boundary layer of many weak-flying organisms generally does not extend above the roughness layer. Once weak-flying and passively transported biota ascend into the surface boundary layer, they can be carried far aloft within eddies that coalesce to form convection currents. Subsequently, they are likely to be returned to the surface layer by subsiding air currents. The birds and insects that use soaring flight (e.g., hawks and monarch butterflies), and spiders that spin silken threads for ballooning, are examples of daytime travelers that rely on the vertical motions of thermals for lift and then glide or float gradually downward through the PBL. On warm, sunny days, some aerobiota can traverse long distances across landscapes using relatively little energy by repeatedly riding thermals upward and gliding slowly downward through the PBL. Passively transported and weak-flying biota that are high in the PBL at sunset can move long distances during the night if they descend into the layer of warm temperatures and strong winds (low-level jet) that often develop at the top of the stratified nocturnal PBL (see Figures 8.8 and 8.9).

CHAPTER 10

The Importance of Microscale Motion Systems and Flight Aptitude to the Ascent of Aphids through the Surface Boundary Layer: A Research Agenda

"If we wish to control the dispersal process, a precise knowledge of the mechanisms involved is preferable to the vague idea that the spores will get there somehow anyway! Success in colonization or fertilization depends on logistics—on getting enough material to the right place at the right time" (Gregory 1973).

In Chapter 4, atmospheric and geographic views of aerial movement were combined into an airscape perspective on the flow of biota within the atmosphere. We believe that this perspective is valuable because it focuses attention on the many ways plants and animals use the atmosphere to flow among habitats. Building on Gregory's lead, precise information and knowledge are necessary to understand and forecast movements of a wide range of important aerobiota to ensure human health and to manage terrestrial ecosystems so they are able to meet human needs in the future.

The airscape perspective has its roots in L. R. Taylor's (1958) application of the term "boundary layer" to flying insects. This concept, which was expounded and illustrated by C. G. Johnson (1969) with examples of the many different ways that insects make use of winds, was fundamental to progress in aerobiology. The insect boundary layer provides a conceptual framework for integrating hypotheses and theories from the meteorological and biological sciences about processes and behaviors that occur in the lower atmosphere. Although Taylor and Johnson applied their ideas to insects, they were aware of their general utility to aerobiology. In his presentation to the Linnaen Society of London, Taylor (1958) proposed that actively flying insects face problems similar to those of passively transported spores during takeoff and ascent. "Spores must cross the thin layer of still air close to the ground before they can enter turbulent air leading into the free circulation necessary for their dispersal" while "for insects the air layer which must be crossed before free circulation is entered is that in which air movement is less than flight speed or within which the insects' sensory mechanisms and behavior permit active orientation to the ground." For example, Taylor (1958) viewed the ejection of ascospores by Ascomycetes as analogous to the vertical takeoff flight of aphids in that "both are responses to current climatic factors and both may lead to dispersal dependent upon air movement."

The strength of the insect boundary layer concept is that it differentiates between a lower atmospheric layer, in which insects have considerable control over their movements, and a zone above, where the speed and direction of the wind govern insect displacement. However, as discussed in Chapter 9, the increase in wind speed with height in the lower atmosphere is highly variable in both space and time because it is strongly influenced by: (1) topography and roughness elements in the terrain that impede airflow, (2) the speed and direction of the wind in the PBL above, and (3) diel changes in the stability of the lower atmosphere. Thus an accurate determination of an insect's boundary layer was regarded as infeasible because its depth not only depends on the physical factors associated with the airflow but also varies with an individual insect's power of flight and from one species to another. Johnson (1969) lamented that "at present it is impossible to describe precisely the responses of flying insects within their boundary layer," and thus he considered the insect boundary layer concept to be "imprecise".

Nevertheless, frictional retardation of the wind by the ground and solar heating of the Earth's surface, create structures within near-surface airflows that are coherent and quantifiable. Consequently, although the physical dimensions of an organism's boundary layer (i.e., depth) may appear imprecise to humans, the layer is amenable to characterization with respect to its structures of motion at the temporal and spatial scales that are important to aerobiota. For example, R. C. Rainey was able to employ his knowledge of atmospheric processes to explain the enormous variation among locust swarms in their density and the altitude of flight of their topmost individuals. In his presentation following Taylor's at the Linnean Society, Rainey (1958) emphasized that "for studying the part played by atmospheric turbulence in these differences, use has been made of the fact that vertical gradients of air temperature not only greatly influence the intensity of turbulence (as has been illustrated by data on the vertical distribution of airborne aphids), but also determine the height to which such turbulence can extend above the ground. . . ." Rainey's application of his considerable knowledge of boundary layer processes to the study of locusts not only contributed substantially to our understanding of patterns of movement exhibited by flying insects, but also provided the capability to rigorously investigate the role of active behavior in the aerial movements of biota within natural environments.

Microscale motion systems in the surface boundary layer (composed of the laminar, roughness, and turbulent surface boundary layers, see Figure 9.4) are perhaps even more critical to the aerial movements of passively transported biota than actively flying organisms. For example, extensive field studies show that only a very small proportion of the tobacco blue mold spores released from lesions on infected plants actually escaped from the canopy of a tobacco field (Aylor and Taylor 1983). Plant canopies reduce the mean wind speed but increase turbulence and present a labyrinth of surfaces that may intercept floating spores. During neutral atmospheric conditions, the proportion of spores that left the crop canopy corresponded directly with wind speed in the surface boundary layer (3, 5, 7% and 1, 2, 3 m s^{-1}, respectively). Greater numbers of spores escape the canopy at higher wind speeds because enhanced turbulence provides more assistance to counteract settling due to gravity than at lower wind speeds. Escape also may be enhanced during unstable atmospheric conditions. Aylor (1978) suggests that spore entrainment in the air may depend more on the motions associated with infrequent peak gusts than the mean wind speed and thus one of the greatest challenges in predicting and managing the spread of airborne plant pathogens concerns the structure of the turbulent airflow within canopies: "Not only do we require knowledge about the frequency of occurrence and spatial variation of these gusts but, importantly, we must know their trajectories."

The ascent of aphids through the surface boundary layer.

Only by studying the structure of the turbulent airflows in the lower atmosphere, and the behavior of a wide range of organisms while moving in them, can generalizations concerning the takeoff and ascent stage (and the descent and landing stage) of bioflow be truly discovered. It is to this end that the general concept of a biota boundary layer was proposed (see Chapters 1 and 4). Here, that theme is developed further through the presentation of a research initiative which would observe and quantify the behaviors of organisms within the microscale atmospheric structures they use to move within and among habitats. Because movement depends first and foremost on the biology of aerobiota, we draw from our experience with aphids to elucidate this research. One important consideration in choosing aphids was the wealth of fundamental studies on basic biology and flight behavior of these insects (e.g., the many works of C. G. Johnson, L. R. Taylor, J. S. Kennedy and their many colleagues). Aphids represent a model system because biological and meteorological factors appear equally critical to the ascent process of this weak-flying insect. In addition, variations in both biological and meteorological factors that can occur over short temporal scales strongly influence the depth of the biota boundary layer for this organism. Finally, many aphids move long distances in the atmosphere, from subtropical overwintering habitats to the interiors of midlatitude continents each spring, transmitting plant viruses that are important to agriculture because of the enormous area and value of the crops they infect and the heavy economic losses that result (Irwin and Thresh 1988). Readers should transcend the particulars of the systems and technologies used to illustrate this research project and perceive the general scientific approach, which is based on evaluating hypotheses concerning the biological and meteorological interactions that govern bioflow. Two AFAR hypotheses (see Table E.1.), "the ascent by organisms into the atmosphere is influenced by environmental conditions" and "initiation of ascent into the atmosphere by organisms that move long distances is biologically controlled," provide the bases for the information presented in this chapter.

Biological and meteorological background.

The influence of environmental factors on the ascent phase of aphid movement is dependent on the aptitude (inclination and capacity) of alatae (winged aphids) for flight. Johnson (1969) compiled evidence from a large number of sources showing a daily bimodal rhythm in aerial densities of aphids and many other insects. Our investigations confirm this for

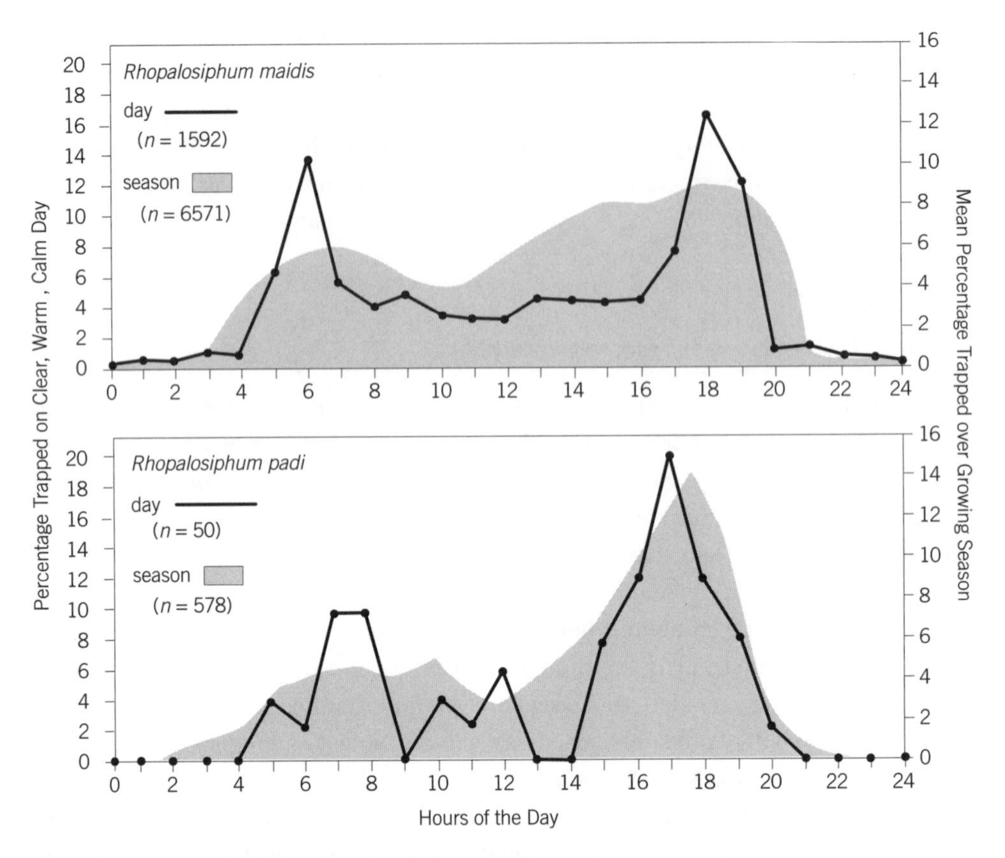

Figure 10.1. Bimodal flight activities of *Rhopalosiphum maidis* and *R. padi.*

Aphid collections were made using Johnson-Taylor suction traps above the canopy of an oat field during the summer of 1978. During a clear day with low wind speeds, warm air temperatures, and high pressure when aphids were particularly abundant, hourly collections of *R. maidis* and *R. padi* peaked during the early morning hours and again in the late afternoon (solid lines). The numbers of aphids captured were low during midday. When the collections are aggregated over the growing season, they display a similar diel pattern, although the bimodal flight periodicy is less pronounced (shaded areas). Cloudy weather, cool air temperatures, blustery winds, and/or precipitation associated with low pressure systems and afternoon convective storms commonly occur during the growing season in Illinois and inhibit takeoff by aphids. Figure adapted from Isard and Irwin 1993.

Rhopalosiphum maidis (Fitch) and *R. padi* (L), the corn leaf aphid and bird cherry-oat aphid respectively, which are important pests of cereals in North America. On a clear day during mid-summer, the number of aphids trapped in flight within the surface boundary layer peaks shortly after dawn, decreases during midday to generally less than 50% of the earlier aerial density, and increases dramatically in the late afternoon, peaking between 1700 and 1900 (Figure 10.1). Flight periodicity in aphids appears to be controlled by the rate of molting into winged aphids, the length of the teneral period from molting to flight, and environmental factors inhibiting flight such as darkness, cold temperatures, strong winds, and precipitation (Johnson and Taylor 1957). On days that are not predominantly sunny, the diel flight curve can assume a limitless number of forms because environmental conditions will delay takeoff by aphids.

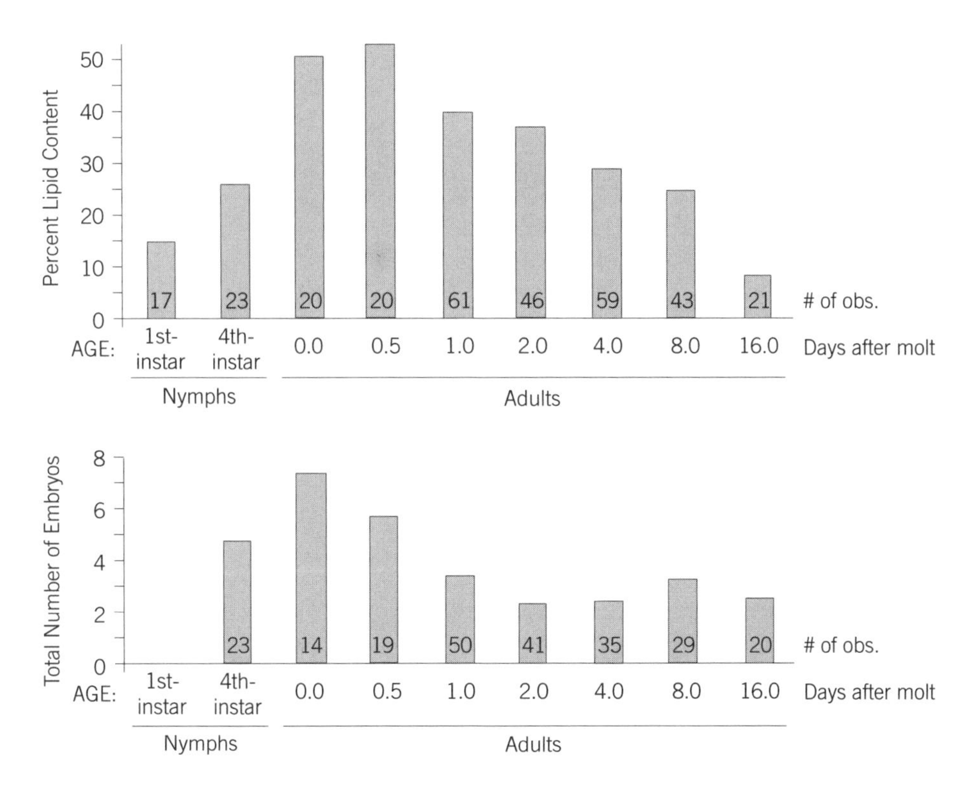

Figure 10.2. Total lipid content and degree of gravidity of nymphs and alates of varying ages of *Rhopalosiphum maidis*.

Total lipid content increases as *R. maidis* progresses through the nymph stage of development and peaks during the first day of adult life. This provides a ready source of energy that may be used for flight and or reproduction. Thereafter, total lipid content gradually declines. The variation in the total number of embryos is similar, reaching a peak immediately after alate emergence. *R. maidis* likely is capable of reproducing before and after long-distance flight, but if flight is delayed by adverse weather conditions, more of the lipids will be utilized for reproduction and fewer will be available for flight (Liquido and Irwin 1986). Figure adapted from Liquido and Irwin 1986.

Laboratory measurements of *R. maidis* indicate a rapid synthesis and storage of lipids prior to alate emergence, ensuring a ready source of energy for flight and/or reproduction. Total lipid content of *R. maidis* reaches a peak among adults approximately 0.5 days after molt, and then decreases. The peak reproductive period of *R. maidis* coincides with its maximum flight aptitude; i.e., during the first day of its adult life (Figure 10.2). The aptitude to sustain tethered flight is also greatest among 0.5-to-1-day-old *R. maidis* alates and decreases thereafter. When cohorts of 0.5 to 1-, 2-, and 4-day-old alates were flown, both the proportion of aphids that initiated flight and the duration of their flight decreased dramatically with age. Eight-day-old alates were incapable of flight (Figure 10.3).

Observations of the free flight of *R. padi* in wind tunnel studies confirm this finding. The mean angle of ascent for 0.5 to 1-day-old, 1-to-2-day-old, and 2-to-3-day-old alates that flew were 34°, 24°, and 17° respectively under a neutral atmospheric stability regime, while the proportion of aphids that initiated flight (64%, 48%, and 15% respectively) decreased equally dramatically with age. Three-to-4-day-old *R. padi* alates were incapable of flight (Figure

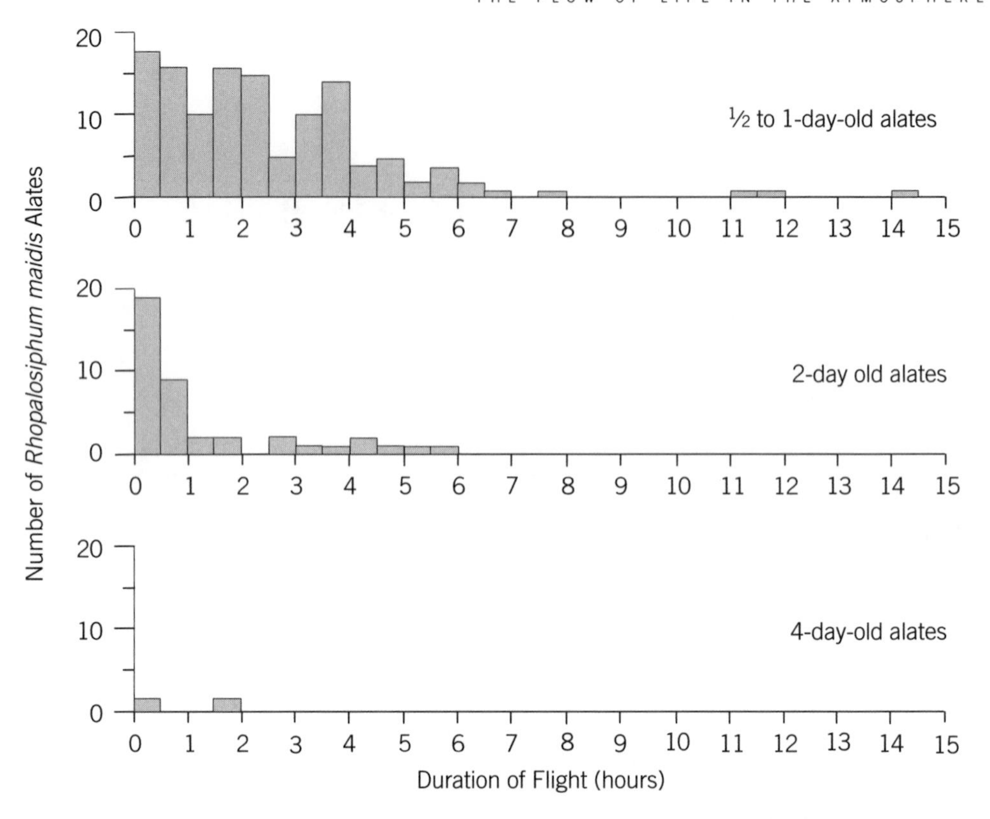

Figure 10.3. Variation with age in the aptitude of *Rhopalosiphum maidis* to sustain tethered flight.

The longest recorded flights of individual tethered *R. maidis* are shown stratified into age classes. The aptitude for sustained flight is greatest immediately after alate emergence and declines rapidly over the next few days. Most 8-day-old *R. maidis* were incapable of flapping their wings and none initiated flight. Although energy use and the motivation to sustain tethered flight is probably much different from free flight, the duration of flight under artificial conditions provides an indication of an aphid's capacity for long-distance flight. Liquido and Irwin (1986) suggest that time to exhaustion during tethered flight may approximate the maximum capacity for free flight because tethered aphids do not need to utilize energy to maintain lift. Figure adapted from Liquido and Irwin 1986.

10.4). These results on differential flight aptitude of *R. maidis* and *R. padi* alate age classes concur with previous findings that the flight muscle in many species of aphids starts to autolyze 2–3 days after adult molt (Johnson 1954).

The structure of the lower atmosphere in the continental interior of North America also experiences a diurnal periodicity in the absence of large-scale low pressure systems. In the growing season, radiative cooling on calm, clear nights usually creates a near-surface temperature inversion. For the first few hours after dawn during summer, the lower atmosphere is generally stable and air parcels that are displaced vertically by mechanical turbulence tend to return to their original level. However, vertical temperature gradients within the surface boundary layer usually increase after a few hours of sunny, calm weather; between midmorning and noon the buoyant forces exerted on air parcels near the Earth are typically large enough to overcome gravity. Air parcels that are displaced vertically by mechanical turbulence tend to continue moving upward as thermals (see Figure 8.10). These unstable

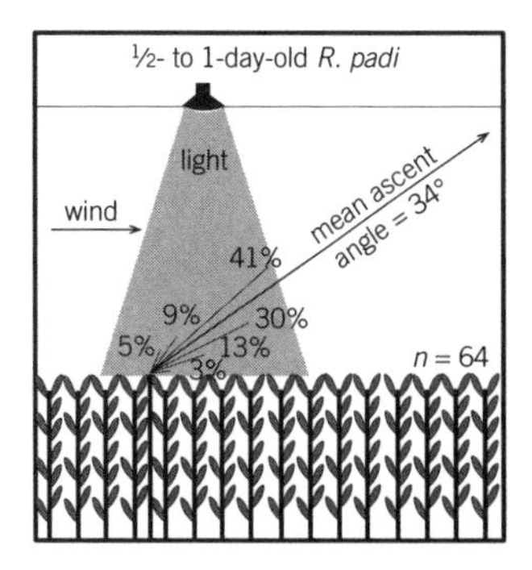

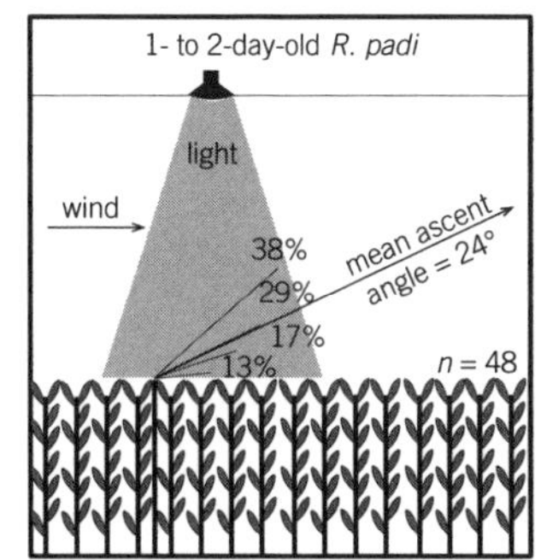

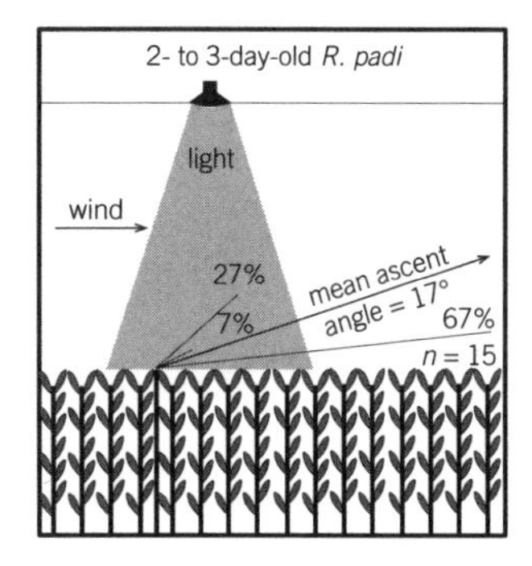

Figure 10.4. Flight trajectories for *Rhopolasiphum padi* alate age classes in wind tunnel experiments.

Trajectories of *R. padi* alates in free flight show a systematic decrease in the angle of ascent with age in wind tunnel studies under conditions of neutral stability. During these experiments, constant temperature and mean wind speed were maintained in the 12.2 m long, 3.7 m wide, and 2.2 m high wind tunnel. The vertical profile was isothermal (26° C). Wind speed at canopy level (0.3 m) was 0.35 ms^{-1} and increased logarithmically to 1.1 ms^{-1} at the 2.1 measurement level. Flight trajectories are displayed as vectors: the angle from horizontal depicts the aphids' mean angles of climb. These ascent angles were stratified into 5 range categories and the length of the vectors represents the percent of aphids initiating flight that exhibited a mean angle of climb in each category. The vectors with arrows represent the summation of the ascent angles for all trajectories in a single experiment. Under still conditions in the wind tunnel, the vast majority of alates that initiated flight from the release vial flew upward in a spiral pattern and reached the light fixture without leaving the cone of illumination. Figure adapted from Isard and Irwin 1996.

conditions often predominate until the onset of radiative cooling and stable conditions around sunset. Atmospheric stability is considered neutral when vertical temperature gradients are small. Neutral conditions can occur during periods of strong winds when mechanical turbulence thoroughly mixes the air or on calm, clear days during the transition periods between stable and unstable regimes (Figure 10.5).

Flight scenarios.

The objective of this research is to determine the biological and environmental factors that influence the ascent of aphids through the surface boundary layer of the atmosphere. On one hand, the success of ascent may depend on the energy status of the alate at takeoff, while on the other hand, the stability of the atmosphere may be equally important to determining an aphid's flight trajectory within the surface boundary layer. Laboratory and wind tunnel experiments (above) suggest that if aphids do not fly during the relatively narrow temporal window (0.5 to 1 day) when their flight aptitude is high ("long-distance flight mode"), they enter a long period during which their flight aptitude is relatively low ("local dispersal flight mode"). Because takeoff of alate aphids from agricultural fields can be delayed by darkness, precipitation, low temperatures, and strong winds, environmental conditions can cause alates to pass from the long-distance to local dispersal flight modes. We postulate that once aphids advance in age and begin to allocate energy to reproduction, they enter the local dispersal flight mode and potentially lose the opportunity for long-distance movement. Throughout calm, clear summer days, the structure of the surface

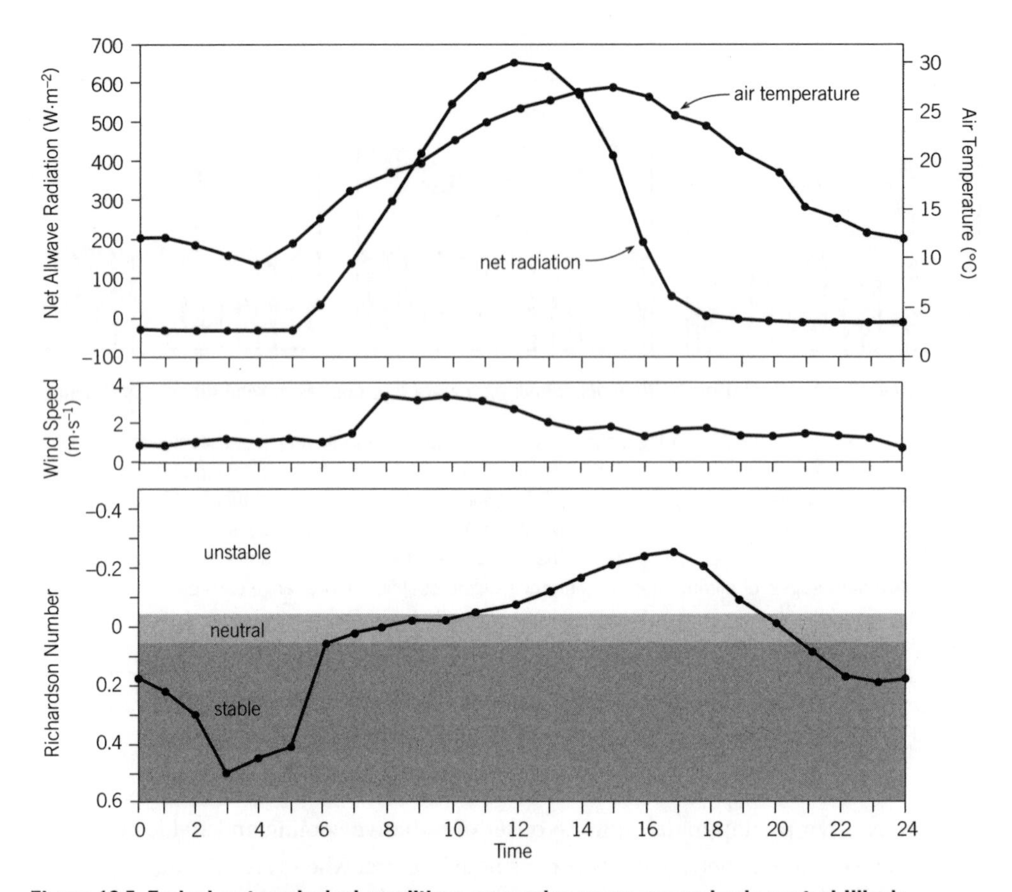

Figure 10.5. Typical meteorological conditions on a calm, warm, sunny day in central Illinois.

Measurements of net radiation, air temperature, and wind speed above a grass field in central Illinois are averaged for six predominantly clear days during August, 1988. The measurements of wind speed and air temperature also are averaged over 7 measurement levels (0.25, 0.35, 0.50, 0.75, 1.0, 1.5, and 2.0 m). The Richardson number represents the ratio of the buoyant to mechanical forces and is a measure of atmospheric stability. Values are stratified into stable, neutral, and unstable categories.

On calm, clear days at this midlatitude continental interior location, conditions in the lower atmosphere are governed by the absorption of net allwave radiation at the ground surface and subsequent heating of the air in the surface boundary layer. A lag of a few hours between the time of maximum radiation and maximum air temperature is common, and a direct relationship between air temperature and atmospheric stability occurs when the mean wind speed remains relatively low throughout the day. Quite often and at any time of day, strong winds associated with convective storms or low pressure systems can enhance mechanical turbulence, thus reducing vertical air temperature gradients and resulting in conditions characterized as neutral stability. However, when neutral conditions are associated with strong winds, aphids generally do not initiate flight. In contrast, aphids can take off during warm, sunny periods during the morning and evening transitions between stable and unstable conditions when the surface boundary layer is calm and neutral. Figure adapted from Isard and Irwin 1993.

boundary layer oscillates through three regimes in a predictable manner: stable (shortly after sunset to shortly after sunrise) and unstable (mid-morning to sunset) with brief transitions or neutral conditions between. The combination of these biological and meteorological factors leads to the six flight scenarios depicted in Figure 10.6 and described in Table 10.1.

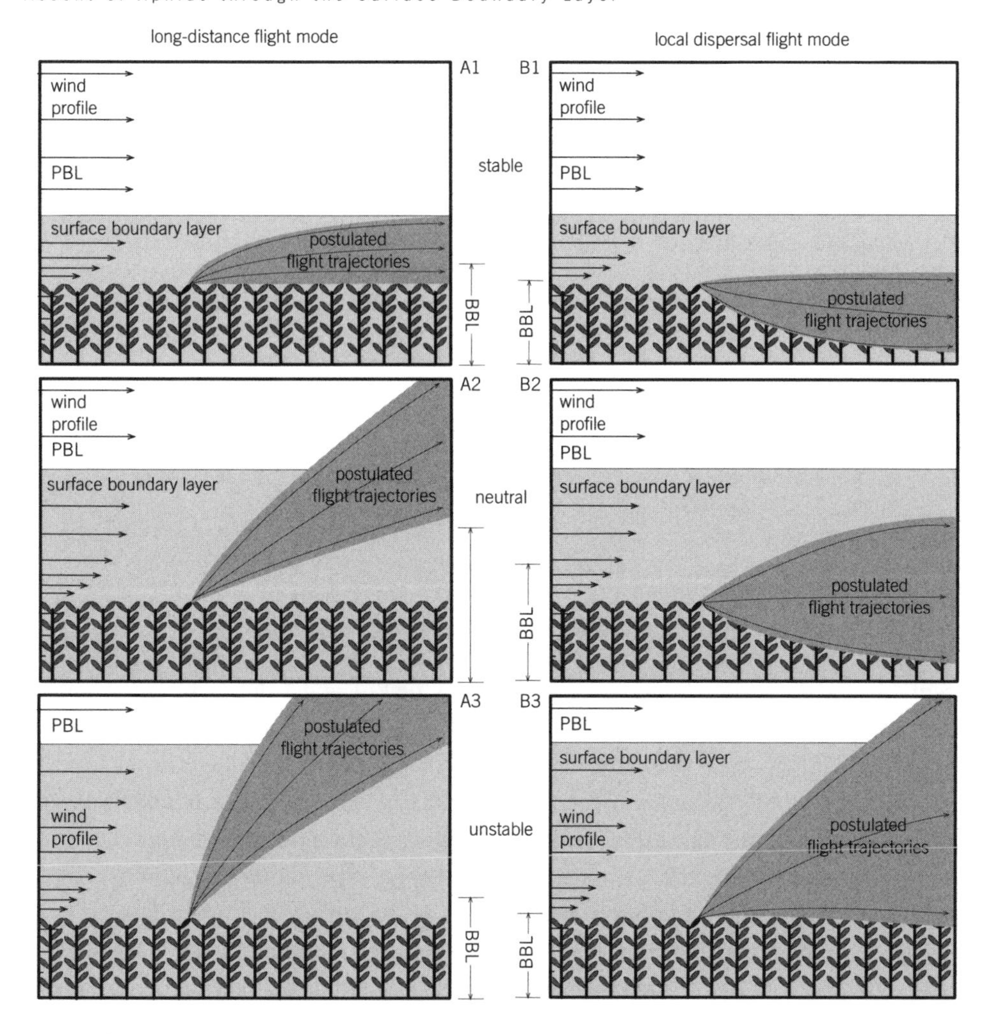

Figure 10.6. Postulated trajectories for aphid ascent flight scenarios.

The dark shaded areas represent postulated flight trajectories. An explanation for each scenario is provided in Table 10.1 and the text. The lightly shaded areas represent the surface boundary layer. The mean wind speed (arrows) in the PBL above the surface boundary layer and the characteristics of the crop are assumed to be similar for all scenarios, thus the surface boundary layer is shallow when the atmosphere is stable and deep when it is unstable. The neutral regime represents a transition between unstable and stable conditions. Generally *R. maidis* and *R. padi* do not initiate flight when air temperatures are cold (< 10–12° C) and/or when the wind speed in their vicinity is strong (> 0.5–1 ms^{-1}). The biota boundary layer (BBL) is deeper during neutral conditions when the buoyant and mechanical forces are about equal than during stable and unstable regimes when the buoyant forces are directed downward and upward respectively. For each atmospheric regime, the BBL is thicker for aphids moving long distances than for those in the local dispersal flight mode.

On calm, warm days, the temporal oscillations of aphid flight activity and atmospheric stability within the surface boundary layer have dissimilar diurnal patterns (compare Figures 10.1 with 10.5). For example, the first peak in aphid flight activity generally occurs between 0600 and 0800 when the surface boundary layer is stable. Thus it is postulated that flight trajectories of aphids in the surface boundary layer are primarily horizontal, and

Table 10.1. Postulated trajectories for aphid ascent flight scenarios.

For aphids that take off when their flight aptitude is high ("long-distance flight mode")

A1. Flight trajectories are primarily horizontal when the buoyant forces are small relative to the mechanical forces within the surface boundary layer (atmospheric stability = stable).

A2. Flight trajectories have an upward component when the buoyant forces are approximately equal to the mechanical forces within the surface boundary layer (atmospheric stability = neutral).

A3. Flight trajectories have a steep upward component when the buoyant forces are large relative to the mechanical forces within the surface boundary layer (atmospheric stability = unstable).

For aphids that take off when their flight aptitude is low ("local dispersal flight mode")

B1. Flight trajectories are downward when the buoyant forces are small relative to the mechanical forces within the surface boundary layer (atmospheric stability = stable).

B2. Flight trajectories are primarily horizontal and short when the buoyant forces are approximately equal to the mechanical forces within the surface boundary layer (atmospheric stability = neutral).

B3. Flight trajectories have an upward component when the buoyant forces are large relative to the mechanical forces within the surface boundary layer (atmospheric stability = unstable).

ascent through the surface boundary layer and into the PBL above during morning hours is unlikely, regardless of whether an aphid is in the long-distance flight mode (scenario A1) or local dispersal flight mode (scenario B1). In contrast, the second (and larger) peak in aphid flight activity generally occurs between 1600 and 2000, when the surface boundary layer is unstable. Consequently, upward movement of aphids within the surface boundary layer is enhanced by atmospheric forces and requires little energy expenditure (scenario A3). In fact, once takeoff has occurred, aphid movement out of the surface boundary layer may be unavoidable during unstable atmospheric conditions, even for aphids in the local dispersal flight mode (scenario B3). During the neutral transition periods between stable and unstable conditions, the buoyant and mechanical forces are approximately balanced, and thus ascent by aphids is probably governed by their flight aptitude rather than atmospheric conditions (scenarios A2 and B2).

Scenario A3 is particularly important because it positions young, vigorous alatae that initiate flight during calm, warm afternoons, to be carried long distances by strong winds that are frequently present in the PBL. In contrast, scenario A1, which is most likely to arise during warm mornings after sunrise, probably results in local aphid movement, even though aphids are in the long-distance flight mode. Under stable and neutral atmospheric conditions, takeoff by aphids in the local dispersal flight mode likely results in intrafield or interfield dispersal (scenarios B1 and B2). Finally scenarios A2 (long-distance flight mode and neutral atmospheric conditions) and B3 (local dispersal flight mode and unstable atmospheric conditions) can lead to either local dispersal or long-distance movement.

Tracking insect flight through the surface boundary layer.

Evaluation of these scenarios requires the capability to obtain precise, high resolution measurements of the position of aphids in flight and the atmospheric structures within the surface boundary layer above a crop. Several technologies are available for tracking insect flight near the surface of the Earth. They include: (1) microwave-based (radar) scanning systems, (2) scanning systems involving visible light sources, (3) passive, narrowband audio receiver (microphone) systems to monitor wingbeats, and (4) phased array ultrasonic systems. Each technology has advantages and disadvantages, and their relative merits are constantly changing as the current systems are improved and new technologies become available. This research will be illustrated using a phased array ultrasonic system to evaluate the ascent flight scenarios, although the technology likely will be dated by the time our ideas are published. [The rapid advance of aerobiological technology is demonstrated by Joe Riley's recent adaption of harmonic radar for tracking insects (Riley et al. 1996). A miniaturized electronic transponder weighing as little as 1 mg attached to target insects allows their movements to be followed within their biota boundary layer. The transponder receives a signal from a radar up to 600 m distant and retransmits some of its energy at a harmonic of the incoming signal (Reynolds and Riley 1997).]

For this application, the disadvantages of the radar, visible light, and acoustic frequency systems outweigh their advantages. Current radar scanning systems are extremely valuable for measuring the movement of aerobiota in the PBL (see Chapter 11) but are excessively complex and costly for surface boundary layer tracking. Harmonic radars are not practical for studying the movements of 100s to 1000s of small insects. Sunlight would cause high background contamination to visible light scanning systems, and, in addition, artificial lights attract aphids. Finally, a narrowband acoustic system would require hundreds of microphones spaced approximately 0.1–0.2 m apart to separate the weak wingbeat signal of aphids from the high background levels (mostly by the wind) of the target frequencies. On the other hand, a phased array ultrasonic system has the spatial and temporal resolution that is comparable to that of radar at close range (1–3 m) while the technology is relatively simple and inexpensive (Figure 10.7). Equally important, flight chamber experiments indicate that the takeoff and ascent of aphids are unaffected by ultrasonic energy at the 40 KHz frequency.

A schematic diagram illustrating the deployment of an ultrasonic sensing system within a tracking volume (5 m long, 3 m across, and 2 m high) above an agricultural crop is shown in Figure 10.8. A phased array of multiple banks of ultrasonic transmitters (a 3-by-5 grid with the long axis oriented downwind from the release vial is shown in Figure 10.8) would emit a pulse at an ultrasonic frequency (40 KHz). The signals emitted from each bank of transmitters (e.g., 3 banks of 5 transmitters oriented downwind or 5 banks of 3 transmitters oriented crosswind) would be phase-delayed to enable the system to "sweep" across the entire tracking volume, alternating between the downwind and crosswind directions. The return echoes from insects flying in the surface boundary layer would be picked up by multiple ultrasonic receivers (8 shown in Figure 10.8), amplified, and passed to a Digital Signal Processor (DSP) and associated microprocessor controller where the return ultrasonic echoes would be converted to Cartesian coordinates (x, y, and z) for each insect, potentially 20 times each second.

These data and associated time tags would be fed from the DSP unit into a portable computer, which would resolve the aphid's rate and direction of movement from consecutive

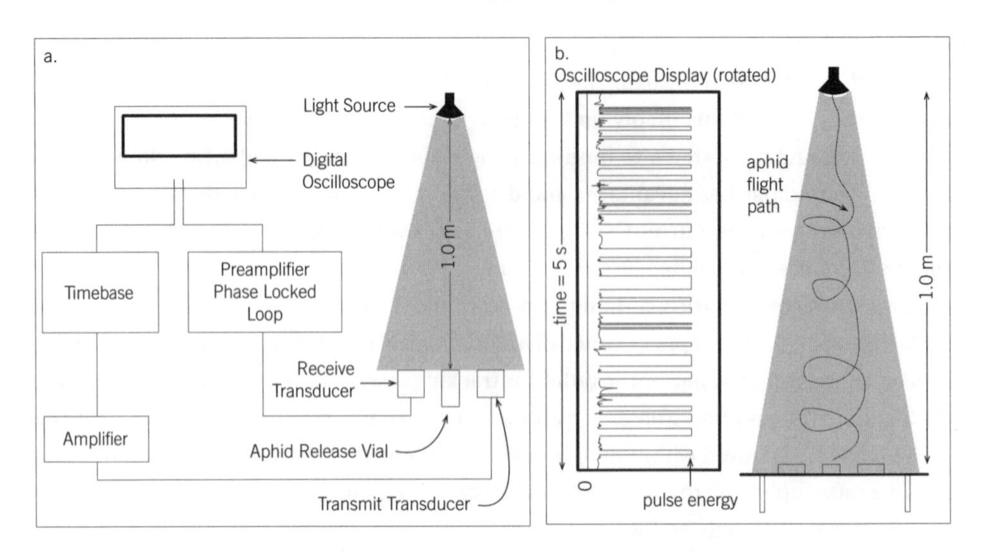

Figure 10.7. Prototype of a phased array ultrasonic insect tracking system.

This prototype system was constructed to assess the feasibility of using an ultrasonic frequency to measure an aphid's position during flight and thus evaluate the six ascent scenarios. The configuration of the components is shown in (a). The system consists of a crystal controlled oscillator and divider chain to generate precise frequencies of 40 Hz and 40 KHz. The 40 Hz pulse has a width of 0.4 ms and is used to pulse a burst of 40 KHz energy to an amplifier that drives a piezo transducer. The energy emitted from the transmit transducer is reflected by the aphid in the beam pattern; the reflected energy is sensed by a separate receive transducer. The receive circuit consists of a preamplifier and phase locked loop detector. By observing the output of the receive circuit on an oscilloscope that is triggered from the transmit 40 Hz timebase, the distance from the reflecting aphid to the ultrasonic transducers can be precisely determined. The ultrasonic transducers were placed on a horizontal table with the beam pattern pointing up. A light source was positioned approximately 1 m above the transducers. A small open vial containing aphids was placed between the two transducers. As aphids flew from the vial toward the light they were "tracked" by the ultrasonic system.

A representative composite plot of digital images from the oscilloscope display showing the location of a single aphid during its flight from the release vial to the overhead light source is shown in (b). Aphids generally flew the 1 m distance from the vial to the light in about 5s. Thus the plot shows the position of an aphid in time and in space. The oscilloscope display is rotated 90 degrees in the diagram. The horizontal bars represent the output of the phase-locked loop detector. The phase loop was set to turn on whenever the receive transducer detected a return echo of a pre-specified magnitude. Potentially the system can detect a return echo from an aphid and display the pulse of energy from the phase loop on the oscilloscope forty times each second, which would have resulted in a vertical line on the composite plot. However, an aphid provides a very small reflecting surface. At certain flight orientations a return echo could not be detected by the prototype ultrasonic tracking system. This resulted in gaps in the composite plot of the oscilloscope images. The thick horizontal bars represent multiple sequential detections. The largest gap in the time series of return echoes is less than 0.5s. A phased array of ultrasonic transducers with its greater number of sensors, higher combined output power, and more sensitive phase loop should enhance the resolution of the measurements.

measurements, and display the aphid's position in near-real time. With the use of a sufficiently fast DSP and appropriate software, tracking multiple insects flying through the volume simultaneously would be possible. The fine spatial resolution of the tracking system should enable determination of exactly where each insect initiated flight within, or entered, the tracking volume. Consequently, the system could be used not only for

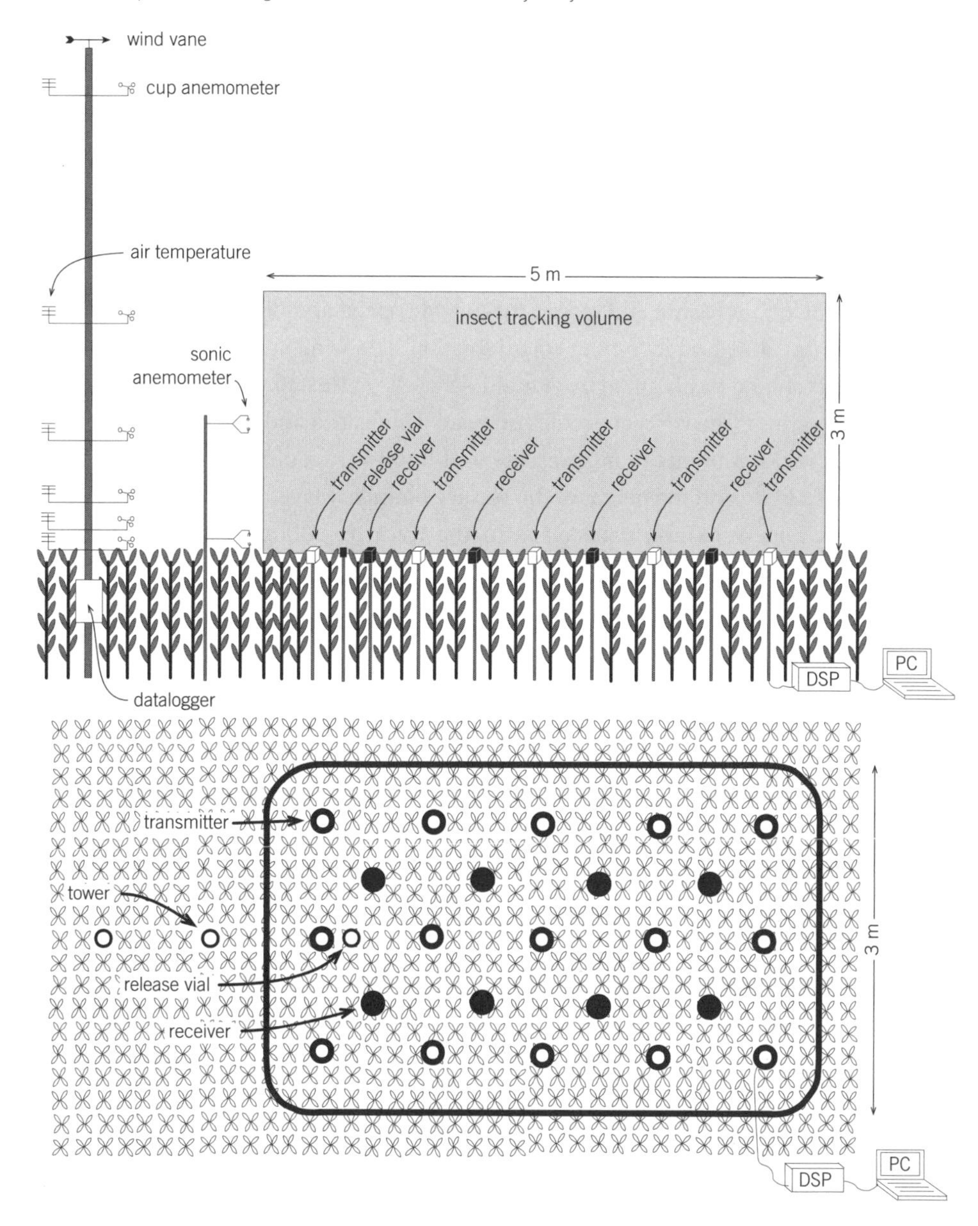

Figure 10.8. Layout and instrumentation for field experiments using the phased array ultrasonic insect tracking system.

The insect tracking volume would be "scanned" by the phased array ultrasonic system 40 times each second. The volume, located above the crop, should be about 5 m long (downwind), 3 m across, and 2 m high. The transmitters and receivers are shown 1 m apart and should be mounted on a rigid steel frame so that the system could be moved easily. The sensors would be connected to a digital signal processor (DSP) and computer (PC). Two towers with meteorological instruments are shown. One should be at least 3 m tall with instruments for measuring air temperature and mean wind speed at multiple levels. Sonic anemometers, mounted on a second tower, would provide precision turbulence measurements above the canopy.

controlled experiments with aphids of known age released from a container, but also to study the ascent of aphids and other arthropods that initiated flight from plants in a natural setting.

Field studies should take place over large plots of corn or small grain crops (at least 50 x 50 m), with the tracking volume located near the downwind edge of the plot to ensure that the vertical profiles of wind speed and air temperature are fully adjusted to the surface roughness and temperature of the crop canopy. A mobile climate station must be established in the field to measure and record mean wind speed and direction, air temperature, humidity, air pressure, and incoming solar radiation. The wind speed and air temperature measurements should be obtained from multiple levels (at least 6) that can be adjusted easily as the crop grows. The vertical profiles of mean wind speed and air temperature and the single height measurements of air pressure and humidity would be needed to compute atmospheric stability and the height of the surface boundary layer. It is important to obtain precise turbulence measurements close to the tracking volume, and currently sonic anemometers are perhaps the most appropriate technology for this application.

Data management and analysis.

Flight speeds for individual aphids and other insects would be resolved by combining the measurements of their rate and direction of movement within the tracking volume with the corresponding vertical wind speed profile measurements from the meteorological data set. Both controlled experiments and natural observations of flight could be conducted simultaneously. First, when aphids from laboratory colonies are introduced into the tracking volume from a vial, the sets of three-dimensional coordinates that constitute each insect's flight path could be stratified by the appropriate biological factors (i.e., aphid's species, age, host plant), meteorological conditions (i.e., atmospheric stability and intensity of turbulence), or some combination of meteorological and biological factors. An analysis of whether the mean angles of ascent from multiple samples differ from each other should be performed using circular statistical procedures. The analysis of flight coordinates of the stratified samples from repeated experiments would enable rigorous evaluation of the six aphid flight scenarios and the research hypothesis that the relative strength of buoyant and mechanical forces in the surface boundary layer and aphids' flight aptitude combine to govern the trajectories of alatae during the ascent phase of flight. During field experiments, the numbers and trajectories of the aerobiota that move through the tracking volume, but do not initiate flight from the release vial, also would be quantified and analyzed with respect to atmospheric stability, intensity of turbulence, wind speed, air temperature, sunlight, humidity, and atmospheric pressure, and the changes of these factors through time.

Outcomes.

The outcomes of this research must be viewed from three different perspectives. First, the research provides a model or strategy for evaluating two AFAR research hypotheses in a manner that could substantially contribute to the formulation of principles concerning the interactive control of biology and meteorology on aerial movement. Second, the results of the research should have important implications to crop protection strategies. Finally, the research would expand our basic understanding of aphid flight behavior and the many ways these weak-flying insects use the winds to move among habitats.

The ultimate goal of scientific research is to discover universal principles about processes and phenomena. Thus, as argued in Chapter 2, aerobiologists should seek generalizations pertaining to the maintenance and stages of the aerial movement process. Most aerobiological processes and phenomena involve interactions between biological and meteorological systems. Consequently, aerobiologists need to focus on the formulation and rigorous testing of scientific hypotheses across a wide range of biological and meteorological systems. The aphid ascent research initiative represents a case study designed to evaluate two of the generic hypotheses developed by Alliance for Aerobiology Research scientists: "initiation and ascent into the atmosphere by organisms that move long distances is biologically mediated" and "ascent by organisms into the atmosphere is influenced by environmental conditions". These hypotheses are linked and further specified to evaluate controls over the ascent stage of long-distance aerial movement in two important aphid systems. However, similar hypotheses can be formulated for a number of biological systems, and the phased array ultrasonic system has the potential for monitoring the flight trajectories of a range of aerobiota (insects and arachnids) in a number of habitats (low, homogeneous vegetation) during a large variety of weather conditions (precipitation-free).

If vertical movements of aphids within the surface boundary layer are governed by the mechanical and buoyant forces that determine atmospheric stability and vertical movement of air parcels, pollutants, and spores, then many aphids that take flight from agricultural fields in the continental interior of North America land in habitats tens to hundreds of kilometers downwind. This scenario holds that aphids taking off during calm, warm afternoons in the growing season are lifted by convection through the surface boundary layer into the PBL. Some are returned to nearby habitats by subsiding airflows while others, particularly those aphids that are in the long-distance flight mode, are kept airborne within thermals and large convective cells until after dusk. When the diurnal period is clear and free from the influence of strong winds associated with low pressure systems, a surface inversion layer frequently develops at nightfall, concentrating aphids and other biota airborne at this time, into the warm layer of strong winds (low-level jet) at the top of the nocturnal PBL (see Chapter 8). Not until shortly after dawn are atmospheric conditions as well as biological and behavioral actions of aphids conductive to flight termination, at which time the aphids may have been transported hundreds of kilometers downwind (Isard et al. 1990). The frequent daily concurrence in the continental interior of North America of (1) calm, warm days which favor the takeoff of aphids and many other insects, (2) unstable atmospheric conditions from mid-morning to sunset in the surface and planetary boundary layers, and (3) the development of surface inversions and low-level jets after dark (see Figure 8.9) has important implications for crop protection strategies. If long-distance aerial movement of aphids and other important pests occurs as frequently as this scenario would suggest, then a greater emphasis must be placed on developing better forecasting capabilities and implementing area-wide preventive pest management programs.

This research also provides the potential to describe precisely the responses of insects to the ever changing conditions within their BBL that C. G. Johnson (1969) believed was fundamental to a general understanding of the movement and dispersal of insects. The collection of biological and meteorological data at comparable scales is prerequisite to this goal. The phased array ultrasonic tracking system would enable quantification of the aerial paths of aphids and many other arthropods in their natural habitats at temporal and spatial scales that correspond to those of the motion systems (turbulent eddies) that they exploit to move in the atmosphere. Equally important, the capability to forecast the mass movements of aphids and other pests, that often occur during calm, warm periods in the growing season, is predicated on discovering principles relating insect behavior to atmospheric conditions within the surface boundary layer. These generalizations would provide the bridge between knowledge of source area distribution and intensity and the concentration of organisms in the PBL, potentially improving the accuracy of bioflow forecast models.

CHAPTER 11

A Decision Support System for Managing the Blue Mold Disease of Tobacco: A Case Study

This chapter draws upon knowledge of a crop-pathogen management program that we believe to be one of the best examples of a decision support system to detect, monitor, and manage a plant pathogen (blue mold) in a commercial crop (tobacco). Tobacco is a seasonal crop in North America and requires warm and humid growing conditions. It is produced in most of the states east of the Mississippi River, Missouri, Kansas, and in southern Ontario where the climate is modified by the Great Lakes. Seedlings are usually propagated in greenhouses and transplanted into fields after the threat of frost is past. Transplanting can occur as early as March in Florida and Georgia and as late as June in southern Ontario. Time of harvest also proceeds from late May in the south to September in Canada. Tobacco plants are susceptible to blue mold from germination to harvest, and there is significant variation in susceptibility to the disease among different tobacco cultivars.

Blue mold is a downy mildew disease caused by *Peronospora tabacina* Adam, an obligate parasite restricted to tobacco (*Nicotiana* spp.). In some years, blue mold epidemics span entire continents while in others the disease occurs in relatively few, highly dispersed areas. Sporangiospores of *P. tabacina* can be transported 100s of km through the atmosphere resulting in costly damage wherever spores are deposited on tobacco plants and the weather is conducive to proliferation of the pathogen. For example, losses at the farm level from blue mold in the Kentucky burley tobacco crop exceeded $200 million in 1996 alone. The spores are too tiny (15–25 micrometers) for growers to detect before the infected plants display disease symptoms. The pathogen attacks the leaves of tobacco plants, producing spots or lesions that reduce crop yield and quality. Often plants develop distinctly cupped leaves with a gray or bluish downy mold on the lower surface—hence the name blue mold. The application of a fungicide, after *P. tabacina* spores have arrived in a field but before the tobacco plants express symptoms of blue mold, is currently the most effective method for disease control. Chemical applications after lesions appear on leaves are less efficacious. Because detection of the presence of spores prior to symptom expression is so difficult, a decision support system that provides growers with a daily assessment of the potential for *P. tabacina* spore deposition in their fields is necessary for effective management of this costly disease.

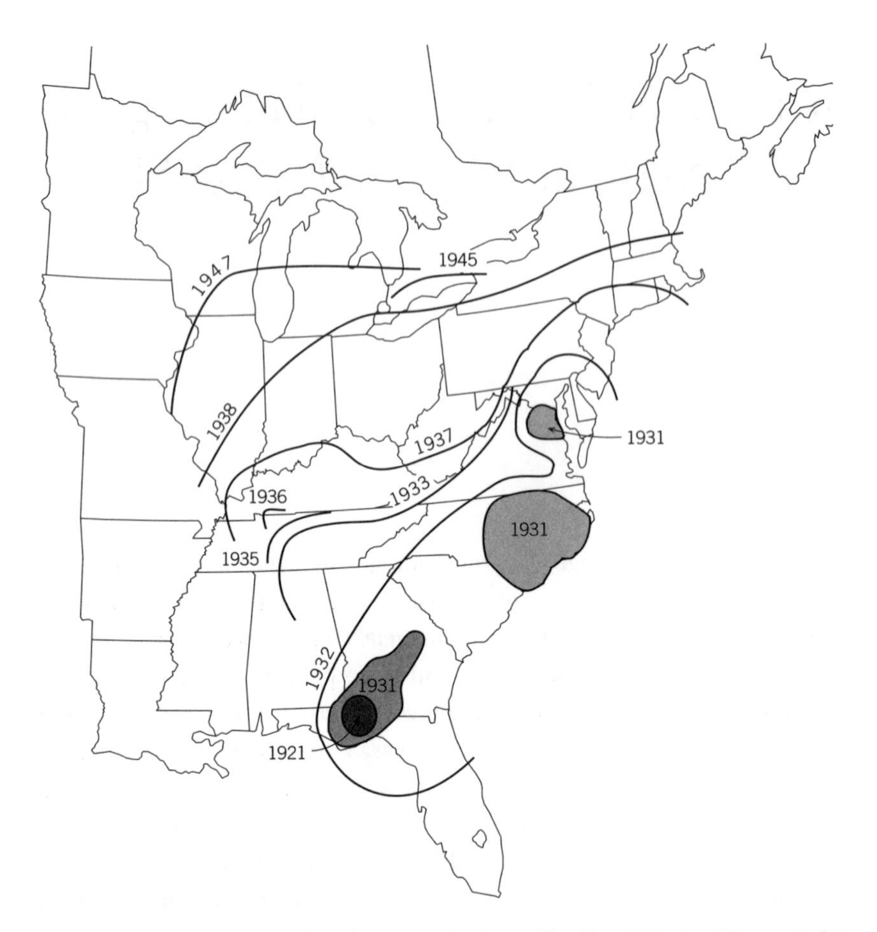

Figure 11.1. The spread of tobacco blue mold in North America between 1921 and 1947.

In North America, the blue mold disease was first discovered on tobacco seedlings during the spring of 1921 along the Florida-Georgia border. Its occurrence was sporadic both in space and time, and the infection centers were effectively eradicated. Further damage by blue mold was not reported in the region until 1931, when the disease reappeared on seedlings in three local centers within the southeastern United States. During the following growing season, blue mold spread throughout the coastal states, and then over the next 15 years it spread slowly to the northwest through Tennessee, Kentucky, and the Ohio River valley, reaching the southern Great Lakes region, including southern Ontario, by the mid- to late 1940s. Throughout this period, *P. tabacina* in North America was unable to tolerate the hot midday temperatures commonly experienced in tobacco fields during summer and thus was essentially a pathogen of young tobacco plants in seed beds. Figure adapted from Weltzien 1981.

The aerial dispersal of *P. tabacina* spores is only one mechanism that spreads the blue mold disease. During spring, tobacco seedlings are grown in greenhouses and transplanted into fields once the likelihood of frost has passed. Young tobacco seedlings can be infected and damaged or killed by blue mold prior to transplanting. The inadvertent shipping of infected seedlings before they express disease symptoms has introduced *P. tabacina* into previously blue mold-free areas 100s of km away, where subsequent aerial dispersal of the pathogen has caused costly epidemics.

Tobacco is not used by humans for food, clothing, or shelter. Smoking and chewing its leaves can cause cancer, and the craving for tobacco by users is irrational. Yet because of the large value of the tobacco crop, the high frequency of damaging blue mold epidemics over the past twenty years, and a handful of dedicated and insightful scientists, much is known about this aerobiological system. Regardless of stigma attached to tobacco due to its detrimental impact on human health, this system is an excellent model for aerobiological studies. The tremendous impact made by scientists and extension personnel associated with the North American Blue Mold Warning System (NABMWS) and the North American Plant Disease Forecast Center (NAPDFC) demonstrates the utility of building decision support systems to help manage agroecosystems impacted by biota that move long distances through the atmosphere.

Brief history of the blue mold disease.

The earliest reports of blue mold on tobacco came from eastern Australia in the 1890s. Over the subsequent 30 years, the disease remained endemic to the Australian continent. Blue mold first was identified in the Northern Hemisphere in 1921, when young seedlings in Florida and Georgia were infected. The disease was not detected in North America again until 1931, when blue mold reappeared in Florida. Over the next decade, the epidemic character of the disease became apparent as the pathogen slowly spread northward, reaching southern Ontario by 1945 (Figure 11.1). In 1935, blue mold was found in Brazil and subsequently spread to tobacco fields in northwest Argentina (1939) and in Chile (1953). Blue mold has been present in Cuba since 1957. Wind transport of *P. tabacina* spores as well as the shipping of infected seeds and oospores in soil have been suggested as possible causes for the abrupt appearance of the disease in North and South America.

In 1958, a particularly virulent pathotype of *P. tabacina* was inadvertently introduced into England from Australia for a company's fungicide trials. The following year, tobacco blue mold appeared on the European continent when plant material was transported from England to Holland and Germany for virological studies. Over the next few years, the disease spread rapidly, and by 1963, only five years after its introduction, blue mold had appeared in tobacco fields throughout the European continent and within the African and Asian countries that border the Mediterranean Basin and the Caspian Sea (Figure 11.2). The Saharan desert appears to be a barrier which restricted the spread of blue mold southward on the African continent. However, the disease continued to spread eastward, and in 1968, it was reported in Cambodia. Each summer since 1963, the disease has spread from its overwintering area near the eastern end of the Mediterranean basin northwestward through Europe. The severity of the resulting losses have been highly variable from year to year and generally confined to local areas that experienced wet and relatively cool summer weather.

The European *P. tabacina* pathotype was particularly virulent because it attacked both young plants in seedbeds and mature plants in the field. At that time, *P. tabacina* in North America was killed by high temperatures (> 30°C), and thus the disease was seldom able to survive field conditions after late May or early June in the southeastern United States. As a result, the blue mold disease in North America was generally confined to young tobacco

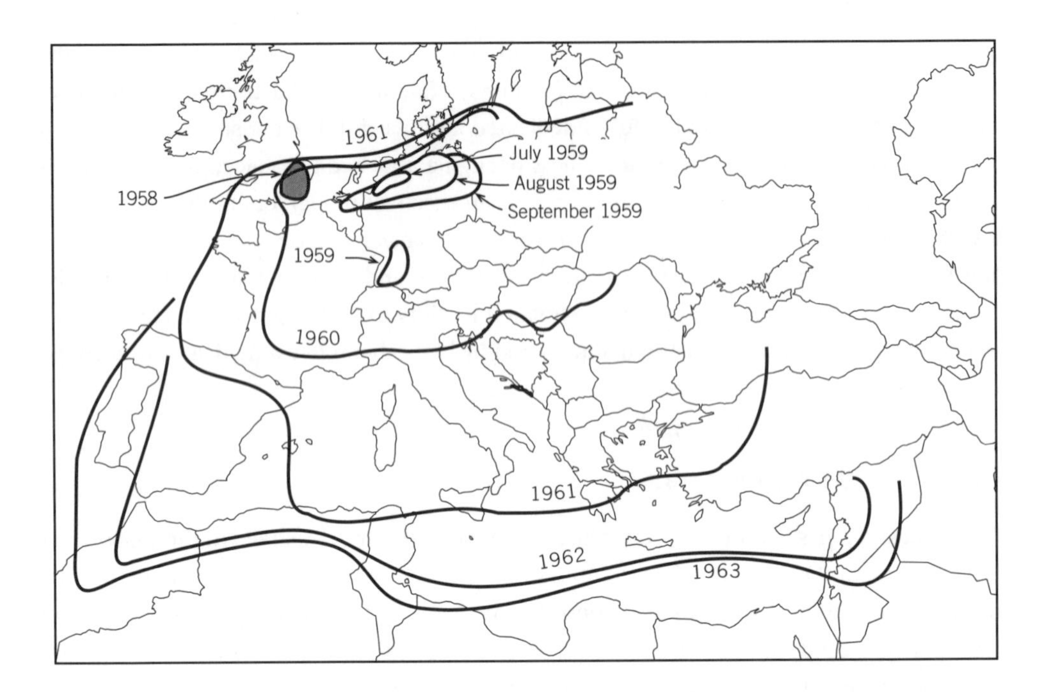

Figure 11.2. The spread of tobacco blue mold disease in Europe between 1958 and 1963.

The blue mold invasion of European tobacco fields illustrates how human-transported biota which subsequently disperse in the air can cause catastrophic damage to production systems in their new environments (see Chapter 1). In 1958, a particularly virulent pathotype of *P. tabacina* was inadvertently transported to England from Australia for fungicide trials. During the initial stages of the resulting epidemic, blue mold developed on ornamental tobacco kept under glass to maintain virus strains. Despite repeated attempts to stop the disease by destroying the *Nicotiana* plants, blue mold appeared in Holland in 1959, following an exchange of plant material for virological studies. Later that year, blue mold was reported in the Dutch, Belgian, and German tobacco fields. Unlike the *P. tabacina* that had invaded North America earlier in the 20th century, this pathotype attacked both young and mature plants, being especially injurious to the latter.

After overwintering in a vegetative state in crop residues or dry leaves, the disease spread to France, Poland, northern Hungary, and Italy during 1960. From these infection centers, blue mold continued to expand its range, invading tobacco fields in Corsica, Algeria, Spain, Sicily, Turkey, Greece, Bulgaria, Romania, Sweden, and the Soviet Republics of Moldavia, Ukraine, Belorussia, and Lithuania the following year. By 1962, five years after its inadvertent introduction from Australia, blue mold had covered the entire European tobacco-growing region and extended into North Africa, Israel, the Crimea, and along the coast of the Caspian Sea in northern Iran.

Losses due to *P. tabacina* were high across Europe in the 1960s. Blue mold infected tobacco leaves, making some unsuitable for manufacture while reducing the quality of others. In some places, tobacco fields were entirely destroyed and many farmers switched to growing other crops. Whereas 64,000 ha of tobacco were planted in the Federal Republic of Germany during 1960, only 4,000 ha were sown the following year. Estimated losses in Flanders, northern France, and Alsace ranged from 80% to 90% of the crop in 1960. The following year, overall losses in the countries affected by blue mold were more than 24% of normal production (Viennot-Bourgin 1981). Figure adapted from Weltzian 1981.

plants in nurseries where it could be managed with a preventive program of fungicide applications. Closing greenhouses on sunny days, and thus exposing diseased seedlings to high temperatures, killed the fungus as well. Between 1931 and the early 1950s, the distribution of the disease in the eastern United States was highly variable in time and space, causing damage in some production areas each year but seldom persisting within a region for more than a few years. By the 1950s, blue mold had ceased to be a major problem. As a result, growers became complacent and stopped using preventive management practices and research on the disease in the United States was for the most part discontinued.

Consequently, the North American tobacco industry was taken by surprise in 1979 when major field epidemics of blue mold suddenly occurred in the United States, Canada, Cuba, and Latin America. Tobacco crop losses in the United States and Canada were estimated at $252 million. For the most part, the disease spread from south to north, impacting all major tobacco-growing areas except those in Missouri and Wisconsin. Weather along the east coast and in the Ohio River valley was unusually cool and wet that summer. Southern growers were caught by surprise without chemical supplies or cultural control practices in place. Northern growers had more warning, but few took action, believing that a change in weather would halt the epidemic. In 1980, growers were better prepared for the disease, although due to favorable weather, blue mold spread from Florida to Connecticut, Massachusetts, and southern Ontario, causing an estimated $94 million in crop losses (Figure 11.3).

Blue mold damage in those two years was even more severe in Cuba, where approximately 33% and 90% of the cigar crop was destroyed, causing the government to lay off 26,000 workers and temporarily closing the country's cigar factories. Tobacco crop failures likely contributed to Fidel Castro's decision to permit the exodus of more than 100,000 Cuban refugees to the United States in 1980. In Canada, tobacco growers brought a major law suit against the government as a result of the losses during 1979, which they blamed on the import of infected tobacco seedlings from the United States. For the first time, growers in North America were confronted with a new temperature-tolerant ecotype of *P. tabacina* that was able to produce viable spores in the field during the hot summer months. Plant pathologists scrambled to discover what had happened to *P. tabacina*, to develop methods to forecast the aerial dispersal of blue mold, and to devise effective management strategies for the disease.

P. tabacina is not known to overwinter north of the 30th parallel of latitude in the Western Hemisphere; however, sporangiospores have been transported each year since 1979 into the southeastern United States from inoculum sources in commercial winter tobacco and/or wild *Nicotiana* species in Mexico, Texas, and the Caribbean islands. Blue mold epidemics are usually progressive in that once blue mold is established, the disease advances as a more or less definable front via windborne spores. However, because the spores can be transported 100s of km by the wind and are extremely weather-sensitive, blue mold epidemics also can be highly variable in time and space. For example, the disease infected almost the entire tobacco-growing region in North America from Florida to southern Ontario during 1980 (Figure 11.3), whereas field epidemics and losses were highly variable in 1981. Kentucky and Tennessee had epidemic conditions with significant losses while central and eastern North Carolina, Indiana, Maryland, New England, Ohio, and southern Ontario, which had been

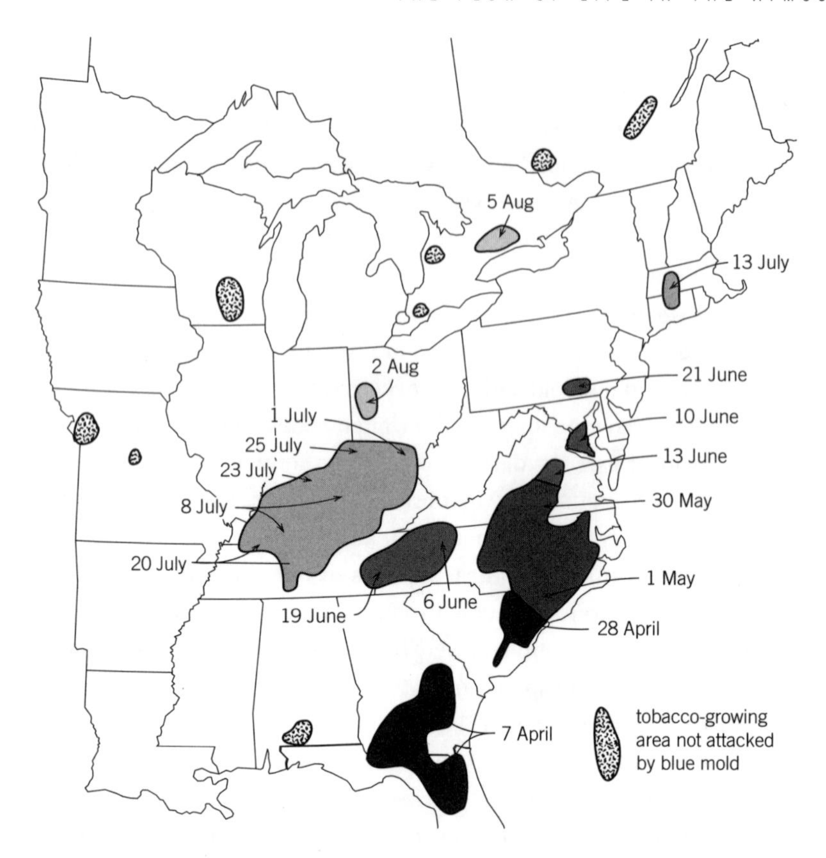

Figure 11.3. The spread of tobacco blue mold disease in North America during 1980.

The blue mold epidemics of 1979 and 1980 raised a series of questions. Why had the disease reappeared in North America after 15 years of latency? Why was blue mold now a bigger problem in tobacco fields than in seed beds? And why did the epidemic spread so rapidly and so extensively in these two years? Researchers posited three plausible hypotheses: (1) a new genetic race of *P. tabacina* had developed, (2) the climate of the tobacco-growing region had changed toward more favorable conditions, and (3) a temperature-tolerant source of inoculum was involved that could produce viable sporangiospores in the field after exposure to temperatures > 30° C.

Experimental evidence using seeds from a special collection of tobacco cultivars of known genetic background grown in nurseries in North Carolina and Puerto Rico revealed that the pathogen population in the United States did not vary in virulence from that in Europe, ruling out the first hypothesis. Climatological analysis indicated that summertime maximum temperatures for the southeastern United States weather stations in 1979 and 1980 did not vary greatly from their thirty-year means, and, therefore, unusually favorable weather was ruled out as a plausible explanation for the resurgence of the disease. In contrast, both field and laboratory evidence provided support for the development of a temperature-tolerant ecotype of *P. tabacina*. Observations from North Carolina during 1980 revealed that spores remained viable in the field until mid-July. This was highly unusual for a pathogen that previously had been unable to survive in the same locations past late May or early June. In 1984, convincing evidence for a temperature-tolerant ecotype of *P. tabacina* was obtained when isolates were tested in phytotron experiments with controlled environmental conditions. In these studies, sporangiospores produced at 34° C and 36° C were shown to be viable, resulting in blue mold lesions (Main et al. 1985). Figure adapted from Davis and Main 1984.

hurt so badly during the previous two years, had very light or no occurrences. For the first time in 1981, blue mold infections were found in several fields in northwest Missouri, but no reports from other areas of the state were received.

Since 1990, blue mold activity has steadily increased with major epidemics in 1996, 1997, and 1998. A number of factors have been implicated as driving recent epidemics, the foremost of which are protracted periods of weather favorable to blue mold and changes in the fungus, especially resistance to fungicides.

The life history of blue mold disease of tobacco.

The blue mold fungus is extremely variable and forms new strains (pathotypes) quickly. Consequently, it is appropriate to include all strains under the name *P. hyoscyami* de Bary, which has priority, and is the name for the pathogen throughout most of the world. However, in this book we use *P. tabacina*, the name given to blue mold in North America. Cool, wet, and overcast summer weather is ideal for multiplication of *P. tabacina* and the spread of blue mold. Under these conditions, tobacco leaves become infected 2 to 4 hours after the sporangiospores arrive. The pathogen grows within the leaf tissue for 5 to 7 days before yellow lesions are expressed on the plant. Sporulation requires a period of darkness with high humidity and is maximized at temperatures between 15 and 23°C. Consequently, the appearance of sporangiophores and their many attached spores is most common on the morning following the first day of symptom expression. Blue mold usually occurs on leaves; however, lesions may occur on buds, flowers, and developing seedpods of the tobacco plant. The intensity of sporulation depends upon the physiological state of the host plant: infections on young, rapidly growing leaves produce more spores than those on older ones. Spore liberation requires an increase in solar radiation and ambient temperature (generally accompanied by a decrease in relative humidity). These factors combine to enhance transpiration from sporangiophores, causing them to desiccate and twist (see Figure 4.5d). As the entangled sporangiophore branches disengage, the resulting spring action mechanically launches the spores into the atmosphere. Generally, the spores are liberated from mid-morning to late afternoon with peak release occurring before noon. One tobacco plant may have 200–400 leaf lesions, and each lesion (≈1.25 cm in diameter) can produce as many as 4,000,000 spores. This means that the number of spores produced by one plant may exceed 1 billion. Consequently, densities on infected leaf surfaces can reach one million sporangiospores/cm^2, and on the order of 10^{11} spores can be released into the air from a single hectare of severely infected tobacco during the late morning when weather conditions are appropriate.

Mature blue mold spores less than a day old have higher germination rates than older ones; however, some can remain viable for weeks. The spores are very sensitive to ultraviolet radiation, and most are killed within an hour when exposed to direct sunlight in summer. Although the length of time airborne spores can remain viable under overcast skies is unknown, it generally is assumed that a small proportion of the sporangiospores released from a field can remain viable for several days when the weather is cloudy. The deposition of airborne spores is influenced by gravitational settling and turbulence; however, spores

Table 11.1. References to publications on tobacco blue mold biology and aerial dispersal.

MONOGRAPH AND EDITED VOLUMES ON BIOLOGY AND DISPERSAL:
Lucus 1975, Spencer 1981, McKeen 1989, Main and Spurr 1990.

MANUSCRIPTS ON AERIAL DISPERSAL:
Lucus 1980, Aylor et al. 1982, Aylor and Taylor 1983, Nesmith 1984, Main et al. 1985, Davis et al. 1985, Davis and Main 1986, Aylor 1986, Davis 1987.

also are washed out of the air by rain to land on plants and the ground below. Wet deposition is especially important for initiating blue mold epidemics because "rainout" can result in the accumulation of many spores in a relatively small area at a time when environmental conditions (e.g., cloud cover, leaf wetness) are conducive to plant infection.

P. tabacina produces sexual oospores as well as asexual sporangiospores. Oospores are sometimes produced in the mesophyll of dead parts of infected tobacco plants. In North America, the asexual cycle of blue mold appears to far outweigh the sexual cycle in epidemiological importance since the oospores cannot overwinter in the field poleward of 30°N latitude. Sexual reproduction potentially could be responsible for the rapid appearance of new strains of blue mold with increased pathogenicity following the release of resistant varieties of tobacco or after the introduction of new fungicides. However, oospores seldom have been found in North America and little is known about the conditions necessary for oospore germination and the role of sexual reproduction in the life history of blue mold. The publications listed in Table 11.1 led to much of the understanding of tobacco blue mold, the aerial dispersal of *P. tabacina* spores, and the on-line DSS that is described below.

The North American Blue Mold Warning System.

No formal system for reporting blue mold was in place to alert growers when the 1979 North American epidemic occurred. An earlier warning system, established by the United States Department of Agriculture (USDA) in 1945, had languished during the previous decade when blue mold was infrequent. During the 1979–80 winter, the Tobacco Disease Council held a special meeting where workers in research, extension, and industry reviewed the new information on blue mold disease and developed plans for coping with the disease in the upcoming year. As a result of this meeting, a preventive program of fungicide applications was designed, and tobacco growers agreed to form a volunteer network to track the spatial extent and severity of the disease in subsequent years. A similar warning system had been operating in Europe by CORESTA (Cooperation Centre for Scientific Research Relative to Tobacco) since the early 1960s. The pattern of south to north spread and the success of previous programs for reporting blue mold epidemics served as the impetus for the North American Blue Mold Warning System (NABMWS). A multinational coordinating committee was formed, composed of local warning system coordinators from each tobacco-producing state/province in the United States and Canada. In the United States, the Cooperative Extension Service (CES) in participating states organized autonomous warning systems involving growers, county agents, agri-business, tobacco specialists, research pathologists, meteorologists, and the media.

During the 1980s, growers used a variety of measures to combat blue mold, including chemical applications, reducing plant populations, good site selection, and early destruction of infected seedbeds and crops. Generally, a systemic fungicide was applied to the soil before planting as insurance against the disease. Tobacco producers would base their decisions about supplemental use of the systemic fungicide at cultivation, and foliar fungicide applications thereafter, on information concerning the status of blue mold in their area. This information was collected, managed, and disseminated by the NABMWS (Figure 11.4). In years when the blue mold disease spread northward, infecting contiguous tobacco-growing areas, the system warned growers when blue mold was active in their vicinity and allowed them to apply preventive fungicides in a timely manner to protect their fields. However, when blue mold "leap-frogged" to northern production areas because *P. tabacina* spores were transported long distances in the atmosphere on southerly winds, farmers were frequently caught by surprise. In these situations, infected tobacco fields would remained undetected until symptoms of blue mold appeared, greatly reducing the effectiveness of fungicide applications for controlling the disease. Obviously, a service was needed to forecast the long-distance aerial dispersal of blue mold.

Further impetus for developing a blue mold forecasting system.

In autumn, 1979, farmers in southern Ontario took legal action against the Canadian government as a result of tobacco losses due to blue mold and lack of compensation from the federal government. The plaintiffs attributed the 1979 epidemic to the importation of infected tobacco seedlings from Florida. They claimed that a truck full of tobacco seedlings had been improperly inspected and that the young plants should have been destroyed or returned to their place of origin. They further asserted that blue mold disease had spread aerially throughout the tobacco-growing region of southern Ontario from the fields managed by farmers who had planted the infected seedlings. The case for the defense was based on the claim that it was unlikely that blue mold was imported with the transplants. They argued that the time between the transplanting of seedlings (6 June) and the expression of blue mold symptoms (11 July) was unreasonably long given the favorable weather conditions for spore production that occurred in southern Ontario. Rather, the defense maintained, the fields became infected with *P. tabacina* spores transported long distances in the atmosphere. Wind transport of spores to southern Ontario from the adjacent American states had likely occurred in 1945 and 1946. However, in 1979, the spores would have had to "leap-frog" across Kentucky and Ohio to Canada, because symptoms of blue mold were not expressed in these states until after they were observed in Ontario. Although the trial dragged on for three years, it was never decided whether the disease originated from wind-borne spores or infected seedlings. However, the controversy seems to have stimulated scientists in North America to develop transport models for quantifying atmospheric movements of plant pathogens.

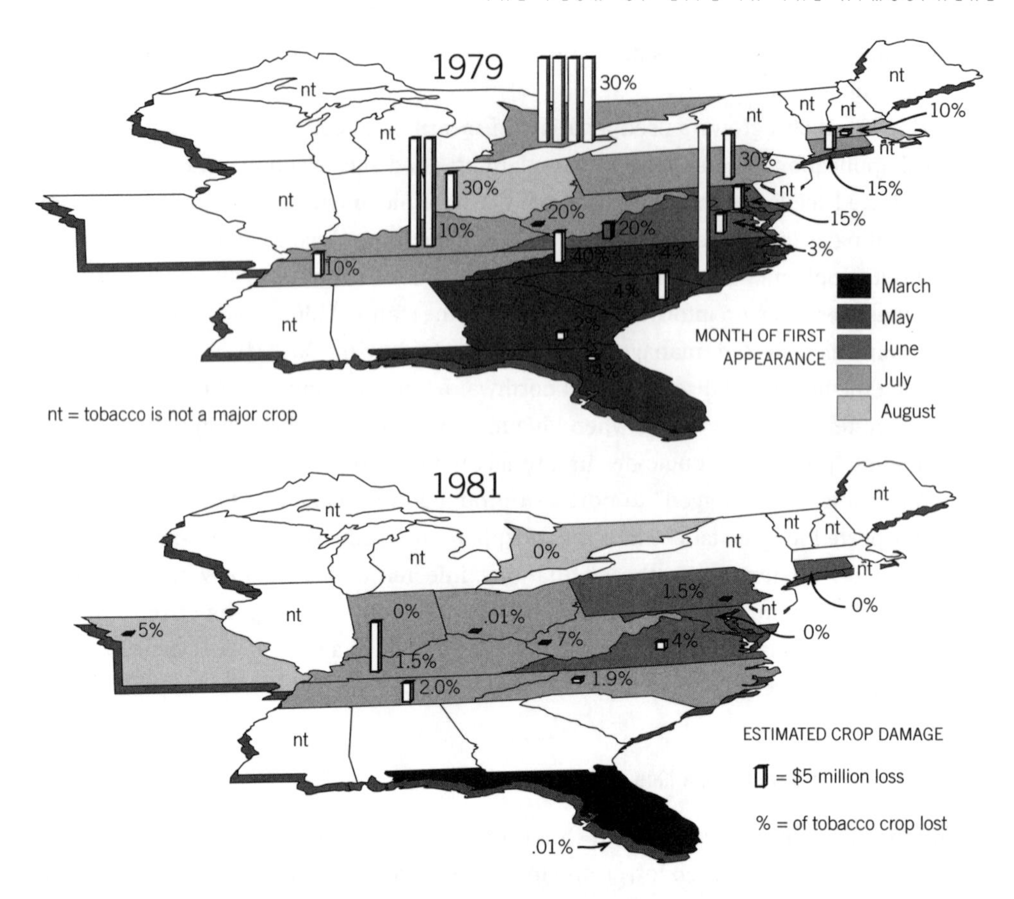

Figure 11.4. First occurrences of blue mold and resulting tobacco crop losses in 1979 and 1981.

The North American Blue Mold Warning System operated by the Tobacco Disease Council in partnership with the United States Department of Agriculture Cooperative Extension Service and its equivalent in Canada was founded in the winter of 1979/80. The system was based on a hierarchical network of coordinators and kept the industry aware of blue mold activity in North America through timely reports. Beginning in spring, 1980, information on the first occurrence of blue mold within individual counties and updates on the progress of the disease were phoned by county extension workers to state coordinators on Monday. The county reports were integrated and forwarded by telephone to the national office by midweek. The national coordinator compiled a warning statement that included the occurrence of blue mold outbreaks and anticipated weather-related changes in disease status. This report was mailed to state coordinators on Friday. State coordinators interpreted the warning statements based on local conditions and used letters and the media to inform county agents and tobacco growers of the disease status and suggested control measures. Generally, weekly reports were first issued in May, and changes in blue mold status were reported until August (Nesmith 1984).

Although many factors, including weather and more effective fungicides, contributed to the decline in blue mold between 1979 and 1981 (above), the North American Blue Mold Warning System was instrumental in convincing growers to implement preventive management programs and providing timely reports of disease status. Crop losses in the U.S. remained relatively light until the 1990s, when weather, *P. tabacina* resistance to fungicides, and production practices changed to favor blue mold epidemics.

The Aylor spore transport model.

Biota transport models can be constructed using either Eulerian or Lagrangian coordinate systems; both can be used to solve the advection-diffusion equation. In a Eulerian framework, the solution is obtained relative to a fixed grid, and both the biota concentrations and meteorological variables are defined at each of the grid points for successive intervals of time. In the Lagrangian frame, the advection-diffusion equation is rewritten such that the change of concentration is computed by following the parcel containing biota as it expands and is advected by the wind. Eulerian approaches are most suitable for biota transport problems when takeoff is highly variable in time and space while Lagrangian models are most applicable when the assumption of a point source for biota is appropriate.

The Aylor Model (see Figure 2.5 and Table 2.3) uses the Lagrangian approach (Aylor 1986). The assumptions employed to operationalize the model require appraisal of the time and space scales relevant to the transport process. Consider the release of a highly concentrated "puff" of spores during a sunny period when the atmosphere is unstable: after the first minute of transport, the spore puff likely has traversed 100s of m and has grown both vertically and horizontally due to the action of small-scale turbulent eddies (see Figure 9.2). After several hours, the center of spore concentration has traveled 10s of km downwind, and the puff has grown into a "cloud" under the influence of thermals and other landscape-scale or larger scale atmospheric motions. Some spores will have reached the top of the planetary boundary layer (PBL), and the cloud now will extend laterally for several km. Over a period of several days, large-scale motion systems will influence spore movement. The spore cloud centroid likely will have traveled downwind 100s to 1000s of km, and the cloud will stretch laterally for 10s to 100s of km. Some spores will have been deposited on the ground, and only a small proportion of those still airborne will be viable.

When computing long-distance movement of spores, it is usually necessary to restrict consideration to larger scale processes. In the Aylor Model, the spores released from a source area during a period of a few hours are considered a single cloud. The initial number of spores in the cloud is equal to the product of the total number of spores released from blue mold lesions in the source area over a specified time period and the fraction of those spores that escape from the tobacco canopy into the air. This calculation represents the takeoff and ascent stage in the aerobiology process model (Figure 2.4). The horizontal transport stage involves consideration of the movement of the spore cloud center downwind, the lateral spread of the cloud, and the survival of the spores in the air. During transport, some spores will be deposited to the Earth (see Figure 4.8), and others will escape from the PBL to the free atmosphere. An atmospheric trajectory model is typically used to calculate the downwind movement of an air parcel representing the spore cloud centroid over subsequent periods of time (e.g., 3-hr intervals). The model produces a cylindrical cloud, extending from the ground to the top of the PBL, with a radius that increases with travel time. The increase in volume occupied by the spore cloud, or equivalently, the decrease in spore concentration (dispersion) within the cloud, is computed in the Aylor Model as a function of the depth of the PBL and travel time. In the absence of precipitation scavenging, the loss of spores from the cloud by dry deposition and escape into the free atmosphere are orders of magnitude

Figure 11.5. Blue mold transport from Texas to Kentucky in 1985.

In early June 1985, blue mold became evident in tobacco seed beds in southcentral Kentucky, suggesting that *P. tabacina* spores had arrived within the previous few weeks. To evaluate the potential source of the inoculum, Davis and Main (1989, 1990) ran back-trajectories using NOAA's Branching Atmospheric Trajectory (BAT) model (Heffter 1983). They determined that storm events of 27 May had characteristics sufficient to cause the observed spatial and temporal patterns of the disease. Air parcel trajectories for the storm implicated southcentral Texas as the probable blue mold source area. Sampling of wild tobacco plants in southcentral Texas established that active sporulation had occurred throughout May 1985. However, given the long time required for movement from southcentral Texas to southcentral Kentucky during the potential aerial transport event (about 50 hrs for the 1500 km trip at 30 $km \cdot hr^{-1}$) and thus the large dilution factor and small likelihood of spore survival, the authors questioned whether Texas could qualify as the source region.

Davis and Main (1990) used the Aylor Model to evaluate the possibility of *P. tabacina* transport from Texas to Kentucky. Their spore production estimate was based on tobacco blue mold sporulation measurements (Aylor and Taylor 1983, Rotem and Aylor 1984), sampling in southcentral Texas (Lemke and Main 1990), and phytotron experiments using wild Texan tobacco plants. They considered only spores that were produced during the late morning on 25 May (⅓ of the daily total), when the BAT model trajectories indicated that the spores deposited in Kentucky seed beds during the 27 May storm had likely initiated movement in Texas. In addition, they assumed that 15% of the spores released by lesions in the source region escaped from the tobacco canopy into the atmosphere above. These assumptions resulted in a cloud above the source region containing roughly 4.68×10^{12} viable spores (N_0).

Following Aylor (1986), a transport function (dilution factor) was computed as:

$$T = (H\pi S_x^2)^{-1} = 1.96 \times 10^{-14} \ (m^{-3}),$$

where H is PBL height (2,000 m) and the radius of the spore cloud, S_x (m) $= 0.5 \times t$, with t equal to the travel time (50 hrs $\times$ 3,600 $s \cdot hr^{-1}$ = 180000 s). Employing a value for spore survivorship ($S = 0.01$) from Aylor (1986) and an estimate of the wet deposition velocity for spores during rain ($v_w = 2.3 \ m \cdot s^{-1}$), they calculated a deposition rate (D) for viable Texan spores that could have infected tobacco seed beds in southcentral Kentucky during the 27 May storm:

$$\begin{aligned} D &= N_0 \times T \times S \times v_w \\ &= 4.68 \times 10^{12} \ (\text{spores}) \times 1.96 \times 10^{-14} \ (m^{-3}) \times 0.01 \times 2.3 \ (m \cdot s^{-1}) \\ &= 0.0021 \ (\text{spores} \cdot m^{-2} \cdot s^{-1}). \end{aligned}$$

Using S_x for the spore cloud length in the downwind direction and the 850 mb-level average wind speed calculated by the BAT model (30 $km \cdot hr^{-1}$), they estimated that the spore cloud would require 3 hrs (90000 m / 30 $km \cdot hr^{-1}$ = 10800 s) to pass over a seed bed in southcentral Kentucky. Integration of the spore deposition rate over the time of cloud passage gives total spore deposition (D_T) in the tobacco seed beds,

$$D_T = \int D \, dt = 0.0021 \ (\text{spores} \cdot m^{-2} \cdot s^{-1}) \times 10{,}800 \ (s) = 23 \ (\text{spores} \cdot m^{-2}).$$

Although Davis and Main (1989) did not know whether a deposition value for *P. tabacina* spores of this magnitude was sufficient to initiate the 1985 Kentucky blue mold epidemic, their analysis demonstrated the utility of the Aylor Model and that viable spores likely were transported from southcentral Texas to southcentral Kentucky in 1985.

smaller than the dilution of spores by the dispersion process, and therefore may be ignored. A survivorship factor, which can be specified as a function of travel time and the weather conditions encountered during transport, allows estimation of the concentration of viable spores at some later time above a potential destination.

Some atmospheric trajectory models allow an air parcel to be pulled apart by vertical wind shear. Branching typically occurs when atmospheric stability changes rapidly near dusk and dawn. When the atmosphere becomes stable at nightfall, NOAA's Branching Atmospheric Trajectory (BAT, Heffter 1983) model separates the spore cloud into three layers with independent trajectories. Mixing due to increased convection is allowed to occur during the transition from night to day as the PBL becomes unstable and grows in height.

Deposition or landing of spores can occur due to sedimentation, impaction, and precipitation scavenging. Precipitation is usually involved in transport events that result in the accumulation of spores in high enough densities to initiate an epidemic in receptor areas. Spore deposition, per unit area, per unit of time, is computed in the Aylor Model as the product of the viable spore concentration in the PBL above the receptor region, and the rate of deposition (deposition velocity). Finally, the total number of spores deposited in a receptor region is calculated by integrating the instantaneous deposition rate over the duration of the spore cloud passage. An application of the Aylor Model to evaluate the possibility that Kentucky tobacco fields were infested by spores from Texas in 1985 is provided in Figure 11.5.

The North America Plant Disease Forecasting Center.

In the mid-1990s, C. E. Main integrated the spore transport modeling framework with the blue mold warning network to create an on-line blue mold DSS for tobacco growers in the United States and Canada. This system has evolved during subsequent years and continues to incorporate new decision support components in response to feedback from growers, scientists, and other users (e.g., school children). The primary components and flows of information in the system are depicted in Figure 11.6. The key component is the Internet User Interface Subsystem which facilitates effective communication among tobacco growers, federal agency personnel, and scientists. Local growers are involved at both ends of the information flow; they monitor their fields for blue mold disease, providing basic biological data, and they are the final users of the risk assessments and management advice. The DSS relies on NOAA for atmospheric monitoring and modeling. The team of plant pathologists and meteorologists at the North American Plant Disease Forecast Center (NAPDFC) maintains the Database Management, Knowledge Interpreter, and User Interface Subsystems.

The on-line DSS is built on the infrastructure for local biological monitoring developed by the North American Blue Mold Warning System (NABMWS). Over the past twenty years, growers have realized the economic benefits of using the warning system to schedule their supplemental fungicide applications and have learned that the effectiveness of the warnings depends directly on their participation. As a result, periodic inspections of fields for blue mold and the immediate reporting of suspected disease has become an integral part of the tobacco farming "culture," i.e., an established management practice. Once blue mold presence is suspected in their fields, most growers seek immediate confirmation of the disease

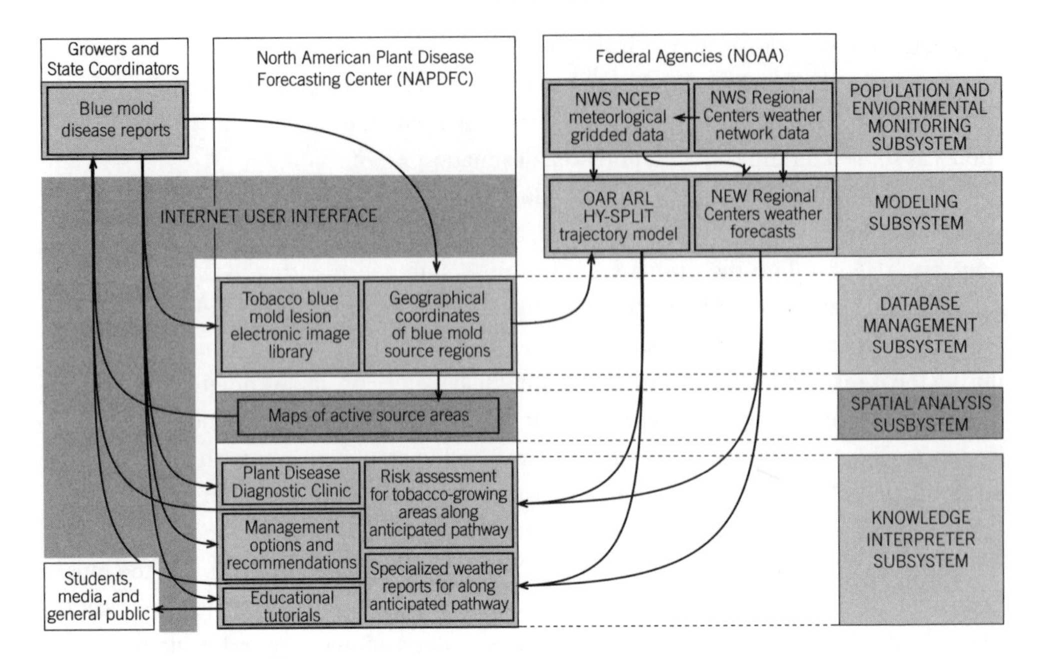

Figure 11.6. Components and information flows within the on-line decision support system for tobacco blue mold in North America.

Tobacco growers and their State Coordinators, NAPDFC scientists, and federal agency personnel participate in the on-line DSS. Growers and State Coordinators monitor fields for blue mold and report incidence and severity to the NAPDFC. Federal agencies collect environmental data and provide daily weather forecasts. The NAPDFC maintains a database and provides maps of active sources and sends parameters to the ARL for the trajectory model runs. NAPDFC scientists interpret the forecasts, assess the risk of blue mold in tobacco-growing areas, and disseminate the information over the Internet to growers, State Coordinators, and the general public.

from CES personnel or other experts, who in turn communicate the information up the hierarchy of participants in the NABMWS.

The on-line DSS has greatly increased the efficiency of the confirmation and reporting process. Growers and CES personnel now can view an image library of blue mold lesions for examples of symptom expression. They are encouraged to electronically transmit digital camera images of the afflicted tobacco foliage to the NAPDFC Plant Disease Diagnostic Clinic for professional examination. A variety of educational tutorials on the life history of *P. tabacina,* detection and epidemiology of blue mold, and aerial movement of the spores are provided. Finally, growers and state coordinators can file disease reports specifying both the geographic location and the severity of infection over the Internet for immediate incorporation into the blue mold source region database. Updates can be transmitted daily as the disease progresses and when locations cease to be sources of inoculum.

The NAPDFC maintains the database on blue mold presence and severity by county for the tobacco production areas in the United States. Supplemental data on the disease are obtained from research pathologists and other experts in Cuba, Mexico, other Latin American countries, and Canada. The on-line DSS uses the atmospheric data, model output, and

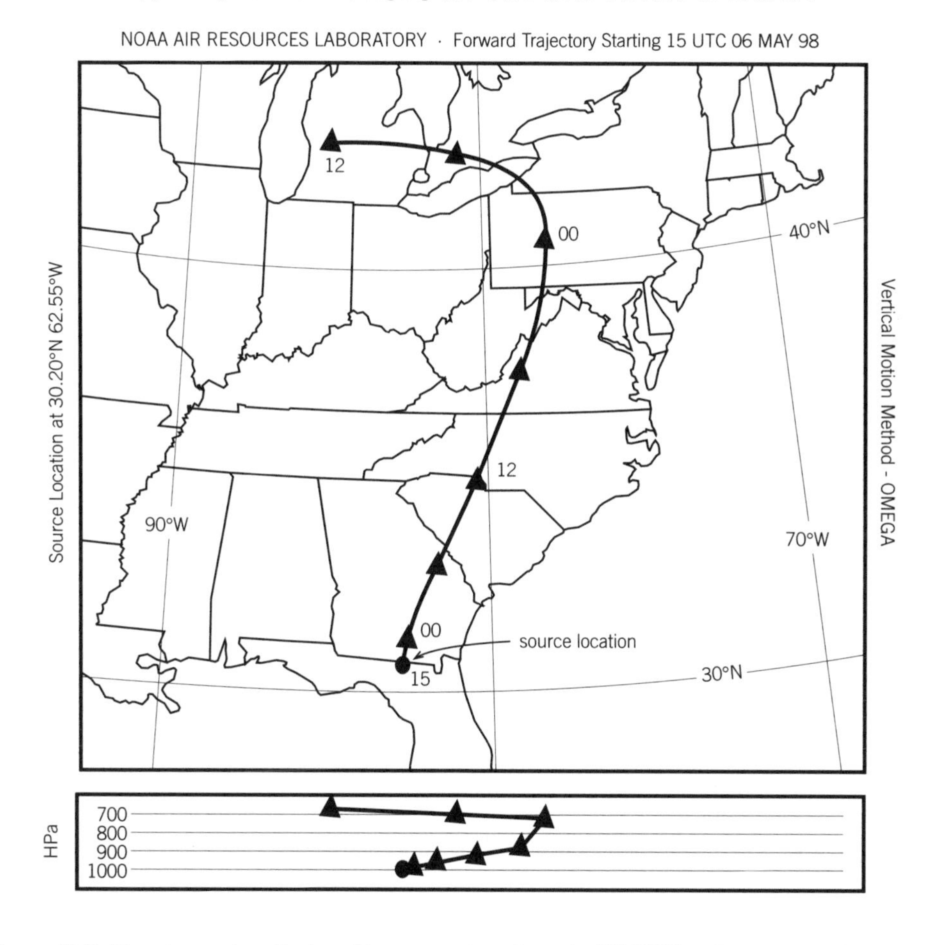

Figure 11.7. Plan view and vertical profile maps output from the HY-SPLIT trajectory model.

The blue mold source is represented on the plan view (top) and profile (bottom) maps by filled circles. The vertical axis on the profile map is pressure in Hectopascals (1 HPa = 1 mb) and can be related to altitude using the hydrostatic equation (see Chapter 5). The triangles represent the forecasted position of the spore cloud centroid at six hour intervals. Source location (latitude and longitude), starting time, and date are printed at the map borders. Time is given in Universal Time Coordinates (UTC) which is 4 hours ahead of Eastern Daylight Time (EDT) in summer.

The source area for the above 6 May forecast was Lake City, Florida. A low pressure center was traversing the Midwest with a trailing cold front. The BAT model output for a spore cloud starting at 1100 EDT shows its centroid reaching North Carolina by 1200 UTC (0800 EDT) on 7 May. By 2000 EDT, the cloud centroid was projected to reach central Pennsylvania, and 45 hrs after release, the spores were predicted to be over southern Ontario. Trajectory confidence was high and the risk of blue mold infection in fields within the northern tobacco-growing region was rated high for 7 May. Moultrie, GA, and Springfield, TN, were also active sporulation areas on 6 May, and trajectories for these source areas predicted northward spore movement as well.

forecasting services routinely provided by NOAA agencies including the National Weather Service (NWS) Southern and Eastern Regional Centers, NWS National Center for Environmental Prediction (NCEP), and the Office of Ocean and Atmospheric Research (OAR) Atmospheric Research Laboratory (ARL). The set of geographic coordinates (latitude and longitude) for active blue mold sporulation areas are transmitted electronically to the ARL.

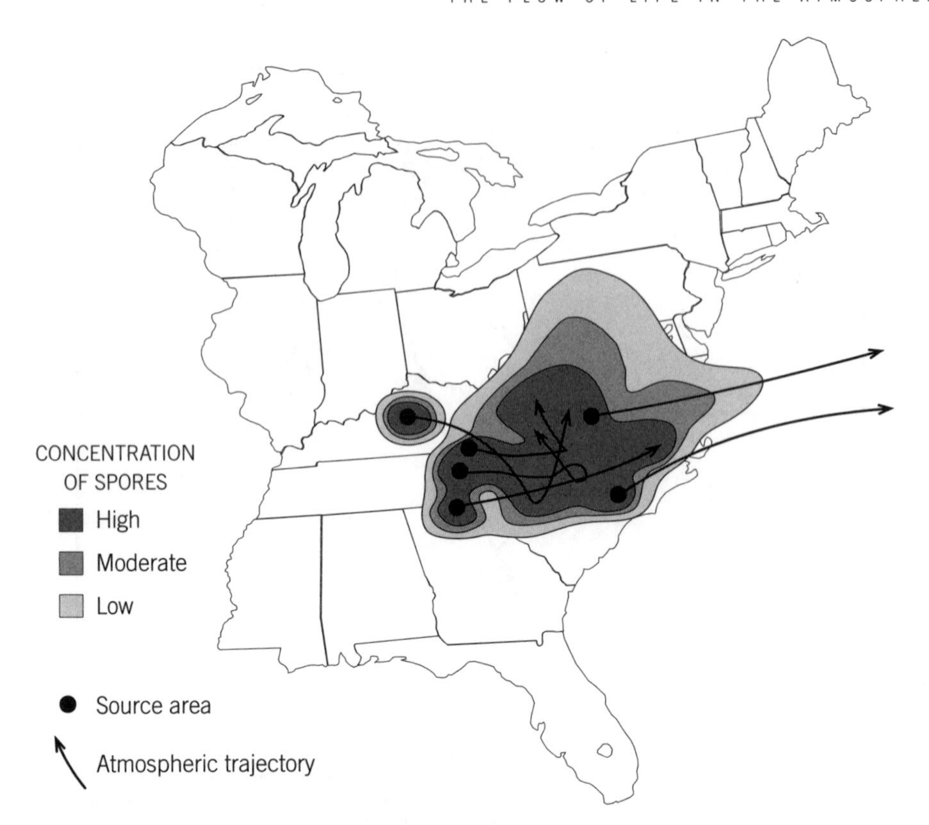

Figure 11.8. Example of near real-time forecasts generated with the BAT and MASS models showing the aerial dispersion of blue mold spores by the wind.

On 26 July 1996 there were six confirmed source areas for *P. tabacina* spores in North America (solid circles). Output from NOAA's HY-SPLIT trajectory model shows the most likely forty-eight-hour trajectory of "parcels" of air containing spores from each source (arrows). The shaded areas represent location and density of the "spore cloud" at the end of the forty-eight-hour period. The cloud concentration was calculated using the Mesoscale Atmospheric Simulation System (MASS) by MESO, Inc., Troy, N.Y., and accommodates multiple sources of spores, the physical features of the underlying terrain, and spores that remain in the atmosphere from previous days. Both sets of forecasts are available in near real-time via the Internet on the North American Plant Disease Forecasting Center Home Page (*http://www.ces.ncsu.edu/depts/pp/bluemold/*). Figure adapted from the North Carolina State University Blue Mold Forecast Home Page that was prepared by T. Keever, C. E. Main, J. M. Davis, and T. A. Melton.

Agency personnel use the HY-SPLIT model to simulate the trajectory of air parcels that initiate movement from the source locations at specified times. A separate run is made for each source area, calculating the position of the air parcel centroid over eight, 6-hr time increments (48 hrs) usually starting at 1000 (onset of peak sporulation). The trajectory model operates using PBL wind fields generated by the NCEP Nested Grid Model, a member of the family of numerical weather prediction models used to forecast weather conditions over North America. In turn, the gridded data is based on weather observations collect by the NWS Southern and Eastern Regional Centers.

The HY-SPLIT model outputs are displayed on plan view and vertical profile maps (Figure 11.7). The scheduling and production of blue mold forecasts during the growing season depends on the spatial distribution of source areas and the severity of the blue mold infestations. Generally forecasts are issued three times each week and more frequently during major blue mold epidemics. The error associated with the forecast trajectories can be substantial and increase with an increase in forecast period. Under the best circumstances, the HY-SPLIT air parcel trajectories deviate from the actual path (calculated using observed atmospheric fields) by about 15% of the transport distance. In most situations, however, the trajectories are off between 25 to 30%, and the error in the forecasted location of the air parcel centroid can exceed 40% of the path length during complex weather. The NAPDFC has worked hard to keep abreast of the most sophisticated atmospheric trajectory models and is currently exploring the Mesoscale Atmospheric Simulation System (MASS) model (MESO, Inc., Troy, NY). This Eulerian model can accommodate multiple sources of spores, the physical features of the underlying terrain, and the spores that remain in the atmosphere from previous days (Figure 11.8).

The plant pathologists and meteorologists at the NAPDFC use their extensive knowledge and experience to integrate the biological data on blue mold incidence and severity, weather data and forecasts, and the output from the blue mold forward trajectory model runs, to assess the risk of disease spread. A separate forecast is made for each active source area and includes maps of the latest trajectory (Figure 11.7). The Trajectory Weather Report describes recent past, present, and near future weather conditions in the blue mold source area and along the anticipated transport pathway. Information on factors important to sporulation in the source region (temperature, precipitation, cloud cover), survival during movement in the atmosphere (cloud cover, which serves as an indicator of ultraviolet irradiation and desiccation), and wet deposition (precipitation) in potential receptor areas are highlighted. Large-scale circulation features (e.g., midlatitude and tropical cyclones, jet streams, and fronts) as well as general weather conditions for the southeastern United States that may influence spore movement, or provide opportunities for blue mold infection, are presented in the Regional Weather Report. The Outlook section of a forecast integrates the biological and meteorological elements, evaluating the likelihood of pathogen spread and the risk of disease over the subsequent 48-hr period. Trajectory Confidence, rated as low, medium, or high, describes the general quality of the forecast pathway (i.e., the complexity of the weather).

Management options and recommendations for blue mold control also are available to growers on the NAPDFC Internet site. A diverse array of strategies and tactics are listed. The current management recommendations are the outcome of a blue mold seminar attended by farmers, industry leaders, and agricultural scientists after the devastating 1996 blue mold epidemic in Kentucky. The document identifies human-induced and natural factors that interacted to drive the recent rash of blue mold epidemics and suggests a number of cultural practices to improve blue mold control. Preventive spray programs for seedlings in greenhouses and plant beds and for adult plants in fields are presented with information on appropriate chemicals, spray volumes, and application schedules. In addition, Internet links are provided to fungicide suppliers for user convenience.

The on-line DSS also provides for the information needs of other sectors of society, especially students, teachers, and the media. All source area maps, trajectory forecasts, and specialized weather reports are archived and easily retrievable for Internet users. In addition, the educational tutorials are designed for students as well as growers with many useful images, a glossary of scientific terms, and self-evaluation components for testing one's understanding of the epidemiology of the blue mold disease of tobacco.

Evaluation of the on-line blue mold decision support system.

The success of the on-line DSS for blue mold is difficult to evaluate. Most standard validation techniques are not appropriate for systems that deliver qualitative advice, such as risk assessments, rather than quantitative results. It is hard to know what the crop losses would have been, both in infected and blue mold-free growing areas, had the producers not used the decision support system to schedule their supplemental fungicide applications.

The NAPDFC began issuing forecasts on 5 March 1996. That year, losses due to blue mold were generally light throughout much of the United States and Canada, except in Kentucky where the burley crop was badly damaged. The majority of the 344 forecasts issued were in the low risk category, especially in the early season. The system successfully predicted the first United States outbreak of blue mold that occurred on 6 May when spores were transported from Cuba to northern Florida. Other successful predictions included the first blue mold outbreaks in southern Ontario (5 June) and in southcentral North Carolina (10 June), by inoculum released in central Kentucky and northern Florida, respectively.

Blue mold occurred in nearly all the tobacco production areas in eastern North America in 1997. A total of 469 forecasts were issued on 75 days between March and early August. The NAPDFC indicates in its summary for the year that tobacco producers' confidence in, and use of, the forecasts increased in 1997 over the previous growing season (see homepage). Twelve of thirteen first reports in affected states could be explained by NAPDFC forecasts.

The 1998 blue mold season was very active with infections in all the major tobacco production areas of North America. The prolonged and widespread epidemic was influenced by two important factors: (1) two month-long periods with favorable weather conditions for sporulation in the southeastern United States (early April–early May and late May–late June) and (2) the inadvertent shipping of infected seedlings into northcentral Tennessee and southcentral Kentucky from southern Florida during mid-April.

Forecasting began on 2 March 1998 when active inoculum sources were reported in western Cuba and Mexico. Over the subsequent six months, 451 forecasts were issued for 72 days. During the height of the epidemic in June and July, trajectories were run for more than ten active and geographically distinct source regions. For the first time since forecasting began in 1996, a transport event in which spores were airborne for two days appeared to have occurred. In this single event on 5–7 May, inoculum was likely blown from northern Florida/southern Georgia to South Carolina, southcentral Virginia, Maryland, southcentral Pennsylvania, and southern Ontario (see 6 May 1998 forecast map in Figure 11.7).

Outcomes of the on-line DSS for tobacco blue mold.

Successful approaches to manipulating agricultural systems to sustain high yields and profits, for the present and in the future, must employ management strategies that treat cropland as a complex system. This is clearly the case for tobacco, where the dynamics of the blue mold disease in a single field, or within a production area, do not depend on *P. tabacina* populations and environmental factors at that location alone. Blue mold epidemics also are influenced strongly by biological and environmental conditions that occur at other locations, some 1000s of km away. The ability to prevent catastrophic blue mold epidemics depends as much on the knowledge of the long-distance movement process as it does on the local factors that influence the rate of spore production and the arsenal of chemicals for controlling the fungus. Consequently, the long-term success of management programs for this important pathogen is contingent on understanding the atmospheric processes that have the potential to link any set of tobacco fields within North America. The NAPDFC uses key components of the current blue mold knowledge base, and the atmospheric monitoring and modeling capabilities of NOAA, to forecast the aerial movement of *P. tabacina*, providing growers with highly valuable information of the risk from this important disease.

It should be noted that the NAPDFC uses essentially the same methodology to forecast the movement of other small aerobiota, including the cucurbit downy mildew spore (see *http://www.ces.ncsu.edu/depts/pp/cucurbit/*) and the pollen of mountain cedar (see *http://pollen.utulsa.edu/mcforecast.html*). The downy mildew disease is caused by *Pseudoperonospora cubensis*, and annually infects squash, cucumbers, pumpkins, and muskmelons grown throughout the southeastern United States. The fungus can kill plants that become severely infected early in the growing season, while later infections result in reduced fruit yield and quality. In contrast, mountain cedar (*Juniperus ashei*) pollen is one of the most potent allergens in the United States. Populations of mountain cedar occur in southcentral Oklahoma and Texas. Pollination of the cedar occurs during December and January, often with the peak occurring the last week of December, and is responsible for mid-winter hay fever problems in the central United States.

As currently configured, the on-line DSS for blue mold satisfies all the requirements of an effective service for managing an organism that moves long distances in the atmosphere (Pedgley 1982, see Chapter 3). It incorporates a disease monitoring network that spans the entire area throughout which tobacco is grown, it has an analysis and forecasting center, it uses the Internet for rapid gathering of information and dissemination of warnings, and its advice is based on a thorough understanding of the host's and pathogen's life histories, the impact of *P. tabacina* in tobacco fields, and available tactics for managing the disease. The frequent forecasts of movement that take into account current environmental conditions in local areas, as well as those at larger scales, make the system very useful from a day-to-day management perspective.

The adoption of an on-line DSS by farmers is contingent on their judgement concerning whether or not it increases crop yields or decreases management costs. Tobacco producers fear blue mold more than any other tobacco disease because it can strike so suddenly and devastate entire seedbeds and fields. Foliar fungicides can provide blue mold control; however, to be effective they must be applied after the *P. tabacina* spores arrive in a field but before lesions are evident on the plants. Consequently, forecasts of likely aerial pathways for the inoculum and associated assessments of risk of blue mold infection are extremely valuable to growers, especially for days when it rains on their farms. For the tobacco blue mold system, reliable forecasting is made possible because *P. tabacina* is an obligate parasite that cannot overwinter in the midlatitudes, and thus active source regions for the spores are relatively easy to identify. For these reasons, tobacco farmers throughout the United States and Canada are willing to provide the timely and well-documented disease reports on which the on-line blue mold DSS is dependent.

Blue mold epidemics likely will continue to occur in North American tobacco production areas during summers wherever the weather remains warm and wet for protracted periods of time. Increased resistance of *P. tabacina* to fungicides and economically-driven alterations in tobacco production practices also have been key factors in recent epidemics. Although it is likely that the disease will never again be of minor importance to tobacco production, as it was in North America during the 1950s through 1970s, blue mold is controllable. Effective management of the pathogen in the future will require changes in production attitudes and activities, in addition to reliable monitoring of the disease in fields and timely forecasts of risks from spore movement in the atmosphere. Frequent long-distance aerial dispersal of *P. tabacina*, current limitations on control options, and the epidemiology of blue mold, dictate that successful management of the disease depends on the tobacco industry operating in community and regional efforts for the benefit of all, rather than growers operating independently. The success of large-area blue mold control programs are contingent on rapid communication and information dissemination for preventive disease management. The on-line DSS operated by the NAPDFC provides an infrastructure for linking tobacco farmers throughout the North American continent making local containment of potentially disastrous blue mold epidemics achievable in the future.

This case study includes the principle components of an integrated pest management program including the integration of environmental monitoring, biological monitoring, modeling, education, and delivery of information and advice to agricultural producers. It also addresses important issues related to the reduction of chemical use during non-outbreak (endemic) disease (pest) phases. Finally, the on-line DSS for the blue mold disease of tobacco provides a template for developing and deploying a complex delivery system over a large geographic area.

CHAPTER 12

A Long-Term Aerobiological Corridor for Characterizing and Forecasting the Aerial Flow of Biota between the Subtropical Area and Continental Interior of North America: A Research Agenda

The movement of organisms from one geographic location across continents to another, in search of food and other life-sustaining resources, is a phenomenon that has fascinated humans for centuries. Movement patterns of large mammals, like buffalo and arctic caribou herds, utilize traditional trails through the landscape to change position in response to seasonal changes in weather and resource availability. Waterfowl and other avian species, guided by environmental cues and aided by atmospheric motion systems, follow traditional airways to move poleward during spring to breeding areas and to fly equatorward during autumn to overwinter. People who use such animal resources for food, comfort, or pleasure have learned to identify the patterns of movement and many of the behavioral characteristics of these organisms. This has helped humans to optimize use of these resources, to understand why populations fluctuate, and to develop laws to protect populations from over-exploitation.

Scientists have long recognized that the lower atmosphere between the Rocky Mountain range to the west, and the Appalachian Mountains to the east, is a very active conduit for the transfer of large quantities of heat and moisture from subtropical to middle and polar latitudes within North America. Much of this transfer is accomplished by near-surface flows of air associated with extratropical cyclones and anticyclones that frequently traverse the area between these north-south oriented cordillera. During the early 20th century, the tracks of cyclones and anticyclones and the seasonal changes in these paths began to be mapped. Analyses of these data revealed preferred tracks for cyclones and anticyclones across the continental interior of North America and a high frequency of large-scale advection of air masses from both the Gulf of Mexico and the Canadian subarctic, along the corridor between the Rocky and Appalachian mountain ranges. As the years passed, researchers in the interior of the North American continent became aware that many plant pathogens, insects, and human allergens that cause serious problems came from afar, usually on southerly winds during the spring and summer (Stakman 1942). Sampling of the air within the PBL using aircraft revealed the tremendous number and diversity of biota that could be found high in the atmosphere over the continent. Light, pressure, temperature, humidity, winds, stability, and their seasonal variations were found to be important factors that contribute to the abundance of aerobiota (e.g., Wellington 1945a, b, c). By the 1940s, sudden appearances of aerobiota in the continental interior began to be associated with the movement of extratropical cyclones and anticyclones (see Figure 7.8). Recent technical advances in ground-based and airborne radars, aerial collection devices, remote sensing platforms, the detection of natural markers, data and information management, and atmospheric and population modeling have greatly increased our capability to investigate and understand the long-distance movements of biota within atmospheric motion systems (see Table 4.3 for references). In 1992, a number of scientists established the Alliance for Aerobiology Research to further advance knowledge of atmospheric transport of organisms by focusing on the study of the interaction of biological and meteorological processes to elucidate the principles of aerial movement of biota (see Chapter 1 and the Epilogue). The following research initiative is a direct outgrowth of the many interactions among AFAR scientists over the intervening years.

The bioflow initiative: objectives and background.

This research initiative focuses on characterizing and quantifying the flow of biota (bioflow) between the subtropical and midlatitude continental interior areas of North America along a pathway that is bounded by the Rocky Mountain range to the west and on the east by the Appalachians. A wide range of biota, including plant and animal viruses, fungi, bacteria, pollen and seeds of higher plants, soil nematodes, arthropods, and birds, use this pathway each year, usually during the spring or early summer, and some, again, in the late summer and autumn (see references in Table 4.3, and especially Johnson 1969, Edmonds 1979, Pedgley 1982, MacKenzie et al 1985, Sparks 1986, Johnson 1995, Kerlinger 1995, and Dingle 1996).

The rationale for the bioflow initiative was presented in Chapter 1; however, the most salient elements of that discussion bear repeating here. It is important to anticipate aerial flows of biota along this pathway and the important consequences of these movements as we begin to develop new strategies to understand and manage the diverse natural and human-modified environments on the North American continent. A sound understanding of the biological and meteorological interactions that govern the movement of organisms in the atmosphere is a prerequisite to developing innovative and successful strategies for protecting human health and for sustaining a large number of terrestrial ecosystems. This is because many biota, some indigenous to North America and others exotic, that are known to afflict human health and production systems can move long distances utilizing atmospheric motion systems. Equally important, the inflows and outflows of organisms to and from habitats can be as important as birth and death rates in regulating the dynamics of populations within both natural and human-built environments.

The general hypothesis, upon which this initiative is based, is that organisms that utilize the atmosphere for translation from one geographic place to another flow within atmospheric motion systems along routes that are mediated by physical and biological features of the Earth's surface. The pathways and timing of the flow of biota are to a large degree regular, and thus the movement of organisms in the atmosphere is predictable. An evaluation of

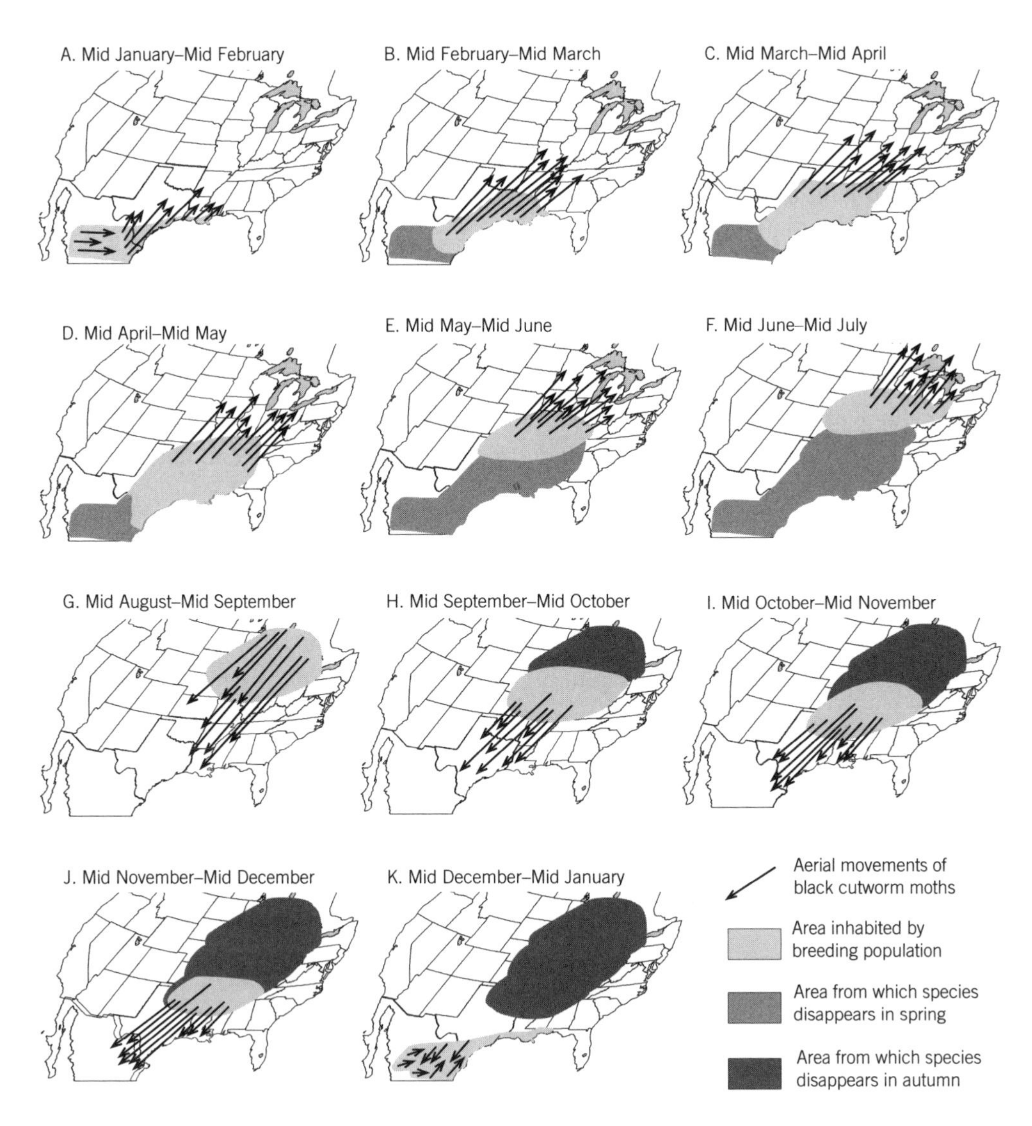

Figure 12.1. Annual long-distance aerial movements of black cutworm moths in North America.

After mid-January, the breeding range of black cutworm moths slowly expands northward, and colonizers establish where temperatures are tolerable and acceptable host plants exist. The species disappears from the south as a zone of intolerance (air temperatures > 36° C) expands. Usually by May, the moths have evacuated the Gulf Coast states and have established breeding populations in the Corn Belt of the United States. During July and August, the breeding range of black cutworm generally extends from about 40° N to 60° N latitude. After mid-August, physiological stress from cold temperatures (<15° C) initiates southward movement in black cutworm. As the zone of cold intolerance expands southward through the midlatitudes during autumn, the zone of breeding slowly shifts to the Gulf Coast and northern Mexico. Figure adapted from Showers 1997.

this hypothesis in the context of the bioflow initiative can be separated into two parts or research objectives: (1) to characterize the biological and meteorological phenomena that influence the flow of biota in the atmosphere, paying particular attention to developing an understanding of the spatial and temporal scales of these phenomena and how their variations, through time and across space, interact to govern the pathways and timing of this flow and (2) to predict the pathways and timing of the long-distance aerial flow of biota from subtropical to midlatitude continental interior areas of North America during spring/summer and the equatorward flow that occurs during the summer/autumn.

Although some understanding exists about a few organisms and their movement through the atmosphere within this North American corridor, an integrated systems approach and design must be set in place to provide a template upon which to overlay important observations about the movement of a wide range of organisms within atmospheric motion systems. As was argued in Chapter 2, this approach is necessitated by the complexity of the meteorological and biological factors that influence long-distance movement, the occurrence of these events and processes at a variety of spatial and temporal scales that are often interconnected, and the intricate interactions of these many biological and meteorological factors that govern movement of biota in the atmosphere. Because of the large scope of this initiative and the extensive resources and expertise involved, simultaneous accommodation of the perspectives and needs of researchers and managers in the agricultural, forestry, human health, wildlife, and environmental arenas is required for improved understanding and prediction of bioflow throughout the North American corridor. Likewise, the knowledge about long-distance movement of organisms in atmospheric motion systems and many of the forecasting tools that are developed and refined during this project should be applicable to understanding and forecasting bioflow within the many other atmospheric corridors in North America (e.g., along the east coast and the corridor between northern Mexico and the southwestern United States) and around the world (see Figure 5.4).

As indicated above, the importance of long-distance aerial movement of organisms as a fundamental component of the dynamics and epidemiology of populations in North America has long been recognized. The sudden and devastating poleward spread of corn leaf blight in 1970 (see Figure 3.1), the seasonal round-trip movements of black cutworm moths (Figure 12.1), the spread of western equine encephalitis by mosquitoes (Figure 12.2), and the long-distance migrations of birds on weather fronts (see Figure 7.10) provide well-documented examples of bioflow along the pathway between the subtropics and continental interior of North America. Although the inventory of organisms that move along this atmospheric pathway has grown rapidly, the quantification and understanding of the causal factors and functional relationships that link these factors and the role of aerial movement in the dynamics of populations have been fragmentary. The quantification and understanding of these phenomena require a concerted multi-disciplinary effort, new perspectives on linking processes and patterns, and a broader world view.

For the first time, the capability exists to take a comprehensive approach to understanding and forecasting the aerial movement of biota that impact natural ecosystems, human health, and agriculture. The dramatic increase in access to information and technology over the last few years has created this opportunity. The technology to monitor large areas of the Earth's surface from remote platforms has been established. The fine temporal and spatial resolutions of these measurements allow the detection of small differences among habitats and their changes through time. These data provide the capability for monitoring and predicting the fluctuations of many populations. Recent advances in radar tech-

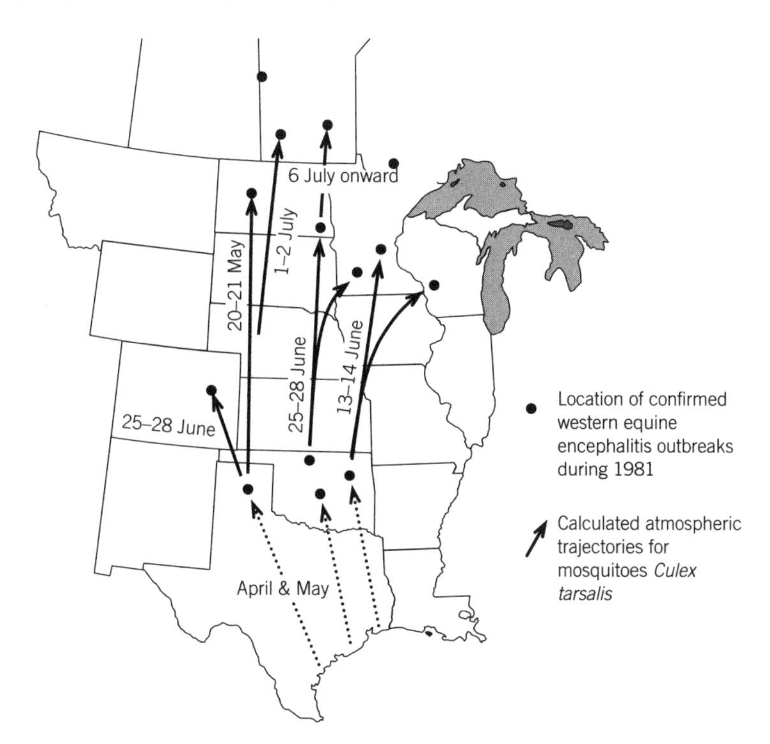

Figure 12.2. Spread of western equine encephalitis by mosquitoes within the North American corridor during 1981.

Western equine encephalitis (WEE) is a serious disease of horses and humans in North America. The virus usually cycles between vertebrate hosts (small mammals, birds, reptiles, and amphibians) and mosquito vectors. Horses and humans are considered dead-end hosts. During 1975, 1977, 1981, and 1983, WEE epidemics occurred within the central prairies (Manitoba, Saskatchewan, Alberta, North Dakota, and Minnesota) of North America. Although WEE is regarded as endemic in these provinces and states, it is thought that the virus is often reintroduced to the area during the spring by migrating birds and/or windborne mosquitoes (Sellers and Maarouf 1988).

During the spring and summer of 1981, at least 390 confirmed outbreaks of WEE in horses occurred throughout the North American corridor (Sellers and Maarouf 1988). Wind trajectory analysis indicated that the WEE virus was most likely carried by *Culex tarsalis* from their overwintering habitats along the Gulf coast to northern Texas and Oklahoma during April and May (dotted arrows). Between May and July, the infected mosquitoes moved on southerly winds to North Dakota, Minnesota, Wisconsin, and Manitoba (solid arrows). The authors suggest that the mosquito vectors may have traveled as far as 1250–1350 km over 18 to 24 hour periods within the PBL and that they were washed out of the atmosphere by precipitation associated with cold fronts. Figure adapted from Sellers and Maarouf 1988.

nology enable real-time detection and continuous observation of many biota during their long-distance movements in the atmosphere. When interfaced with the existing weather radar network in North America, this technology provides the opportunity to monitor the flows of these biota over large areas. Finally, fast parallel digital processors and computer networks provide the capability to connect and manage large and diverse data sets to discover new knowledge and to make these data sets and knowledge easily accessible to users around the world. Together, these advances provide the tools that enable a comprehensive approach to forecasting the movement of biota along the North American corridor.

However, as argued in Chapter 2, the ability to accurately forecast the flow of biota along this pathway is predicated on an extensive knowledge-base about the biological and meteorological factors and their interactions that govern the movement process in each of its stages (see Figure 2.4). The atmospheric corridor between the Rocky and Appalachian mountain ranges is almost as wide as it is long. The winds in different areas of the pathway can be more conducive to long-distance movement than those in others parts, and these airflow patterns often change very quickly. Likewise, habitats that sustain the abundance of biota that may be ready to move along this atmospheric pathway vary tremendously with space and time. The abundance of biota in source areas and the size and location of source areas vary on temporal scales that can range from weeks to years. In addition, the life histories of organisms and the characteristics of source areas differ greatly among aerobiota. Consequently, the keys to predicting the routes and timing of bioflow through the North American corridor are to (1) monitor the abundance of populations in their source habitats and (2) understand the biological and meteorological events and processes that interact to govern aerial movement. Special emphasis must be placed on investigating the spatial and temporal scales of these phenomena and how their variations, through time and across space, interact to govern the flow of biota within the corridor.

Sources of spatial and temporal variations of bioflow.

The temporal scales of the biological and meteorological phenomena that are important to biota while they move in the atmosphere range from fractions of seconds to weeks. The spatial scales range from leaf parts to continents. To identify and analyze the biological and meteorological phenomena that influence aerial movement, it is convenient to group the events and processes by scale (Table 12.1). The bioflow initiative focuses on sources of variation in events and processes that occur at the large- and mesoscales, and influence bioflow within the entire North American corridor. Global scale processes that dictate climate zones and the evolution and distribution of biotic domains and their divisions on Earth are beyond the scope of the bioflow initiative. We recognize that atmospheric and biological phenomena which occur at small scales (local and micro) are nested within those that occur at larger scales and at times can exert considerable control over larger scale events and processes. Also, any program to monitor biota within a landscape must be acutely aware of local- and microscale processes and events that cause spatial and temporal variations in populations and the accuracy of population measurements. Examples of research initiatives designed to illuminate some of the smaller scale aerial movement processes and events are presented in Chapter 10.

In Table 12.1, phenology is used to characterize the time sequences of biological events and to help define areal units with common biological attributes. The life histories of organisms can be viewed as a sequence of processes and stages. The phenological stages that most plants and animals transition through include emergence, early development, late development, maturation, reproduction, and senescence. Many taxa within a spatial unit at any moment in time are at similar stages in their life histories. This general correspondence through time of phenological stages is caused by both the seasonal progression of climate and

Table 12.1. Sources of variation in the routes and timing of bioflow in the North American corridor.

TEMPORAL & SPATIAL SCALES	METEOROLOGICAL EVENTS & PROCESSES	BIOLOGICAL EVENTS & PROCESSES
Global scale (centuries; planetary)	Global climate system that governs the distribution of climate zones.	Evolution and movements of plants and animals determine the distribution of biotic domains and their divisions. Climate, landforms, and soils primarily determine the types of vegetation and associated species of animals that inhabit a biotic division.
Large-scale (weeks–months; continents)	Mid-tropospheric patterns of airflow (e.g., the presence and position of Rossby waves and the polar jet stream over the corridor.)	Phenological development of the assemblage of plants and animals as it is influenced by the progression of the seasons within biotic divisions.
Mesoscale (days; landscapes)	Presence and location of extratropical cyclones and anticyclones within the corridor that influence the large-scale advection of air within the PBL.	Phenological development of the assemblage of plants and animals within landscapes. A landscape is a geographic assemblage of spatially contiguous ecosystems which are interconnected by corridors and superimposed upon a matrix of associated physical features.
Local-scale (hours; subregions)	Topographic features and the configuration of aerial units with different types of surfaces that influence PBL airflows (e.g., drainage winds and lake breezes).	Phenological development of plants and animals within ecosystems. An ecosystem is a biological community and its associated physical environment. Nutrients pass among different organisms in specific pathways and cycle through the organisms in hierarchical interdependent processes.
Microscale (seconds–minutes; patches)	Types and configuration of plants and human-built structures within a landscape unit that influence surface layer airflows (e.g., channeled flow within row crops and standing eddies to the lee of buildings).	Phenological development of plants and animals within habitats. A habitat is a place in which an organism lives, characterized by physical and biological features.

extreme weather conditions which can cause dramatic changes in phenological progression and thus create convergence of the development stages of a wide range of organisms. For example, an early spring will act to hasten the development of many plants and animals within an area while a severe drought or an early frost may result in widespread senescence.

Meteorological factors that operate at the large scale and cause variations in bioflow.

The transport of energy, moisture, and many biota, from the subtropical to higher latitudes in North America, is influenced by large-scale horizontal wave-like motions called Rossby waves. Chapter 6 describes these large scale circulation mechanisms that cause temporal and spatial variation of bioflow within the corridor, and only a few elements of that discussion bear repeating here. Rossby waves are embedded in a global scale vortex of winds that blow from west to east aloft around the North Pole and are always in a state of flux. However, the position of these long waves over North America is often anchored by the Rocky Mountains (see Figure 6.5). Because the strength of the airflow aloft, and the position of the polar front, are governed by the horizontal temperature gradient between the subtropical and polar latitudes, the latitudinal position of the Rossby waves varies with the seasons. During winter and early spring, the latitudinal temperature gradient is large, winds in the upper troposphere are very strong, and the Rossby waves can extend to the subtropical latitudes of North America. Springtime emergence and the development of many plants and animals are triggered when the circumpolar vortex begins to contract, allowing warm tropical air masses to penetrate into the continent. In summer, when the temperature gradient is much weaker, the winds aloft are weaker and the circumpolar pattern of airflow is over the northern United States and Canada. Finally, senescence of many biota occurs in autumn when cold air masses penetrate equatorward beneath upper air troughs in the Rossby waves as the circumpolar vortex expands over the North American corridor.

Meteorological factors that operate at the mesoscale and cause variations in bioflow.

In the North American subtropics, the weather during much of the year is dominated by subtropical high pressure (STHP) cells that are centered around the Tropic of Cancer over the North Atlantic and North Pacific oceans. The high surface pressure, created by subsiding flows of tropical air from aloft, produces diverging winds at the Earth's surface. As a result, winds in the subtropical portion of the North American corridor (i.e., east of the Rocky Mountains in the southern United States and Mexico) are predominantly southeasterly during much of the year. In summer, these warm moist airflows from the Gulf of Mexico penetrate far into the continental interior as the polar front moves north and the circumpolar vortex of winds aloft contracts.

The seasonal shifts of the location of the circumpolar vortex, the presence and position of Rossby waves in these mid-tropospheric winds, and the penetration of subtropical air, cause important temporal and spatial variations in the opportunities for bioflow within the North American corridor. These large scale flow patterns are most variable in late spring and autumn, when many biota are motivated to move within the corridor due to the changing availability of food and other resources.

Weather in the North American corridor between the Gulf of Mexico and Hudson Bay is characterized by itinerant extratropical cyclones and anticyclones that are nested in the larger-scale circumpolar airflow aloft. The structure and genesis of these mesoscale features that cause temporal and spatial variations in bioflow in the corridor are discussed in Chapter 7. These eddies often result in the advection of air for 1000s of km within the PBL.

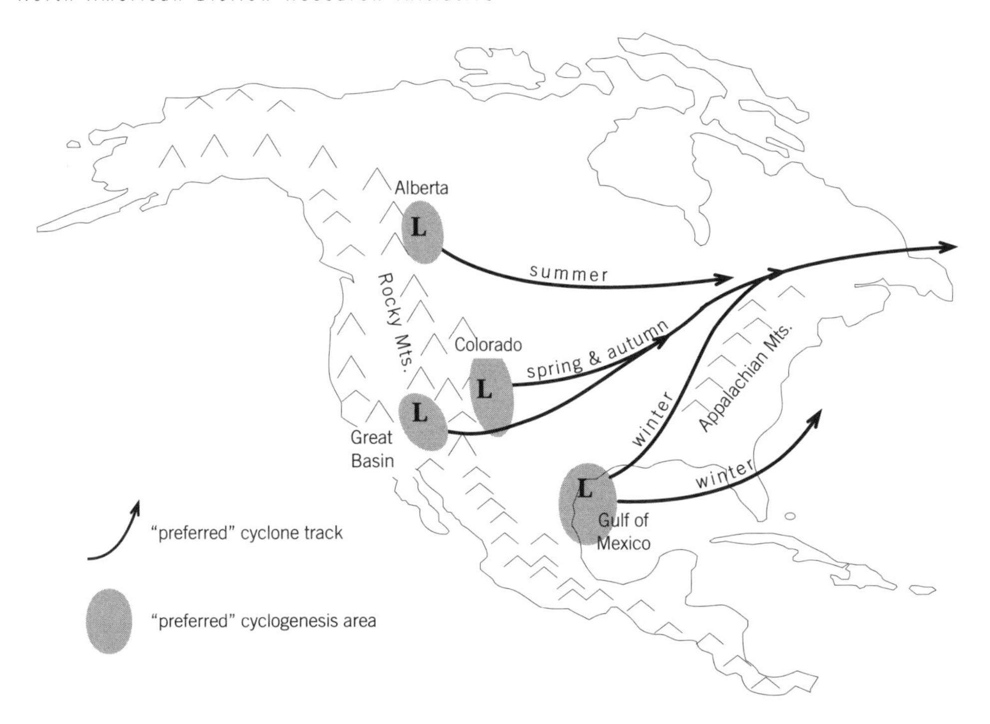

Figure 12.3. Common extratropical cyclone tracks across the continental interior of North America.

The principle cyclogenesis areas in the North American corridor are the Gulf of Mexico and the lee side of the Rocky Mountains although cyclogenesis can and does occur almost anywhere in the corridor. Cyclogenesis areas are characterized by: (1) steep lapse rates, a nearby source of warm humid air, and/or large horizontal temperature gradients. A steep lapse rate favors rising motions along a front. Warm, humid air allows condensation to occur at low altitudes increasing the buoyancy of the air. Large horizontal temperature gradients are associated with the presence of fronts and jet maxima in the area. Cyclone movement is generally from west to east. Cyclongenesis in the Gulf is typically in the winter and spring. Many of the Gulf cyclones remain fairly weak as they move across northern Florida or southern Georgia and then intensify as they move northward along the east coast of North America. Alberta is an active cyclogenesis area throughout the year although the role of these cyclones in transporting aerobiota is greatly diminished in the cold seasons. Most of these cyclones move along the United States-Canadian border with some remaining well north in Canada and others dropping south of the Great Lakes. Colorado cyclones generally form in spring and autumn and track through the Great Lakes region. Lee-side cyclones can also develop in Nevada to the east of the Sierra Nevada Mountains. These cyclones often intensify once they move into the corridor and incorporate warm, moist tropical air from the Gulf of Mexico.

Because extratropical cyclones generally form along the polar front where there are divergent airflows aloft and track beneath the polar jet stream that forms within the downwind side of a Rossby wave, preferred areas of cyclogenesis are found within the North American corridor, and these cyclogenesis areas vary with the seasons (Figure 12.3). Likewise, the tracks of the itinerant anticyclones are governed by the larger-scale Rossby waves. Because these mesoscale features and their associated surface airflows within the corridor are embedded in the larger-scale mid-tropospheric flows, they vary in strength and position with the season and produce substantial temporal and spatial variations in opportunities for long-distance aerial movement of biota within the North American corridor.

Biological factors that operate at the large scale and cause variations in bioflow.

During the 19th century, scientists developed classification schemes for dividing the world into biogeographical realms to characterize the distribution of plants and animals (e.g., Wallace 1876, Merriam 1898, Shelford 1913). The early classifications resulted in six global ecoregions: Neoarctic, Paleoarctic, Neotropical, Ethiopian, Oriental, and Australian. The North American continent falls within the Neoarctic realm and was divided into biomes using plant formations. It is generally assumed that the distribution of animals is closely associated with that of the plant communities, and thus each biome consists of a distinct assemblage of plants and animals.

Recently, the United States Forest Service has adopted a classification scheme based on the concept of an ecosystem. This worldwide classification system of "ecoregions" was developed by Bailey (1983, 1995, 1998) so that management of natural resources could be accomplished in a more effective manner. The classification defines an ecosystem as a biological community and its associated physical environment so organized that a change in any one of its components results in changes in the others and thus alters the operation of the total system. The ecoregion classification includes four domains (Polar, Humid Temperate, Dry, Humid Tropical) with each domain partitioned into divisions. There are eight biotic divisions within the North American corridor (Subarctic, Prairie, Warm Continental, Hot Continental, Subtropical, Temperate Steppe, Tropical/Subtropical Steppe, and Savanna). Each of these divisions has an associated high elevation regime, although only the high elevation regime of the Subtropical division (Ozark Mts.) is included within the North American corridor (Figure 12.4). In many respects, the pattern of divisions and their associated elevation regimes is similar to the pattern of biomes for North America. However, because the ecoregion classification is based on the concept of an ecosystem as an interrelated system, and was designed for the purpose of managing ecosystems, its use is very appropriate for characterizing the distribution of plants and animals in North America, with respect to atmospheric movements of biota. Each biotic division and its associated regime can be partitioned into landscapes—geographical assemblages of spatially contiguous ecosystems that are interconnected by corridors and superimposed upon associated physical features. At the smallest spatial scale, an ecosystem is considered to be composed of a variety of habitats—places in which organisms live that are characterized by physical and biological features.

The large-scale synchrony in temperature and moisture that is associated with the progression of the seasons generally places the plants and animals within the same biotic division in similar phenological states (e.g., a warm spring will advance development while a cool spring will delay development). Changes in growing conditions can be experienced on a weekly basis and at the large spatial scale are primarily influenced by latitude, position on the continent, and elevation. The seasonal progression of temperature and moisture conditions within a biotic division induces cohort synchrony in both plant and animal populations, setting the stage for synchronous long-distance movement by a large number of individuals and species once they enter a "dispersal-ready" state. Large-scale human modification of biotic divisions has increased the synchrony of phenological development

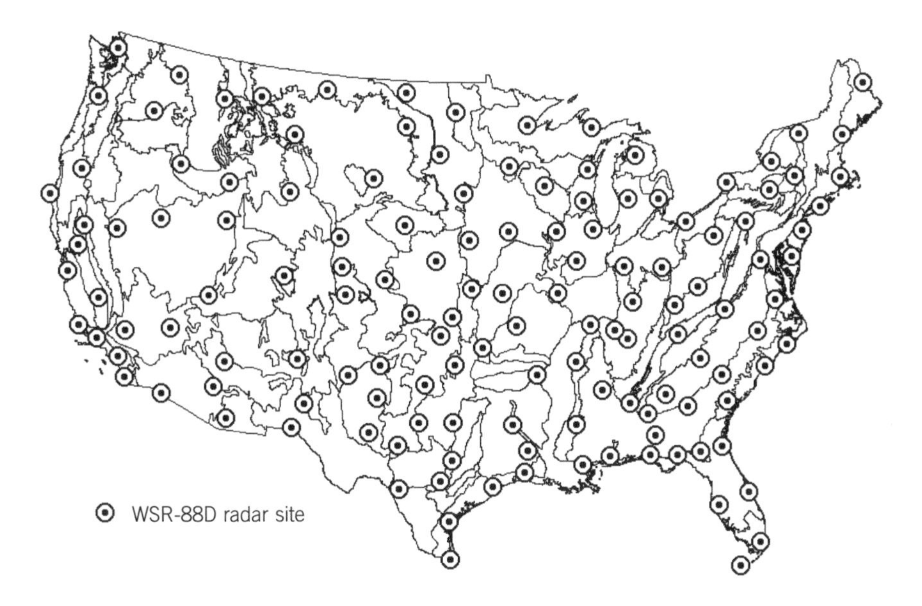

Figure 12.4. WSR-88D radar facilities superimposed on Omernik ecoregion classification for the United States.

The WSR-88D (NEXRAD) radar was designed to protect against severe weather events that threaten life and property through measurements of winds and rainfall. The Doppler technology allows computation of both the speed and direction of motion of hydrometeors in the air. Approximately 160 WSR-88D radars will be deployed by the onset of the 21st century in the United States. Users of radar data have access to WSR-88D products via the NEXRAD Information Dissemination Service. The products include reflectivity values, radial velocity measurements, rainfall accumulations, and wind profiles. Radars work by transmitting short pulses of electromagnetic energy. Because biota contain large amounts of water, they are good reflectors of radar signals. A small fraction of the emitted waves are scattered by an object and returned to the radar (reflectivity). The Doppler effect is the change in the frequency of the return signal that is produced by the motion of the reflecting object. When the object is moving away from the radar, the frequency decreases. The return frequency increases, if the reflecting object is moving toward the radar. Thus Doppler radars measure the radial velocity (speed toward or away from the radar) of hydrometeors and aerobiota.

in a wide range of organisms. For example, the simultaneous planting of crops throughout large regions of the continental interior of North America has synchronized the life stages of many agricultural pests and their predators. Many organisms move long distances in the atmosphere to exploit these abundant but ephemeral resources, and to survive, some must move again after harvest to overwintering habitats.

Extreme events that occur at large temporal scales can induce unusual and unexpected mass movements of organisms in the atmosphere. Many populations of animals do not move from their habitats until the ecosystem is under severe stress. For example, the migratory grasshoppers in the grasslands of North America have not entered a migratory phase since the 1930s. Conditions within the grasslands have not allowed their populations to reach high enough levels to cause sufficient ecological stress to induce long-range movements of grasshoppers.

Biological factors that operate at the mesoscale and cause variations in bioflow. Landscapes are the matrices of spatially contiguous ecosystems within biotic divisions. Natural landscapes are commonly delineated by landforms (e.g., river basins and mountains) and soils. They are composed of ecosystems that are connected by corridors through which biota freely move. Generally, the topography and the types and arrangement of surface cover units within a landscape exert considerable influence on the impact of the seasonal variations in temperature and precipitation on the landscape's ecosystems and populations. Consequently, patterns of landforms and surface cover cause mesoscale temporal and spatial variations in bioflow within the North American corridor. Within this vast domain, large portions of each of the biotic divisions have been substantially modified by human use. Today, landscapes within the corridor are more often delineated by land development and management practices (e.g., urban, suburban, agricultural, and recreational) than by topography and natural resources.

Population pressures within a landscape, including predation, disease, resource shortages, and other processes that may induce stress, can increase the propensity of organisms to move. Ecosystem diversity, variations in soils, surface water abundance and quality, and the quantity and distribution of nutrients can modify an organism's responses to these population pressures as well as to adverse weather conditions, thus affecting an organism's phenology and dispersal potential. A relatively homogeneous landscape may provide a single resource in great abundance (e.g., food); however, it may not constitute a suitable habitat for many organisms even though they can exploit the plentiful resource, because the landscape does not supply the variety of resources that the organisms require. It follows that organisms that can move long distances in the atmosphere are better able to exploit homogeneous landscapes with abundant resources. In contrast, the presence of adjacent ecosystems within a landscape that can supply organisms with resources sufficient for development during the growing season and habitats suitable for overwintering may make long-distance movement for a species unnecessary.

From the perspective of organisms that have short life spans, the spatial distribution of resources in natural landscapes changes with time in a manner that is usually predictable. However, in human-managed ecosystems and landscapes, changes in vegetation patterns may vary from season to season and from year to year, depending on economic and land management factors. For example, in some years crops may be rotated according to strategies that can sustain long-term production (e.g., a corn-soybean-wheat rotation) whereas in other years economic forces may require an agricultural community to grow continuous crops (e.g., corn after corn). These changes in vegetation patterns from year to year influence the local distribution of organisms within a landscape and may enhance the potential of organisms to enter a population outbreak phase. The degree to which humans increase the homogeneity of landscapes (e.g., urbanization and modern agriculture) directly influences the distance that many organisms must move to find the variety of resources they need to survive. From this perspective, current human land use practices within landscapes are increasing the importance to many biota of being able to move long distances.

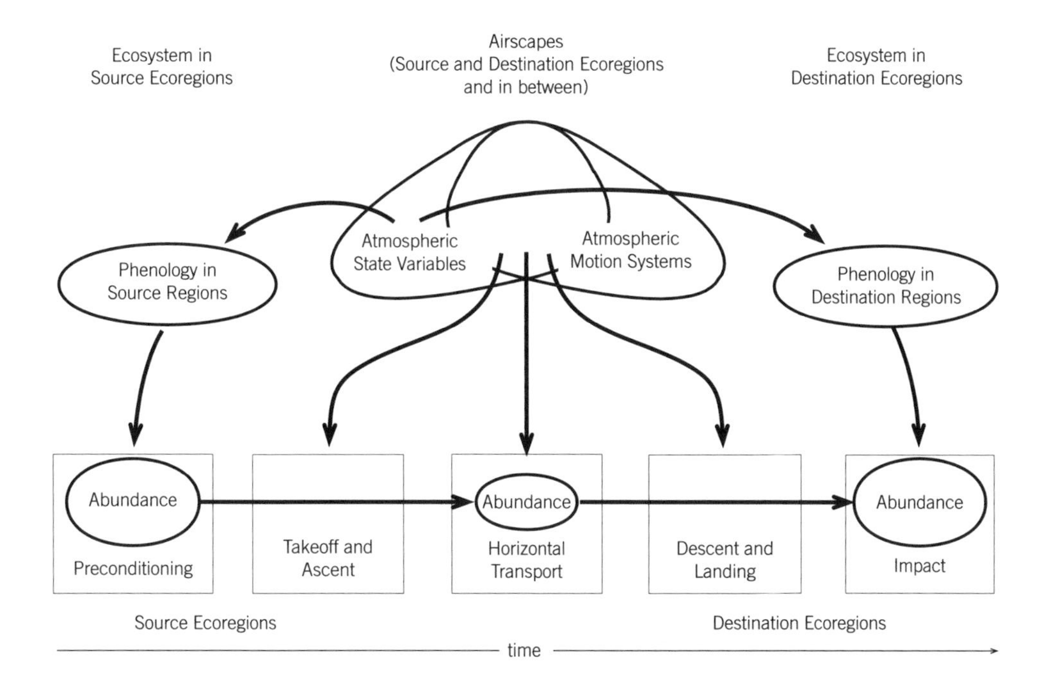

Figure 12.5. Operational form of the aerobiology process model for the bioflow research initiative.

The general aerobiology process model (see Figure 4.2) has been simplified to forecast the large-scale movement of biota within the North American corridor. The components of the model that are encircled with heavy lines would be monitored during the project. Atmospheric state variables and motion systems, vegetation phenology, and the abundance of key aerobiota should be monitored throughout the corridor on a near-continuous basis. Periodically, the abundance of biota in the atmosphere must be measured directly. Because the focus of the bioflow research agenda is to forecast long-distance movement, emphasis is placed on the influence of the atmosphere on the survival of biota in the air and on the distance and direction of their aerial transport. In this application, the biological components of the general model (life history of organisms, current vitality of organisms, and the environment and ecology of their terrestrial ecosystems) are represented by vegetation phenology. To a large extent, the life history of organisms in their source ecoregions, and thus their readiness for aerial movement, is linked to vegetation phenology. The abundance of biota in the air during takeoff & ascent and descent & landing would not be monitored because these aerobiology process stages occur very quickly (seconds to minutes). Nonetheless, the inclusion of the control of temperature and wind on takeoff and ascent and the influence of precipitation on descent and landing may be important.

Conceptual framework for understanding and predicting bioflow.

Figure 12.5 shows an operational form of the general aerobiology process model (see Figure 4.2) that has been adapted for predicting large-scale flows of biota within the North American corridor. In this very simple conceptual model, phenology is used to characterize the time sequences of biological events and to predict abundance and incidence of organisms in biotic divisions that represent potential source areas (preconditioning). As discussed above, the many taxa within a biotic division at any moment in time are at different stages in their life histories. However, the rates at which they progress through their life history sequences are linked because variations from the usual rates of phenological progression are governed to a large extent by atmospheric state variables (i.e., the seasonal progression of weather). As argued in the previous chapters, atmospheric state variables and motion systems greatly influence the takeoff and ascent, horizontal transport, and descent and landing stages in the movement process. Because the focus here is long-distance bioflow, emphasis is placed on the influence of the atmosphere on survival in the air and the distance and direction of horizontal transport. Nonetheless, it is important to account for the regulation of temperature and wind on the takeoff and ascent of many aerobiota and the influence of precipitation over their descent and landing. Finally, vegetation phenology in potential destination areas (i.e., landscapes with the requisite types of resources) influences the immigrants' success at colonization and their subsequent impacts on receptor landscapes.

The bioflow initiative seeks to study the pathways and timing of aerial movement of organisms within the North American corridor by monitoring the atmospheric state variables and motion systems, vegetation phenology, and abundance of key aerobiota at sites throughout the entire corridor during a long series of growing seasons. The abundance of aerobiota in the atmosphere must also be measured directly at a few locations intermittently throughout the study period. Details of the measurement program are provided in the Plan of Action below. The initiative would quantify the relationships represented by arrows in Figure 12.5, and in so doing, characterize the biological and meteorological phenomena that influence the flow of biota in the atmosphere. This knowledge can then be used to forecast the pathways and timing of bioflow within the North American corridor.

The bioflow initiative: plan of action.

Many classes of organisms have evolved to utilize the atmosphere to move long distances. Animals that commonly use atmospheric motion systems for transport include birds, arthropods, and nematodes. Many plants, some considered by humans to be beneficial and others to be weeds, use the atmosphere for dispersing their seeds, pollen, and spores. Microbial organisms transported by atmospheric motion systems include bacteria and viruses. Some of these aerobiota travel together, with the larger organisms serving as vectors for seeds, pollens, nematodes, spores, and microbes. The vast diversity of these classes of organisms provides an opportunity to integrate the different methods that are currently used to measure atmospheric flows of biota. The consolidation of methodologies and technologies that are useful for measuring and interpreting the atmospheric flows of a broad range of biota under a single observation system will greatly enhance the utility of aerobiological observations at reduced cost.

The primary objective in the bioflow initiative, which is to characterize the biological and meteorological factors that influence the atmospheric flow of biota within the North American corridor, can be approached from two perspectives simultaneously. On one hand, correlations of spatial and temporal patterns of aerobiota with patterns of important meteorological and phenological phenomena throughout the entire corridor, will increase the understanding of the aerial movement process by inference. On the other hand, direct observations of types and concentrations of aerobiota, and the associated atmospheric motion systems, using entomological and weather radars will increase the knowledge base about aerobiology process controls. Integration of the direct measurements of the bioflow process, to be obtained over small spatial and temporal scales, with the large-scale inferred information relating pattern to process, is required to characterize the biological and meteorological interactions that influence bioflow at large spatial scales.

A biological monitoring program based upon standardized measurement principles is a fundamental component of the bioflow initiative. Its design must provide the capability to compare observations of specific aerobiota over time and at different geographic locations. Permanent locations for measuring incidence, abundance, and occurrence of selected organisms need to be established to characterize the temporal dimension of bioflow. Standardized monitoring principles must be implemented to ensure that observations made at one site will be comparable to those made at all other sites in the network.

The components of a measurement and analysis program for the bioflow initiative include: strategic biological sampling landscapes, vegetation phenology measurements, radar measurements of biota, information management and exchange, and computational analyses. Each of these components is discussed below as it was envisaged by the authors when they presented the bioflow initiative as a proposal to the AFAR Governing Board. Since that time, the initiative has been improved in many areas by suggestions from the AFAR membership. Also, a number of the specifics of the methodology have been updated as technologies and knowledge have advanced. Nonetheless, many of the details of the bioflow initiative methodology are included here to provide an example of how we believe modern technology should be harnessed to further the understanding of large-scale aerobiological processes.

Strategic biological sampling landscapes.

Accurate determinations of atmospheric bioflow across a large geographic area require that measurements be made (1) during the time when the organisms are being transported and (2) throughout the area where the organisms are flowing. Because the North American corridor is very large, and due to financial limitations on the number of measurement sites, the strategic sampling landscapes must be carefully selected. At least one landscape in each of the eight different biotic divisions in the North American corridor must be represented. As more resources become available to implement the bioflow initiative, other sampling landscapes should be added.

Establishment of strategic sampling landscapes requires an analysis of the geography of the ecoregion by integrating climatological, physical, and biological information within a GIS. The climate analysis includes patterns of upper airflows, air temperatures, and precipitation amounts. Topography, watershed characteristics, soils, and land use patterns would be encompassed in the analysis of the physical geography. The biological characteristics of the ecoregion includes vegetation types, agronomic factors, and animal species distributions. Most of this information is currently available in digital form. An analysis of the resulting digital map library of climatology, biology, and physical features will guide the selection of representative landscapes to be designated as strategic sampling landscapes.

Each landscape selected for strategic biological sampling must contain a National

Weather Service recording station with a long time series of climatological measurements. Current measurements of important atmospheric factors, including solar radiation, air temperature and humidity, soil temperature, air pressure, wind speed and direction, and precipitation, must be accessible in near real-time via the Internet. In addition, the landscape must be located near a WSR-88D ("NEXRAD") Doppler weather radar measurement installation (see Figure 12.4). Only by co-locating strategic biological sampling landscapes with weather and climate measurement facilities can the specific research objectives be achieved.

The populations of at least ten aerobiota within each of the landscapes chosen for sampling should be monitored. The species chosen should represent the wide diversity of organisms that move in the atmosphere and are important to humans. Examples might include a plant pathogen (e.g., wheat rust), human pathogen (e.g., virus or bacteria), human allergen (e.g., ragweed), plant seed (e.g., weed), weak-flying insect (e.g., leafhopper or aphid), strong day-flying insect (e.g., monarch butterfly), strong night-flying insect (e.g., Helicoverpa spp.), day-flying bird (e.g., hawk or crane), night-flying bird (e.g., Swainson's thrush), and strong flying waterfowl (e.g., Canada goose). A wide array of population measurement technologies can be deployed including: spore traps for spores, pollen grains, and plant pathogens; virus filter traps for viruses; sticky traps for weed seeds and insects; light, water pan, and pheromone traps for insects; infrared cameras and mist nets for birds; and visual observations. These measurement technologies require positioning within the ecosystems across the landscape at standardized locations that are appropriate for measuring the various biota. The coordinates of each measurement location can be determined by a Global Position System (GPS) unit.

During important periods in the year, daily measurements must be made within each of the landscapes by a trained observer. Protocols for specific organisms should be developed by specialists and standardized across the network. Data collected using these protocols can be managed according to the specific methods prescribed. Organisms should be identified *in situ* if possible, and where this is not appropriate, sent to a specialist for identification. After identification and enumeration, the observations would be entered using a computer-linked form. Data then can be made accessible to project personnel and administrators. Automated mapping of organism distribution and abundance will be incorporated into the Internet system.

Vegetation phenology measurements.

Vegetation base maps displaying the distribution of vegetation types can be acquired for the study area based on the works of Kuchler, (1964, 1970), Omernik and Griffith (1991), and Bailey (1995) and converted to digital format. Cropping distribution patterns can be derived from National Agricultural Statistical Service data. Satellite data on vegetation growth patterns are available on a biweekly basis. One of the more useful satellite sensors for observing changes in seasonal vegetation patterns is the Advanced Very High Resolution Radiometer (AVHRR) aboard the NOAA Polar Orbiting Satellite, which collects daily observations over the lower 48 states and parts of Mexico and Canada. Two week composites of daily satellite imagery at a 1 km scale provide information about vegetation growth patterns based on a Normalized Difference Vegetation Index (Eidershink 1992). Current AVHRR satellite imagery can be obtained from the EROS Data Center in Sioux Falls, South Dakota. The remotely sensed data should be frequently calibrated with field measurements of vegetation phenology. A sampling grid must be established to standardize the phenological measurements from each landscape that is chosen as a strategic biological monitoring landscape. Regular observations of the growth stages of key plant species at multiple grid points within the various ecosystems in the landscape, using prescribed protocols, should be digitally recorded using standard methods. The coordinates of each of these measurement locations can also be determined by a GPS unit.

Analytical techniques for forecasting bioflow.

Two analytical techniques are currently used to study and forecast the atmospheric movement of biota in the North American corridor (see Chapter 11). Trajectory analysis is commonly employed to describe the mean displacement of biota in the atmosphere between successive measurements and to predict movement over subsequent intervals (e.g., Figure 11.7). Atmospheric diffusion models are also utilized to calculate the dispersion of biota along their concentration gradient within the PBL. Both of these analytical techniques can forecast the aerial movements of biota from single or multiple points in space and time (Figure 11.8). They differ in that trajectory models compute the most likely path that biota will travel while diffusion models treat the biota as a "cloud" that spreads out, becoming less dense with distance in the downwind direction. A number of atmospheric trajectory and diffusion models are currently available, and they continually are being improved. The most advanced of these models should be used and the computational analysis procedures updated throughout the project as new analytical procedures are developed and more sophisticated computer hardware becomes accessible. Although it has long been acknowledged that any significant improvement in the capability to forecast bioflow requires validation of these analytical techniques through coincident atmospheric measurements and biological observations, heretofore the technology and expertise to address this important issue was unavailable. However, this is no longer the case, and these models should be evaluated by direct measurements of the bioflow process and the associated atmospheric motion systems using state-of-the-art entomological and weather radars.

Radar measurements of biota.

Two unique radar systems have the capability to remotely detect aerobiota: Doppler weather radars and insect monitoring radars (see Table 12.2). Birds, bats, large insects, and some smaller aerobiota (e.g., "clouds" of small insects, pollen, and spores) can be detected by the network of WSR-88D ("NEXRAD") Doppler weather radars operated by the United States Departments of Commerce, Defense, and Transportation. Approximately sixty WSR-88D installations operate continuously within the North American corridor (see Figure 12.4), and this number is growing. The Doppler radar has the capacity to measure the direction and speed of bioflow in addition to reflectivity. The resolution of the WSR-88D reflectivity and radial velocity measurements are 0.95° by 1 km and 0.95° by 0.25 km within a range of 460 and 230 km respectively. At altitudes below 3 km, biota are the principle source of "clear-air

Table 12.2. References to works on WSR-88D and IMR radars.

WSR-88D ("NEXRAD") DOPPLER WEATHER RADARS:
Larkin 1991, Klazura and Imy 1993, Jain et al. 1993, Westbrook et al. 1998.

INSECT MONITORING RADARS (IMRS):
Riley 1992, Drake 1993, Riley 1993, Smith et al. 1993, Drake et al. 1998.

echo" and are used routinely to estimate wind speed and direction. Radar coverage within the PBL extends sufficiently far to allow these weather radars to be used as areal surveillance tools for bioflow. However, the density of the radar installations is insufficient to continuously track "clouds" of aerobiota as they move through the North American corridor. Currently, aerobiological measurements are confounded by precipitation when the WSR-88D radars are operating in the "precipitation mode" to detect nearby storms. This significant limitation needs to be addressed because much bioflow occurs during stormy weather within prefrontal airflows. Radar operators are interested in measuring biota concentrations in the atmosphere because aerobiota (as well as thermal stability and turbulence) can degrade or contaminate WSR-88D wind velocity information. Wind velocity bias due to aerobiota is typically most pronounced when actively flying organisms are collectively aligned and concentrated in layers of the atmosphere.

Russell and Wilson (1997) provide an excellent example of the utility of using sensitive Doppler radar to study the aerial movement of biota. Calculations based on multiwavelength dual-polarization radar data and theoretical models of radar backscattering, as well as a small number of aerial collections using nets attached to kites, indicate that clear-air radar echoes in the PBL primarily result from biota that range in length from 1 to <10 mm. These findings lead to the speculation that clear-air echoes largely result from small insects rather than microbiota such as pollen and spores or large strong-flying aerobiota such as moths, grasshoppers, and birds. The authors note that a common characteristic of the spatial distribution of weak-flying aerobiota is their concentration in near-linear convergence zones or "fine-lines" in the PBL. These convergence lines are frequently present during warm afternoons and are often associated with the development of convective storms. In contrast, Achtemeier (1992) provides Doppler radar observations of apparent active flight within a cloud of grasshoppers and/or other insects. A change in flight orientation appeared to be a response to an environmental stimuli—a turning downward within a zone of strong updrafts associated with a gust flow from a convective storm. The insects either may have folded their wings to fall or else actively flown downward through the updraft to avoid being carried aloft. Table 12.3 from Russell and Wilson (1997) lists other circumstances of strong correspondence between the spatial and temporal variations in clear-air radar echoes and activity patterns of aerobiota.

Data from Insect Monitoring Radars (IMRs) provide information about the size of aerobiota and allow separation of birds, insects, and other aerobiota into broad taxonomic groups. IMRs are vertically looking 8.8 mm wavelength radars that can detect individual aerobiota weighing as little as 2 mg at a range in excess of 1 km. They are considered sampling

Table 12.3. Characteristics of diurnal clear-air radar echoes compared to spatial and temporal patterns in the ecology of aerial insects.

RADAR OBSERVATIONS	AERIAL INSECT ECOLOGY
Clear-air return over land is widespread and intense during summer, much scarcer and weaker during winter.	Aerial insect populations are much larger during the summer than during the winter in temperate and subtropical areas.
Intensity of clear-air return during winter is temperature-dependent, with echoes almost nonexistent below 10°C, but present and increasing in intensity between 10°C and 20°C.	Flight thresholds of most small insects are between 10–20°C.
Clear-air return is rare over the ocean except for short distances offshore in conditions of strong offshore winds.	Aerial densities of small insects are much lower at sea than over land.
Depth of the clear-air scattering layer varies geographically, but is usually identical to the height of the convective boundary layer.	Small insects depend on convective updrafts (i.e., thermals) for ascending, and are therefore unlikely to occur above the top of the [planetary] boundary layer.
Exceptions to [the above] pattern occur only on some days when the sun is obscured by extensive high cloud cover; in these instances, depth of the particulate scattering layer is much less than the height of the boundary layer.	Thermals are weaker when the incidence of solar radiation on the ground is reduced. Insects appear to have trouble ascending in conditions of extensive cloud cover.
Reflectivity factor decreases with increasing altitude in the [planetary] boundary layer.	Aerial insect density decreases with increasing altitude in the [planetary] boundary layer.
Average size of radar scatterers decreases with increasing altitude in the [planetary] boundary layer.	Average size of aerial insects decreases with increasing altitude in the [planetary] boundary layer.

From Russell and Wilson 1997. The authors provide citation for each of the characteristics.

devices because they can measure bioflow only in a small conical volume above the radar. They are highly automated and can be configured to disseminate their data rapidly across the computer network. IMRs are mobile and allow study of the flow of various target aerobiota during appropriate time periods in the biotic divisions. Generally, the IMRs should be operated near WSR-88D installations to allow simultaneous use of the surveillance and sampling capabilities of these radars. Short case studies that include aerial sampling of biota will validate WSR-88D and IMR data and relate vertical zonation of biota to atmospheric stability and turbulence. Concurrent WSR-88D and IMR measurements provide valuable information in three areas. First, they allow identification of aerobiota (broad taxonomic groups) and provide estimates of the density, orientation, flight speed, and direction of these aerobiota. Second, they provide information for analyzing relationships between aerial movement of biota and atmospheric motion systems that can validate and refine (improve the parameters of) numerical models that forecast bioflow. Finally, they provide information that could lead to the development of algorithms for removing the undesirable effects of biological scatterers on WSR-88D products such as the wind velocity errors that can be pronounced when biota are concentrated in atmospheric layers.

Information management and exchange.

The success of the proposed research requires a multi-investigator team composed of participants with a wide range of expertise from locations dispersed throughout the corridor. The complexities of communication require a coordinated approach. Fortunately, the Internet provides a medium and a methodology for information communication unprecedented in history. A system designed to use the Internet allows: (1) easy communication between field observers and laboratory staff, (2) taxonomic support using image capture and image transmission, (3) transmission of biological and meteorological data collected in the field, (4) access to field data for cooperating investigators, (5) extraction and management of regional weather observations associated with biotic flow observations, and (6) a mechanism to report findings to colleagues, the larger scientific community, and the general public.

Computational analysis.

The computational analysis can be divided into three components. The first component of the computational analysis, the compilation of a digital atlas of climate, biology, elevation and physical landforms for the corridor, was described above with respect to the selection of strategic biological sampling landscapes. The second component, the measurements of patterns of incidence, occurrence, and abundance of the target organisms and their association with temporal and spatial patterns of vegetation phenology and atmospheric flows, would be used to improve the parameters of quantitative models of bioflow. Because data will be collected at all sites in the sampling network during the same time windows, using similar methods of observation for each organism, interlocation comparisons can be made to develop correlations among observations of organism occurrence, atmospheric processes and events, and vegetation type, growth, and phenology. To keep the records updated, the multi-location observations must be linked to the digital atlas of climate, vegetation, elevation, and landforms and to the biweekly digital maps of the phenological development of the vegetation of the corridor derived from satellite imagery. Measurements of the spatial and temporal distributions of biota in the atmosphere and associated motion systems by entomological and weather radars and tropospheric soundings would be included at this stage. Application of geostatistics within a GIS is a critical component of this spatial analysis. Regression and gradient analyses, paired comparisons between different types of organisms, and organism diversity computations, will serve to determine parameters for models of the incidence, occurrence, and abundance of aerobiota in a landscape.

Trajectory and diffusion models would be used to link weather, atmospheric motion systems, and landscape characteristics to predict changes in the distribution of each target aerobiota across the landscapes in the biotic divisions. These models can be validated using observed patterns of biota from the strategic sampling landscape network and radar measurements of bioflow. Once validated, these models will be refined and generalized, using information on patterns of vegetation phenology to evaluate temporal and spatial variations in opportunities for long-distance aerial movement within the corridor. Experiments should be conducted to test specific hypotheses which evolve from the quantitative analysis of patterns of aerobiota revealed by observations and by the computer models. Subsequent spatial and temporal measurements of aerobiota from the strategic biological sampling network and direct observations of the aerial movement of biota from WSR-88D and IMR radar measurements and aerial collections would facilitate evaluation of these hypotheses.

The final phase of the computational analysis addresses the second objective of the initiative, which is to develop a regional-scale model to predict the pathways and timing of the long-distance aerial flow of biota from subtropical to midlatitude continental interior areas of North America during spring/summer and the equatorward flow that occurs during the summer/autumn. The validation of trajectory and diffusion models for predicting specific bioflows would lead to the development of an interactive model characterizing the flow of organisms within the atmosphere throughout the North American corridor. The goal is the development of a bioflow model for the entire North American corridor that can be operated in near real-time by a variety of users. This large-scale forecasting model would be driven in part by satellite imagery and field measurements (from the strategic sampling landscape network) that track patterns of vegetation phenology in source areas. Wind fields generated by the National Center for Environmental Prediction (NCEP) Nested Grid Model can be used to run the atmospheric trajectory and diffusion models. In addition, the resulting large-scale model could be calibrated (updated on a daily basis) using near real-time WSR-88D measurements of aerobiota density and movement as well as the measurements of populations in the destination areas within the corridor. User specified parameters would be combined with this data to run the bioflow transport model. The output from the model, characterizing the ascent, transport, and descent of organisms within the North American corridor should be displayed visually in three dimensions using a daily time step. Optimization of model output and visualization integration requires a model be developed using high end visualization software linked to a relational database management system.

The bioflow initiative: products/outcomes.

This research initiative transitions through three components: (1) compilation of a digital atlas of climate, biology, elevation, and physical landforms for the corridor and analysis of this historical data to help locate strategic sampling landscapes, (2) analysis of data: field measurements of populations of selected aerobiota and vegetation phenology within representative landscapes, satellite imagery of vegetation phenology throughout the corridor, radar measurements of bioflow and concurrent atmospheric motion systems, aerial collections of biota, current weather measurements and maps of airflows, and the digital atlas of climate, biology, elevation, and landforms to increase our understanding of bioflow in the corridor, and (3) evaluation with bioflow measurements of the existing models (trajectory and diffusion types) for predicting aerial movements of individual taxa and development of a large-scale model of the atmospheric flow of biota over the North American corridor.

In addition to improving our understanding of the interactions among biological and meteorological processes that govern bioflow, the bioflow initiative provides three primary products and a large number of indirect outcomes. The direct products will be available to users via the Internet. First, a digital atlas of climate, biology, elevation, and physical landforms within the corridor will be updated annually with phenological and climatic data from the strategic sampling landscape network. Second, maps will be produced showing the extent and abundance of target aerobiota in their source areas within the North American corridor. These maps will be created by extrapolating population data from the strategic sampling landscapes to the scale of biotic divisions using the vegetation phenology measurements derived from satellite imagery and calibrated by weekly phenology measurements from the strategic sampling landscapes. Finally, forecast maps of the aerobiological pathways for target species (e.g., Figures 11.7 and 11.8) will be produced daily. These maps will show: (1) wind vectors, (2) the mean trajectory and the location and concentration of the biota "cloud" for targeted biota from each source area, and (3) a composite of the trajectory and diffusion analyses for each target biota. The indirect products and outcomes of the bioflow initiative will likely be numerous. For example, synergism hopefully will be created between the data-gathering activities associated with the strategic biological sampling landscape network and studies involving local- and microscale population measurements within the selected landscapes by using local biological scientists to operate the sampling stations. Also, in the short case studies to validate WSR-88D and IMR data with aerial sampling of biota, scientists from the United States Departments of Commerce, Defense, and Transportation will be included. These field measurement projects could provide valuable information leading to the development of algorithms for removing the undesirable effects of biological scatterers on WSR-88D products. In addition, because the initiative embraces researchers from a large number of disciplines and institutions, it enhances communication, education, and the sharing and pooling of information and technology among participants to create a "climate" that will propel the science of aerobiology forward into the 21st century.

EPILOGUE

The Alliance for Aerobiology Research (AFAR): A Grassroots Mechanism for Enhancing Understanding of Aerobiological Systems

Origins of the Alliance for Aerobiology Research.

The idea of forming a grassroots organization to coordinate research in aerobiology evolved from a meeting of the North Central Regional Committee on Migration and Dispersal of Insects and Other Biota (NCR-148) in 1990. NCR-148 had been organized in the mid-1980s by agricultural entomologists from the United States who needed to pool knowledge and expertise about the vast array of new tools that were becoming increasingly available for studying long-distance aerial movement of biota. These tools included mobile entomological and ornithological radars, powerful weather radars, genetic marking techniques, atomic absorption spectrophotometry, atmospheric diffusion and trajectory computer algorithms, geographic information systems, and systems approaches to modeling aerial insect movement. The primary objectives of the group were to share their research findings and to foster collaboration. Equally important was the need to expand NCR-148 to include scientists from other disciplines, to learn more about and share new and exciting technologies and research opportunities. By the early 1990s, substantial collaboration between aerobiologists with biological and atmospheric expertise had developed, and the NCR-148 annual meetings were attended regularly by scientists from entomology, plant pathology, meteorology, engineering, geography, and systems science. The majority of the participants were from the United States Department of Agriculture (USDA) Agricultural Experiment Stations, the Agricultural Research Service, and/or the Cooperative Extension Service.

Despite the emphasis on interdisciplinary and multi-institutional cooperation within NCR-148, a strong bias toward entomological research remained, with little collaboration among biologists studying different taxonomic groups (e.g., entomologists and plant pathologists) and no participation from scientists in the human health or environmental arenas. Realizing that aerobiologists, regardless of their specializations, were united in the pursuit of principles that govern aerial movement of biota, NCR-148 participants drew up a set of generic hypotheses to elucidate these principles and embarked on a campaign to obtain funding for an international workshop to enhance interactions among scientists from a wide range of disciplines. The workshop organizers promoted their efforts through the Pan-American Aerobiology Association (PAAA) and the American Meteorological

Society (AMS) and sought support from the USDA, United States Agency for International Development (USAID), the National Institute of Health, and the National Science Foundation. In October, 1992, approximately seventy scientists and outreach specialists from North America and Europe attended the Alliance for Aerobiology Research (AFAR) workshop where they deliberated over a wide range of important aspects and issues in aerobiology. At the conclusion of this workshop, the AFAR was formalized.

Summary of the proceedings of the Alliance for Aerobiology Research workshop.

The AFAR workshop brought together scientists and outreach specialists from agriculture, climatology, engineering, entomology, environmental sciences, forestry, geography, human health, meteorology, ornithology, palynology, plant pathology, systems science, and weed science who shared a common interest in aerobiology. Participants were asked to discuss the generic hypotheses of long-distance atmospheric transport of biota and to formulate strategies to share technologies and information systems. To launch the AFAR meeting, the workshop organizers presented a conceptual model of AFAR to workshop participants for their input and modifications. The subsequent discussions established the functional basis for AFAR (Figure E.1). The alliance was formalized as a set of research projects. Each project was to focus on an important organism and/or meteorological process and evaluate a subset of the generic hypotheses (see below). During the workshop, three key components of the alliance were debated: research foci and hypotheses, information systems and diagnostic technologies, and organization and administrative structure. The consensus of participants on these subjects, as summarized in the AFAR Workshop Report (Isard 1993), are presented below.

Research foci and hypotheses.

"Allergens, pests, and other biota in the atmosphere affect our daily lives by influencing [human] health, [the] environment, and agricultural productivity. [These] biota rely on atmospheric processes to disperse, and biota that die en route may still be important as allergens. Thus, the quality of life and the economic health of all humanity are affected by long-range movement of biota. Major gaps in our knowledge of the relationship between atmospheric processes and the aerial transport of biota impede the resolution of problems caused by aerobiota."

"To understand long-distance movement processes holistically, one must understand what happens at each end of the movement process. Dynamics of source area populations are crucial to the initiation of movement and the status of the biota during that movement. [After landing], if organisms descend into suitable habitats, the dynamics of the fallout population largely determine its impact. Many of the questions of population dynamics have been successfully addressed using established research practices. However, the descent of aerobiota to the Earth's surface is not well understood. Organisms may fall out in unsuitable habitats, then may or may not move again. Finally, the time of transport and atmospheric conditions during transport may affect the organism's mortality, viability, fecundity, and fallout location. These can all play crucial roles in determining the impact of the translated

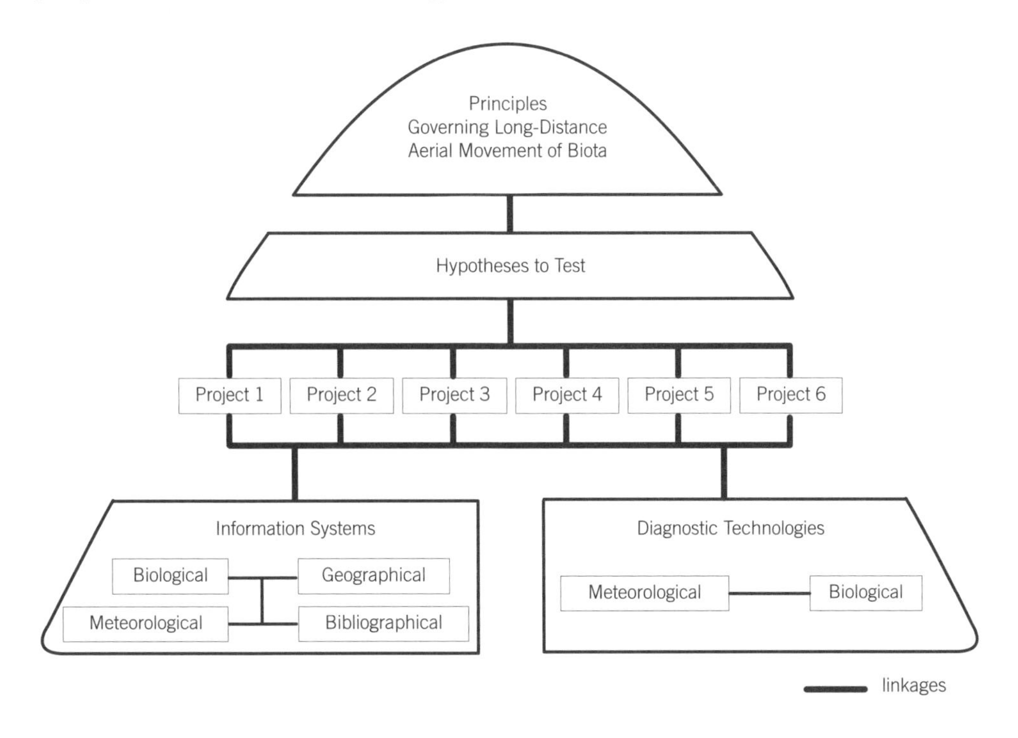

Figure E.1. Conceptual basis of the Alliance for Aerobiology Research.

The functional basis of AFAR as envisioned by workshop participants was a set of research projects, each focusing on an important organism and/or meteorological process. The design called for individual and multi-institutional projects to study the interactions between biological and meteorological systems by evaluating specific hypotheses governing the long-distance aerial movement of biota. Aerobiological systems include pests and natural enemies that range from passively transported biota (such as viruses, bacteria, fungi, nematodes, mites, arachnids, and the pollen and seeds of higher plants) to more actively flying insects and vertebrates. Aerial movement involving atmospheric motion systems of all spatial and temporal scales from eddies to global wind belts constitute part of the aerobiology system. Research projects were to formulate general principles of long-distance aerial movement of biota. The goal was to integrate the research projects comprising AFAR through a "state-of-the-art" information system with geographical, biological, meteorological, and bibliographical elements. Details of this system are provided in chapter 12 on the bioflow project. A direct communications link among research projects through a computer network was designed to provide the capability of sharing vast arrays of information pertaining to habitats, climates, temporal and spatial distribution of biota, and real-time meteorological forecasts. This computer-based information system was envisioned to have a decision support component to be accessed by farmers, veterinarians, health officials, and government agencies throughout the world so that effective pest management and human health protection strategies could be implemented. AFAR provided a mechanism for sharing advanced technologies among participating institutions and could facilitate the development of new diagnostic technologies in aerobiology.

population, be it an impact on health, the environment, or agricultural/forest production. The problems caused by long-distance movement have long been recognized, and the major impediment to resolving them is an understanding of the long-range movement process itself. Study in this area will not only advance our understanding and appreciation of the science of aerobiology but will pave the way for the resolution of a broad range of practical problems."

"For any organism involved in long-range movement, an important question is how that movement is maintained from generation to generation, whether it is strictly environmentally mediated or maintained through selection. The broader issue of gene flow in populations is heavily affected by long-distance movement. An increased understanding of these areas will shed much light on issues such as forecasting the atmospheric pathways and understanding the impact of immigrants. Many questions of aerial movement that relate to takeoff, horizontal displacement, and descent have yet to be answered, both in terms of the biology of the organisms involved and the large- and small-scale atmospheric processes that affect them. The aerobiology surrounding severe weather events is poorly understood. Recent advances in technology in both atmospheric and biological sciences, with the linkages provided by AFAR, can now enable us to address these perplexing aerobiology questions. We propose the following as guidelines, posed as generic foci [see below] and hypotheses [Table E.1], to conceptualize and direct research in aerobiology. . . ." Workshop participants suggested that aerobiology research should be organized around three general questions: (1) "What are the causes and consequences of long-distance movement?," (2) "What are the relative contributions of round-trip and one-way movement among major types of organisms?," and (3) "What are the relative contributions of genes, the environment, and the interaction of genes and environment, to takeoff and ascent, translation, and descent and landing?"

Information systems and diagnostic technologies.

"The Alliance for Aerobiology Research must be a well-organized umbrella that facilitates the coordination of research projects that focus on aerial movement of biota. AFAR should include individuals from a wide range of disciplines and outreach groups to ensure a broad perspective on aerobiology. Diverse biological organisms are recognized, including arthropods; plant viruses, fungi, and bacteria; birds; human allergens; and other airborne biota. These biota move within atmospheric motion systems that range from the microscale to macroscale."

"Existing biological, atmospheric, and other environmental monitoring networks will be identified, and they should be coordinated and enhanced for application in analyzing and forecasting aerobiological events. Networks should be encouraged that can accommodate appropriate new technologies. Existing biological sampling technologies should be used, including insect pheromone traps, aerial netting, particulate samplers, and meteorological instrumentation such as surface observing stations, rawinsonde stations, and satellite radiometers. Volatile, state-of-the-art technology such as next-generation radar (NEXRAD), frequency-modulated/continuous-wave (FM/CW) radar, wind profilers, vertical-looking radar, and lidar must be evaluated for aerobiological research applications before expensive technology and unique scientific expertise is lost due to program reductions within federal agencies. The diagnostic technologies should monitor basic atmospheric properties and biological activity that affect biota during ascent, aerial transport, and descent. Aerobiological information should forecast the time, occurrence, and flux of airborne biota for advisories of environmental contamination, public health, and agricultural production. Furthermore, the aerobiological information should advance the science of aerobiology by

Table E.1. General purpose hypotheses governing the atmospheric flow of biota.

Maintenance of the Movement Process

1. Long-distance movement is a one-way process.
2. Long-distance movement is a two-way process.
    - 2a. Return movement is ancillary to long-distance aerial transport.
    - 2b. Return movement reinforces the genetic control over long-distance transport.
    - 2c. Return movement is driven by existing environmental conditions.

Components of the Movement Process

A. Takeoff and Ascent

3. Takeoff and ascent into the atmosphere by organisms that move long distances is biologically mediated.
    - 3a. The phenological state that invokes initiation of ascent is genetically controlled.
    - 3b. Environmental preconditioning induces a physiological state that causes initiation of ascent.
    - 3c. Intraspecific/interspecific interactions influence the initiation of ascent.
    - 3d. Ascent may be influenced by aerodynamic properties.
4. Ascent by organisms into the atmosphere is influenced by environmental conditions.
    - 4a. Ascent is governed by convection within the lower atmosphere.
    - 4b. Thresholds of important atmospheric factors limit the tendency to take off and the success of ascent.
    - 4c. Ascent may be caused by hydrometeors.

B. Horizontal Transport

5. Organisms are concentrated within atmospheric layers during long-distance aerial movement.
    - 5a. Organism behavior and aerodynamic properties govern the vertical distribution of organisms during long-distance aerial movement.
    - 5b. Atmospheric factors govern the vertical distribution of organisms during long-distance aerial movement.
6. Horizontal transport of organisms within the atmosphere is predictable.
    - 6a. The duration and direction of movement are determined by the organism.
    - 6b. The duration and direction of movement are affected or driven by atmospheric processes.
    - 6c. The duration and direction of movement are influenced by environmental preconditioning.

C. Descent and Landing

7. Organisms actively descend from the atmosphere.
    - 7a. Environmental cues induce descent.
    - 7b. Physiological states govern descent.
    - 7c. Intraspecific/interspecific interactions influence the initiation of descent.
    - 7d. Descent may be influenced by aerodynamic properties.
8. The descent of organisms from the atmosphere is governed by meteorological factors.
    - 8a. Descent is caused by hydrometeors.
    - 8b. Descent is caused by downdrafts.
    - 8c. Descent is caused by changes in stability/turbulence.
    - 8d. Descent is caused by gravity.

Developed by participants at the Alliance for Aerobiology Research Workshop (Isard 1993).

defining the vertical distribution and identity of airborne biota to indicate the appropriate altitudes and times for aerial sampling. Aerobiological measurement needs should be communicated to foster the development and application of technology to meet these needs."

"An existing electronic network, such as the Internet, should be used to accommodate appropriate new technologies and provide linkage among agencies and institutions with similar or associated interests in, and needs for, aerobiology knowledge and tools. The Committee on Biometeorology and Aerobiology of the American Meteorological Society, the International Aerobiological Association, the Pan-American Aerobiological Association, and other professional societies with similar interests should be identified and linked to provide a forum for knowledge transfer. The communication network should facilitate scientific dialogue within and among disciplines associated with aerobiology, inform the public of aerobiological issues, and provide information that can influence research funding policy. Additional forms of communication may be required to provide private sector enterprises and developing countries with access to aerobiological information."

"Information systems should be established on the electronic communications network to facilitate access to aerobiology information. A distributed information delivery system should identify the appropriate contacts for requesting information, rather than actually contain vast arrays of aerobiological data. The distributed information delivery system should contain information about: (1) scientists involved in aerobiology research, (2) physical hardware used for aerobiology research, (3) software used for aerobiological data analysis and modeling, (4) data resources appropriate for aerobiology research, (5) plans for field research, and (6) bibliographies. The distributed information delivery system should help aerobiologists locate meteorological information that is temporally and spatially appropriate for the biological system being studied. Furthermore, the availability and utility of geographic information systems and other graphic analysis software should be identified. International standards should be accepted or established for sampling formats, geographic reference, and protocols for data use."

Organization and administrative structure.

Participants decided that an elected governing board was required to serve the membership and forge a successful alliance. To ensure balance, they specified that at least one representative from eight categories of scientists and outreach specialists should serve on the board: human health, environmental sciences, agriculture/forestry, meteorology, physical sciences/engineering, human health outreach, environmental outreach and agriculture/forestry outreach. It was also decided that a secretariat to provide the functional and coordinating support needed to maintain a viable alliance should be established as soon as possible.

The AFAR Governing Board.

The AFAR Governing Board has met annually since the initial workshop and has provided the alliance with strategic and operational plans. These meetings, open to all interested individuals, have been held in conjunction with conferences sponsored by related professional societies. At their inaugural meeting, the board established the following mission for AFAR:

> The Alliance for Aerobiology Research is an interdisciplinary organization to advance the understanding of atmospheric transport of organisms and biological particles important to agriculture, forestry, human health, wildlife, and the environment.

The goals for AFAR, established by the Governing Board, are to support aerobiology by: (1) enhancing communication and education, (2) sharing and pooling information and technology, (3) developing strategic partnerships, and (4) identifying and acquiring resources for collaborative aerobiology research.

Over the past five years, AFAR has made substantial impacts on aerobiology research within North America on two notable broad fronts. First, AFAR has dramatically increased the frequency and quality of communications and collaborations among aerobiologists from diverse regions, disciplines, and institutions. Second, alliance members, through their interactions with colleagues from the scientific community, have heightened the awareness of the importance of aerobiology as an integral component of environmental science and management.

Shortly after the inaugural AFAR governing board meeting, a Movement and Dispersal homepage was constructed to enhance internal communications among the membership. Over the intervening years, many aerobiology groups in the United States have posted the results of their ongoing research activities on the Movement and Dispersal homepage each year. These reports have provided an important avenue for information exchange among the AFAR membership, and in some cases, have led to collaborative research projects among scientists from different disciplines, regions, and institutions. In addition, the Movement and Dispersal homepage provides a window through which scientists, students, and administrators, external to the aerobiology research community, can view the activities of AFAR and locate contacts for further inquiries about aerobiology research projects. The homepage provides a directory of (e-mail) addresses and phone numbers for aerobiologists in North America, links to professional aerobiology associations, notices of aerobiology symposia and meetings of related professional societies, minutes from aerobiology meetings, an aerobiology newsletter, and short descriptions and schedules for upcoming aerobiology field projects and programs. Interested readers should visit the Movement and Dispersal homepage: *http://www.inhs.uiuc.edu/cee/movement/migr_dis.html*

A continuing series of symposia in association with AFAR has regularly assembled aerobiologists from across North America to share and discuss their ideas and research findings. These symposia have been held concurrently with meetings of the AMS, American Phytopathology Society, Entomological Society of America, International Congress of Biometeorology, PAAA, and the USDA National IPM Workshops (Table E.2). Rotating the AFAR symposia among the conferences of these larger professional societies and organizations has proven to be a successful strategy by: (1) increasing the attendance of aerobiologists at these professional meetings, (2) enhancing the exposure of aerobiologists to research emphases and methodologies in disciplines other than their own, and (3) stimulating a number of scientists with little or no previous awareness of aerobiology to participate in AFAR.

Substantial indirect outcomes from the alliance also have resulted. Perhaps the most notable is the impact of discussions about AFAR on the scientific community in North

Table E.2. AFAR Workshops and Symposia.

TITLE (TYPE OF MEETING)	*SPONSOR, LOCATION, AND DATE*
Movement and dispersal of biotic agents (workshop).	*USDA ESCOP-PMSS, National IPM Symposium/Workshop, Las Vegas, NV; April 1989.*
Insect biometeorology (symposium).	*AMS, 10th Conference on Biometeorology and Aerobiology, Salt Lake City, UT; Sept. 1989.*
Formulating generic hypotheses on aerial movement of biota (workshop).	*AMS, 10th Conference on Biometeorology and Aerobiology, Salt Lake City, UT; Sept. 1989.*
Migration and dispersal of arthropods (symposium).	*Entomological Society of America, Reno, NV; Dec. 1991.*
Alliance for Aerobiology Research Workshop (workshop).	*Sponsored by NCR-148 and supported by the USDA, Experiment Station Committee on Organization and Policy, Pest Management Strategies Subcommittee (ESCOP-PMSS), Agricultural Research Service, and Cooperative State Research Service and Michigan State University; Oct. 1992.*
Takeoff and ascent component of aerial movement of biota (symposium).	*13th International Congress of Biometeorology, International Society of Biometeorology, Calgary, Canada; Sept. 1993.*
Translation component of movement (symposium).	*AMS, 11th Conference on Biometeorology and Aerobiology, San Diego, CA; March 1994.*
Movement and dispersal and IPM (symposium/workshop).	*USDA, ESCOP-PMSS and Extension Service-IPM Task Force, Second National IPM Symposium/Workshop, Las Vegas, NV; April 1994.*
Predicting the aerial movement of biota: accuracy, precision, resolution, and limitations (symposium).	*PAAA, Tulsa, OK; June 1995.*
Formulation and evaluation of hypotheses for ascent, transport, and descent of airborne biota (symposium).	*AMS, 12th Conference on Biometeorology and Aerobiology, Atlanta, GA; Jan. 1996.*
Aerobiology networks and information systems (symposium).	*PAAA, Boston, MA; June, 1997.*
Aerial dispersal of pests, diseases, and their natural enemies (symposium).	*Joint conference of the Entomological Society of America and the American Phytopathology Society, Las Vegas, NV; Dec. 1988.*
Aerobiology in the 21st century (panel discussion).	*PAAA, Tucson, AZ; June 1999.*

America. The interactions that have resulted from discussions about AFAR between aerobiologists and other scientists/administrators from the agricultural, atmospheric science, environmental, and human health disciplines have enlarged the "world view" of many of these colleagues to include aerobiology as an element in their scientific domain. On one hand, the alliance has provided a platform to launch and formulate discussions between aerobiologists and administrators in government agencies concerning the importance of bioflow as a research area. One example was the recent formation of the Movement and Dispersal Working Group by the USDA, Experiment Station Committee on Organization and Policy, Pest Management Strategies Subcommittee. On the other hand, AFAR has stimulated scientists to work in teams to incorporate aerobiology into research proposals that they have submitted to grant programs in the natural resource management and integrated pest management arenas.

Many of the ideas on aerial movement presented in this book owe their genesis to AFAR aerobiologists. Through reading their numerous scientific works and sharing provocative discussions on a myriad of diverse aerobiology topics at the AFAR Workshop and subsequent AFAR symposia, our appreciation and knowledge base of aerobiology have been considerably expanded. To these sagacious founders of AFAR, who contributed so immensely to this airscape perspective, we dedicate this book.

Glossary

The technical terms listed below are defined with respect to their usage in this book. Most of the words have broader and/or additional meanings in other contexts.

abundance—a measure of population density. Abundance is used in a technical way as a relative measure of population, the number of organisms measured with a specific sampling method over a given time period.

active transport—movements of biota in the atmosphere that are influenced by the behavior of the organism. Organisms that engage in active transport have a degree of control over their direction and speed of travel and consequently their destination. Contrast passive transport.

adiabatic process—a process in which there is a change in temperature without heat entering or leaving the system. Whenever an air parcel enters a higher or lower pressure environment its volume changes, and a corresponding change in temperature occurs. Thus a rising air parcel cools by adiabatic expansion and subsiding air warms by adiabatic compression.

advection—the horizontal transport of atmospheric constituents (including heat, moisture, and biota) by the wind. Advection of air masses causes much of the variations in weather and can assist the transport of biota over long distances.

aerobiology—the study of factors and processes that influence the movement of biota in the atmosphere.

aerobiology process model—a "soft" systems model of the sequence of stages through which each organism that moves in the atmosphere proceeds. The purpose of the model is to focus research on the many factors that affect biota in each stage of the process and the coupling of the various stages in the life cycles of aerobiota.

aerobiota—organisms that use the atmosphere to move, including plant and animal viruses, fungi, bacteria, pollen and seeds of higher plants, soil nematodes, arthropods, and birds.

African trypanosomiasis—a tropical disease of humans caused by a protozoal parasite that infects the muscle, heart, and brain. The parasite is transmitted among humans by the tsetse fly.

agroecosystem—an ecosystem in which some form of agricultural activity is taking place. *See* ecosystem.

air mass—a large body of air that is characterized by relatively homogeneous air temperature and moisture characteristics in the horizontal plane. Air masses can move great distances across the Earth's surface carrying aerobiota.

air parcel—a small volume of air.

airscape perspective on the flow of biota—a perspective on the aerial movement of biota that emphasizes how plants and animals use the atmosphere to flow among habitats and how knowledge and information about the movement phenomena can help humans manipulate terrestrial ecosystems.

anabatic wind—a local wind that blows up a mountain slope. During sunny weather, the air over the slope is heated, and the difference in density between the warm air in proximity to the slope, and the cooler air at a corresponding altitude away from the slope, initiates an upslope airflow.

anemometer—an instrument for measuring wind speed.

anthropocentric—having a human-centered approach or perspective.

anticyclone—a large scale closed circulation or air mass characterized by high atmospheric pressure. In the Northern Hemisphere winds diverge clockwise from anticyclones while in the Southern Hemisphere they diverge in a counterclockwise pattern. Also called a high. Contrast extratropical cyclone.

aphid—any member of numerous species of small insect of the family Aphidae. Many aphids are economically significant because they infest a wide variety of food crops and ornamental plants and are vectors for many plant disease viruses.

aptitude for movement—a term used in conjunction with the aerobiology process model to denote the combination of vitality and life history components (influenced by intrinsic factors, extrinsic ecology, and the terrestrial environment) that determines an organism's inclination and capacity for aerial movement.

arbovirus—short for arthropod-borne virus. Viruses such as viral encephalitis and yellow fever that can affect humans and are carried by mosquitoes, ticks, and other bloodsucking arthropods.

arthropod—any members of the animal phylum Arthropoda, which includes species adapted to almost every type of habitat and food source. Arthropods that move in the atmosphere include spiders, mites, and insects.

ascent—the stage in the aerobiology process model involving upwardly directed movement through the biota boundary layer in which an organism may exert considerable control over its flight direction and/or speed.

atmosphere—the envelope of gas surrounding planet Earth that is held in place by gravity.

atmospheric circulation—any flow of air in the atmosphere. In many circumstances, airflows can form closed circulations.

atmospheric diffusion models—methods for calculating the dispersion of atmospheric constituents. These models treat biota as a cloud moving downwind that spreads out and becomes less dense with time and distance away from a source.

atmospheric dispersion—the spread of atmospheric constituents, including biota, by convection and diffusion within the air.

atmospheric motion systems—flows of air that range in size from tiny eddies to global wind belts and assist the takeoff and ascent, horizontal transport, and descent and landing of aerobiota.

atmospheric pressure—the downwardly directed force on an area that results from the weight of the overlying column of air. Pressure decreases exponentially with increasing altitude because air is compressible.

atmospheric stability—a measure of the behavior of air pertaining to vertical motion that is intimately connected to the lapse rate of temperature. *See* stable, neutral, and unstable atmospheric conditions.

atmospheric state variables—elements of weather, including air temperature, pressure, moisture, and solar radiation.

atmospheric trajectory—a path in the atmosphere tracing the points successively occupied by an air parcel (or "cloud" of biota) in motion. The path is commonly computed by an advection-diffusion equation in the Lagrangian or Eulerian coordinate system, using gridded fields of atmospheric data.

bacteria—microorganisms that reproduce either by fission or by forming spores and generally lack chlorophyll or a distinct nucleus surrounded by a membrane. Some bacteria are parasitic and cause such human diseases as pneumonia and typhoid fever.

beneficial—a name often applied to a species that helps humans compete with other species (pests) for food and fiber resources. In a broader sense, most species are beneficials in that they are integral to the food web and contribute to the diversity and stability of life on Earth.

bioflow—the flow of biota in the atmosphere.

biota—(*see* organism).

biota boundary layer (BBL)—the zone in the lower atmosphere within which an organism has some degree of control over its flight direction and consequently its destination. The BBL concept generalizes the insect boundary layer concept to all aerobiota.

biotic division—a large geographical area that circumscribes the distribution of a major assemblage of plants and animals. Primarily climate, landforms, and soil determine the types of vegetation and animals which inhabit a biotic division. Biotic divisions are composed of ecosystems in the ecoregion classification scheme.

birth rate—the ratio of the number of births to the total population during a specific time period.

boundary layer—a relatively thin zone of air near the Earth's surface where friction between the airflow and the Earth slows the wind and causes mechanical turbulence.

buoyancy—the property of an object that allows it to float in a fluid. If the temperature of an air parcel increases, its density decreases, and the parcel becomes buoyant within its colder atmospheric surroundings.

buoyant forces—forces caused by vertical air density gradients that result in turbulent mixing of the air and its constituents.

centroid—a point that represents the center of a multidimensional object, such as a spore cloud.

chinook wind—a warm dry downslope (katabatic) wind on the leeward (eastern) side of the Rocky Mountains. The air warms by adiabatic compression as it flows downslope. Equivalent to a foehn.

circumpolar vortex—the large-scale westerly vortex of winds in the middle and upper troposphere, roughly centered on the poles and often extending over the polar and middle latitudes.

climate—the long-term aggregate of weather (including average and extreme conditions of temperature, humidity, precipitation, wind speed, and cloud cover) within a large area. Climate establishes the limits of potential weather conditions for an ecoregion during a particular temporal period.

cloud—a dense concentration of suspended water droplets or tiny ice crystals in the atmosphere; formed by condensation.

cohort—plants or animals that form a group of the same age and as a result are subject to the same environmental and ecological factors that may differ from those experienced by older or younger biota.

cohort synchrony—refers to the correspondence in the life history stages of taxa within an area. This general correspondence through time is usually produced by extreme weather conditions which can cause dramatic changes in phenological progression and thus create convergence of the development stages of a wide range of organisms (e.g., a killing frost or a severe summer drought).

cold front—the leading edge of a cold air mass as it pushes warmer air upward within an extratropical cyclone. The cold front slopes steeply upward through the lower troposphere and is usually characterized by cumulus clouds and intense precipitation.

conceptual model—an abstract representation of a system.

condensation—the process by which water vapor becomes a liquid. Condensation releases latent heat from the water molecule to warm the surrounding environment.

continentality—the degree to which the climate at an interior continental location is subject to the influence of the land mass. Continental climates in the middle and high latitudes are characterized by large annual temperature ranges.

convection—heat transfer by mass movement of molecules in the atmosphere. Also refers to atmospheric motions that are primarily vertical and produced by air density differences.

convection current (thermal)—upward moving flow that results from density differences caused by the heating and expansion of air near the ground surface. Cooler, denser air from above replaces the rising, warmer air at the Earth's surface.

convective storm—any storm with cumulonimbus clouds and precipitation that results from convective currents. If lightning and thunder are associated with the storm it may be called a thunderstorm.

convergence—the net horizontal inflow of air into a specified zone. Atmospheric convergence can occur whenever airflows from different directions merge or where wind speed decreases over distance. The "excess" air will ascend and/or descend within the atmosphere. Contrast divergence.

Coriolis deflection—an apparent force on a moving object in a rotating system. In the atmosphere, the Coriolis deflection results from the rotation of the Earth and causes moving air parcels to deflect to the right in the Northern Hemisphere and to the left in the Southern Hemisphere. The effect of the Coriolis deflection on the wind increases with latitude from the equator to the poles.

cost-benefit analysis—estimates and comparisons of the long- and short-term economic, environmental, and social costs and benefits from a management option.

crop rotation—a system of agriculture in which different crops are grown in the same field in alternating years to reduce the depletion of nutrients in the soil, to increase yields, and/or to reduce losses from plant diseases and pests. The rotation can involve two or more crops.

cultivar—a distinct subspecies of a plant that does not occur naturally but has been produced by selective breeding.

cumulonimbus—a very dense, dark, and vertically developed cloud that may have an anvil head and produces precipitation. The precipitation is often accompanied by lightning, thunder, and hail.

cumulus—a vertically developed cloud.

cyclogenesis—the formation of an extratropical or tropical cyclone.

cyclone—a term for a large-scale closed circulation characterized by low pressure. Winds inspiral in a counterclockwise direction around cyclones in the Northern Hemisphere and converge in a clockwise manner in the Southern Hemisphere. Also called a low. Contrast anticyclone.

cyclone track—the path commonly followed by an extratropical or tropical cyclone.

death rate—the ratio of the number of deaths to the total population during a specific time period.

decision support system—a computer-based methodology and tool for integrating different types of information to help ecosystem managers make the best decisions possible about manipulating populations and their environments.

dengue—a tropical viral disease of humans resulting in high fever, joint pains, headaches, and rash. Dengue is transmitted among humans by mosquitoes.

density—the ratio of the mass of an object to its volume. Also, the number of organisms in a given area of land or volume of air.

density current—the intrusion of cold air underneath warmer air such as occurs with a sea breeze.

density dependent—a factor whose impact on a population is proportional to the density of the population.

deposition—the settling out of atmospheric constituents, including biota, onto the Earth's surface from the air. *See* dry deposition and wet deposition.

deposition velocity—the rate at which particulates, including biota, are deposited from the atmosphere to the ground. Deposition velocity is strongly influenced by the vertical wind speed gradient in and above a plant canopy.

descent—the stage in the aerobiology process model involving movement of organisms from the planetary boundary layer into their biota boundary layer after horizontal transport.

destination area—the ecosystem in which aerobiota land and subsequently impact. The end point in the aerobiology process model. Contrast source area.

diapause—a period of reduced activity entered into by some insects to survive adverse environmental conditions (e.g., cold temperatures).

diffusion—the transport of atmospheric constituents, including biota, by either the random movements of molecules (molecular diffusion) or by turbulent eddies (eddy diffusion). Molecular diffusion moves atmospheric constituents along their concentration gradients and is important in the laminar layer of the atmosphere. Above the laminar layer, atmospheric constituents and biota move as a cloud downwind, spreading out and becomes less dense with time and distance away from a source, due to the random movements of turbulent eddies within the mean flow.

disease—an illness or pathological condition in which normal physiological functioning is impaired.

dispersal—the passive or active spreading of organisms away from each other within the atmosphere.

diurnal (diel)—daily. Usually refers to a phenomena or process that occurs over 24-hr periods.

divergence—a horizontal spreading of an airstream, resulting in a net outflow of air from a zone in the atmosphere. Divergence can occur as a response to the inflow of air from above or below, or where airflows accelerate due to either increasing Coriolis deflection or as a result of a decrease in friction with the ground surface. Divergence in the middle and upper troposphere contributes to the genesis and maintenance of extratropical cyclones at the Earth's surface. Contrast convergence.

division (*see* biotic division).

domain—a group of biotic divisions that have common large-scale climate characteristics. Four domains (Polar, Humid Temperate, Dry, and Humid Tropical) are described in the ecoregion classification scheme.

doppler weather radar—an electronic instrument that can determine the velocity of an object by accounting for changes in the frequency of waves that are reflected by the object (the doppler shift). Doppler weather radars have the capacity to measure the direction and speed of biota in addition to their concentration. These large scanning radars can be used as arial surveillance tools for bioflow.

downdraft—a downward flow of cool air from a convective storm. Downdrafts are initiated by precipitation within the cumulus cloud and often can be felt before the onset of precipitation.

drift (*see* horizontal transport).

dry adiabatic lapse rate—the lapse rate that would exist in a column of air that is thoroughly mixed with its heat content evenly distributed throughout and in the absence of condensation, or equivalently, the change in temperature with height of a dry (humidity less than saturated) air parcel which is forced to ascend or descend in the atmosphere ($\approx$ 9.8°C/km).

dry deposition—fallout of atmospheric constituents, including aerobiota, from the atmosphere to the ground due to the attraction of gravity (*see* gravitational settling). Contrast wet deposition.

dynamics of populations—fluctuations in population levels.

easterly wave—a trough of low pressure that develops in the trade winds and migrates from east to west. Easterly waves often are associated with convective storms that can develop into tropical cyclones.

ecoregion—a general term for a geographical unit composed of a mosaic of terrestrial ecosystems. Contrast region.

ecosystem—the interacting biological, environmental, economic, and social components that comprise a plant and animal community. Ecosystems are composed of many habitats with boundaries that are demarcated by population processes and patterns. Nutrients pass between different organisms in specific pathways and cycle through the organisms in hierarchical interdependent processes. Ecosystems are considered local scale and parts of landscapes.

ecosystem diversity—a measure of the complexity of an ecosystem that takes into account the number of inhabitants belonging to different species.

ecosystem integrity—the capability of an ecosystem to maintain a balanced, integrated, and adaptive community of organisms. Important aspects of ecosystem integrity include species composition, diversity, and their functional organization.

ecosystem monitoring program—a program for observing and measuring changes in relevant environmental and economic factors and sampling important populations in the field on a regular basis and at fixed locations.

ecosystem stability—the ability of an ecosystem to resist change when confronted by a disturbance.

ecotype—a locally adapted genetic variant within a species that has adapted to a particular set of environmental conditions in an area. Ecotypes of the same species differ from one another less than pathotypes or strains.

eddy (*see* turbulent eddy).

Ekman spiral—the turning and acceleration of the wind with altitude in the planetary boundary layer.

emigration—the aerial movement of organisms from a source area.

environment—the sum of the surrounding biotic and abiotic conditions within which an organism or community exists.

ephemeral environment—an area that experiences dramatic seasonal fluctuations in the quantity and quality of resources available to biota.

epidemic—an outbreak of a disease that spreads rapidly and affects a large proportion of potential hosts in an area.

epidemiology of a population—the geographic spread of a population. Usually refers to the outbreak of a plant, human, or other animal disease.

equatorial trough—a nearly continuous belt of low pressure that lies between the subtropical high pressure cells of the Northern and Southern Hemispheres.

equilibrium—a state of balance in a system.

eradication—the total elimination of a pest.

Eulerian model—any representation of a system within a 2- or 3-dimensional array of grid cells. Properties of a fluid are assigned to points in space at each given time. Within each grid cell, advection, diffusion, transformation, deposition, and emission are modeled, typically using gradient transport theories. There is no attempt to identify individual fluid parcels (or "clouds" of biota) from one time to the next.

evaporation—the change of state from liquid water to water vapor. Evaporation is a cooling process for the surrounding environment.

evolution—the development of organisms through changes in inherited characteristics over many generations.

exotic species—a species that is not native to an area.

extrapolation—to extend knowledge into an unknown area or time as with a prediction or forecast.

extratropical cyclone (midlatitude cyclone)—a large mid- or high latitude low pressure system with weather fronts. Extratropical cyclones generally move from west to east embedded in the mid-tropospheric flow. Also called a low or depression.

extrinsic ecology—a term used in conjunction with the aerobiology process model to denote the components or characteristics of the life history of organisms that include (1) resources for nutrition, reproduction, and protection, (2) predators, parasites and pathogens, and (3) mutualists and commensals. Contrast terrestrial environment.

foehn—a warm dry downslope (katabatic) wind usually occurring on leeward sides of the Alps. The air warms by adiabatic compression as it flows downslope. Equivalent to a chinook.

food web—the interconnected pattern formed by multiple series of organisms dependent on each other for food (i.e., complex web of food chains).

forecast—a statement of the anticipated conditions (e.g., weather or populations) in a given area for a specific time in advance. As the time in advance for which the prediction is intended increases, the forecast generally becomes less reliable.

forecast period—the time interval for which a forecast is made.

forced convection—vertical motions in the atmosphere that are caused by mechanical forces. As faster airstreams move over slower air layers, the friction among the layers results in turbulent mixing of the airstreams and their constituents. At larger scales, forced convection can be caused by the lifting of warm air at a weather front or by the upward deflection of air by a topographic obstruction. Contrast free convection.

free atmosphere—the layer of air above the planetary boundary layer in which the effect on motion of friction between the atmosphere and Earth is negligible.

free convection—vertical motions in the atmosphere that are caused by buoyant forces. Heating or cooling of air causes density differences that result in vertical motions. Contrast forced convection.

front—the boundary or transition between air masses. Typically a zone of relatively low pressure and storminess.

fungi—a member of the kingdom of organisms called Fungi, consisting of eukaryotes that absorb their food, reproduce through spore production, have chitinous cell walls, and lack flagella at all stages of their life cycle. The spores of fungi often move long distances in the atmosphere.

fungicide—any substance that kills fungi.

genetic engineering—the altering of genes or genetic material in an organism to produce desirable new traits or to eliminate undesirable characteristics.

geographic information system (GIS)—a set of statistical tools for quantifying spatial relationships between map layers or data themes, each of which is composed of a single type of data that is spatially referenced. A GIS usually includes capabilities for data capture, database management, and map display.

geostatistics—a family of statistical methods for describing temporal and spatial correlations among phenomena. Geostatistics can be used for quantifying and modeling correlation at a spectrum of spatial scales and for interpolating between, and extrapolating beyond, sample points.

geostrophic wind—the wind above the planetary boundary layer where the Coriolis deflection is balanced by the pressure gradient force resulting in flow parallel to isobars or contours.

germination—the process by which a seed or spore begins to grow and develop.

global circulation—the general circulation of the atmosphere.

global positioning system (GPS)—a system for precisely determining the position of a place on Earth. The geographic coordinates for a location are calculated by an electronic receiver that simultaneously tracks the signals emitted by several of a network of satellites.

global scale—spanning the Earth and/or lasting centuries, millennium, or more. The largest scale of atmospheric motion.

global wind belts—areas of persistent winds that extend over large portions of the Earth (e.g., trade winds, midlatitude westerlies, and polar easterlies).

gravitational force—the attractive force between all masses. Gravity pulls all atmospheric constituents, including airborne organisms, towards the Earth's surface.

gravitational settling—the descent and landing of particulates, including biota, from the atmosphere due to gravitational force. *See* dry deposition.

growth rate—the change in a population over a period of time. Also the ratio of the number of births minus the number of deaths to the total population during a specific time period.

habitat—a place in which an organism lives within an ecosystem. Habitats are generally characterized by physical features and dominant plants.

Hadley cell—a large thermally driven circulation that consists of warm rising air near the equator (equatorial trough), a poleward flow of air in the middle and upper troposphere, sinking air near 30° latitude (subtropical high pressure cells), and an equatorward near-surface flow of air (trade winds) that completes the circulation.

hail—an ice pellet that has concentric shells formed by the successive freezing of layers of water in a large convective storm cell.

heterogeneous—made up of many different elements.

high (*see* anticyclone).

high latitudes—a geographic reference to polar and subpolar areas.

highly mobile organisms—a term applied to biota that have adapted to moving long distances in the atmosphere as part of their life histories.

holistic perspective—a focus on the whole system rather than on its component parts.

horizontal transport—movement of biota that is assisted by strong winds that govern the direction and speed of travel in the planetary boundary layer. Horizontal transport is used in conjunction with the aerobiology process model to denote aerial movement above the biota boundary layer. Drift is often used for the movement of microorganisms in the PBL.

host—an organism that supports another organism at its own expense.

humidity—a general term for the air's water vapor content.

hurricane (typhoon)—a tropical cyclone exhibiting wind speeds greater than ≈ 33 m s^{-1} and heavy precipitation.

hypothesis—a conceptualization to explain a phenomena or process.

ideal gas law—the relationship of pressure, temperature, and density that governs the behavior of air. The ideal gas law states that the pressure exerted by a gas is proportional to its density and temperature.

immigration—the movement of organisms into a destination area.

impact—a forcible impinging of a process or phenomena on something else. Also the final stage in the aerobiology process model where immigrants impinge upon the dynamics of an ecosystem in a destination area.

incidence—a measure of occurrence. Used in a technical manner as a rating of abundance of microorganisms and other small biota that are not directly counted.

infection—the condition of disease in plants or animals due to invasion by pathogenic organisms such as fungi, bacteria, protozoa, viruses, or animal parasites.

inoculum—a spore or other microorganism that is introduced to a host suited to promote its growth.

insects—a numerous and diverse class of arthropods with enormous structural and habitat diversity. Insects are the pollinators of more than half of the flowering plants on Earth, major pests of most crops, and vectors of numerous plant and animal diseases.

insect boundary layer—the zone in the lower atmosphere within which an insect can control its flight direction and consequently its destination. Also called the insect flight boundary layer.

insect monitoring radar (IMR)—a vertically looking very short wavelength radar that can detect individual birds and large insects at a range of approximately 1 km. These electronic instruments are considered sampling devices because they measure bioflow only within a small conical-shaped volume above the radar.

integrated pest management (IPM)—a strategy for controlling pests in agriculture that incorporates a variety of diverse tactics. An important goal of IPM is to decrease the economic cost of pest management and environmental pollution by reducing the application of synthetic chemicals to human managed ecosystems.

intrinsic factor—a term used in conjunction with the aerobiology process model to denote a component or characteristic in the life history of an organism that through its genetic underpinnings is in large part common to a species (e.g., size, weight, morphology or physiology).

Internet—the world-wide electronic computer network.

intertropical convergence zone (ITCZ)—the area at the surface of the Earth that marks the boundary between the northeast trade winds from the Northern Hemisphere and the southeast trade winds from the Southern Hemisphere. The ITCZ represents the equatorward wing of the Hadley cell circulation and is typically characterized by low pressure and rising motions stretching around the Earth.

intertropical front (ITF)—the boundary separating airflows from the Northern and Southern Hemispheres within the intertropical convergence zone.

inversion—a layer in the atmosphere with an inverted lapse rate. The warmer air near the top of the layer traps the cooler, denser air below. Consequently an inversion layer is characterized by pronounced stability and little turbulent mixing.

inverted lapse rate—an increase in temperature with increasing altitude in the atmosphere.

isobar—a line drawn on weather maps joining locations with equal atmospheric pressure.

isotropic—having physical properties that are uniform in all directions.

jet stream (jet)—a relatively strong horizontal wind concentrated within a narrow zone in the atmosphere.

katabatic wind (downslope wind)—a local slope wind that develops when radiative cooling at night in high elevations causes air density to increase. The cold air forms a density current that flows down the slope.

kinetic energy—the energy associated with a mass in motion.

Lagrangian model—any representation of a system that simulates diffusion, transformation, and deposition in a fluid within a moving-reference frame that is centered on a given parcel. Langrangian models can be used to track air parcels containing biota that are moving in a larger scale wind field in a numerically simple way.

laminar flow—smooth, nonturbulent flow in the atmosphere.

laminar boundary layer—a shallow air layer in very close proximity to roughness elements on the Earth's surface. Flow in this layer is straight and horizontal, and vertical transport of atmospheric constituents through this layer is by molecular diffusion.

land breeze (*see* sea-land breeze circulation).

landing—a term used in conjunction with the aerobiology process model to denote the act of terminating aerial movement in a destination area. Landing must involve movement from an organism's biota boundary layer to the Earth's surface, so as to avoid physical injury. Deposition is often used for the landing of microorganisms.

landscape—a geographic assemblage of ecosystems that are interconnected by passageways through which biota can freely move. Groups of landscapes compose biotic divisions. Patterns in landscapes correlate with processes within the landscapes.

lapse rate—the change of temperature with increase in altitude in the atmosphere. The average lapse rate in the troposphere is approximately –6.5°C/km.

large scale—spanning continents and/or lasting weeks to months. Also used in a more qualitative and relative manner to indicate vast distances and/or long time periods as opposed to small scale.

latent heat flux—the transfer of heat from the Earth's surface to the atmosphere by evaporation and transpiration or from the atmosphere to the Earth's surface by condensation. Contrast sensible heat flux.

leeward—facing away from the wind or being downwind of an obstruction to airflow. Contrast windward.

lesion—a localized, defined area of diseased tissue on a plant or animal.

liberation (*see* takeoff).

life history of organisms (life cycle)—the growth stages and life processes through which each species continuously cycles, typically including birth, growth, maturation, reproduction, movement, and death.

lightning—a sudden flash of light produced by the flow of electrons between oppositely charged parts of a cloud or between a cloud and the ground.

local aerial movement—a change of position by an organism usually within a habitat or to nearby habitats that results from aerial movement within its biota boundary layer. In accordance with the airscape perspective, an organism for the most part can control its direction and speed of movement and destination during local aerial movement.

local scale—spanning ecosystems and/or lasting hours.

local winds—small scale winds that differ from those which would be appropriate to the larger scale pressure distribution. Local winds are often due to effects of the landscape features (e.g., mountains and large water bodies) on air density.

long-distance aerial movement—a change in position by an organism, usually between distant habitats, that results from aerial movement within the planetary boundary layer above its biota boundary layer. In the planetary boundary layer, strong airflows dictate the direction and rate of movement by an organism and thus strongly influence its destination.

low (*see* extratropical cyclone).

low latitudes—a geographic reference to equatorial or subtropical areas.

low-level jet stream—a narrow, strong wind or jet stream in the planetary boundary layer.

lymphatic filariases—a tropical disease of humans caused by a nematode transmitted by arthropods which can cause elephantiasis.

malaria—a common tropical disease of humans caused by a parasite that lives in red blood cells and is transmitted by mosquitoes.

management strategy—a plan of action (involving multiple tactics) for manipulating components of an ecosystem.

management tactic—a single method or approach for manipulating components of an ecosystem.

mean wind—the speed of horizontal airflow in the downwind direction averaged over periods of 10 minutes or more far from any obstructions and at a height where the surface roughness elements have only moderate impact on the airflow.

mechanical forces—forces caused by friction in the air as faster airstreams move over slower air layers. This results in turbulent mixing of the airflows and their constituents. Contrast buoyant forces.

mechanical turbulence—movements of air characterized by small-scale changes in direction and speed caused by mechanical forces. Contrast thermal turbulence.

meridional flow—a large-scale pattern of tropospheric circulation over the middle latitudes where the north to south component of the wind is pronounced. Contrast zonal flow.

mesoscale—spanning landscapes and/or lasting days.

microorganism—an organism that is too small to be seen by the human eye. Bacteria, viruses, protozoans, and some fungi and algae are commonly considered microorganisms.

microscale—spanning geographical areas equal to or smaller than habitats and/or lasting for time periods of minutes, seconds, or shorter.

midlatitude cyclone (*see* extratropical cyclone).

midlatitudes (middle latitudes)—a geographic reference to the latitudes between approximately 30° and 50° N and S.

migration—the seasonal movement of a population from one area to another and back again. Migrants often traverse tremendous distances within the atmosphere.

mites—members of an extremely diverse subclass, Acarina, of the class Arachnida. Many are parasitic and some are vectors of human and animal diseases such as Lyme disease.

model—a simplified representation of a system that depicts the relationships among its component parts. A soft systems model is conceptual while the relationships in a computer simulation, or hard systems model, are quantified for simulating the behavior of the system.

momentum—the amount of motion of a body calculated as the product of an object's mass and its velocity.

monsoon wind system—a system that displays a seasonal reversal of the prevailing wind direction between winter and summer. Typically the wind blows from sea to land and is accompanied by abundant precipitation in the summer and is a dry, land-to-sea wind in the winter.

mountain and valley circulation—a system of local winds that commonly develops during calm, clear weather in mountainous areas and is driven by air temperature differences. The flow is uphill during the day and downhill at night.

natural enemy—a predator or pathogen of an organism.

natural resource—any biotic or abiotic material that is exploited by organisms.

negative feedback—a flow of information, mass, or energy that causes an effect which counteracts the stimulus. Contrast positive feedback.

nematode—round worms that belong to the phylum Nematoda whose species are distributed widely as free-living forms in water and soils or as animal and plant parasites.

nested scale concept—an abstraction that focuses on the reciprocal relationship between phenomena and processes that are produced at different scales but in the same space and/or time. The nested scale concept emphasizes linkages and continuity of events and places.

neutral atmospheric conditions—a state of the atmosphere with respect to stability that represents the boundary between the enhancement of vertical motion and its suppression. A condition of neutral stability exists when the lapse rate is close to the dry adiabatic lapse rate, either during a transition between stable and unstable conditions or when the wind speed is great enough to thoroughly mix heat within the lower atmosphere.

NEXRAD—an acronym for Next Generation Weather Radar. More technically called the WSR-88D Doppler radar. *See* doppler weather radar.

nocturnal low-level jet—a strong wind that develops near the top of the planetary boundary layer at night in many locations on Earth. Clear skies and a variety of landscape characteristics contribute to the formation of nocturnal low-level jets.

North American corridor—the "ecoregions" and the lower layer of atmosphere (planetary boundary layer) that extends from the subtropics into the continental interior of North America. This corridor, within which many aerobiota move long distances, is bounded by the Rocky Mountains to the west and the Appalachian Mountains to the east.

object of control—environment dichotomy—a framework for quantitatively evaluating the interdependence of organisms and their environment. The dichotomy separates the components of a system into a set of important organisms and the web of their interactions that one desires to manage (object of control) and the other important biotic and abiotic factors that influence the object of control's resource pool.

oceanality (oceanicity or maritime influence)—the degree to which the climate at a location is subjected to the influence of the ocean. A maritime climate is generally characterized by a relatively small annual air temperature range.

occluded front—the boundary between air masses that is formed when a cold front overtakes a warm front within an extratropical cyclone. The air in the warm sector of the cyclone is lifted upward within the troposphere as the fronts merge, usually causing extensive precipitation.

occurrence—an event or incident. Used in a technical manner as the frequency at which an organism is found at a sample location.

on-line decision support system—an ecosystem management decision support system organized around a program which continuously monitors populations and their environments. The frequent measurements provide continuous information about the current state of the system and a basis to validate and adjust models to forecast future population levels.

onchocerciasis—a tropical skin or eye disease of humans caused by nematodes.

oospore—the fertilized female cell in fungi and some algae.

organism (biota)—any single or multicellular living body whose components work together to carry out life processes.

orographic wave—a wave to the lee of a mountain whose form and extent is determined by the disturbing effects of topography and the speed and direction of the passing airflow.

over wintering habitat—usually a tropical or subtropical area where a migrant spends the winter months of the year.

parameter—a variable in a system that plays a role in determining the behavior of the system.

passive transport (passive dispersal)—movements of biota in the atmosphere that are generally independent of the behavior of the organism. However, organisms that rely on the atmosphere for transport are never totally passive. Contrast active transport.

patch—a small ecological unit that is relatively homogeneous in its vegetation and physical features.

pathogen—an organism (e.g., parasite, bacterium, or virus) that causes disease.

pathogenicity—the disease-producing capacity of a microorganism or spore.

pathotype—a subpopulation of a pathogen. Analogous to strain.

pest—a species that successfully competes with humans for food and/or fiber resources. Pathogens that are detrimental to human health are also considered pests.

pesticide—a chemical that is used to kill pests. Often a relatively stable organic compound synthesized by humans and differentiated according to the taxa of their targets (e.g., herbicides, insecticides, and fungicides).

pesticide resistance—an inheritable capacity developed in a population to survive exposure to pesticides. It derives from the selective effect of exposures that kill or disable a portion of the population, the subsequent generation originating from the survivors. Among the survivors are those which carry pre-adaptations for resistance to that pesticide. An example of human-induced evolution.

phenology—the life history stages of an organism as they relate to seasonal variations in the physical environment. These transitions include emergence, early development, late development, maturation, reproduction, and senescence.

physiological stress—a strong negative impact on an organism's ability to function, brought about by a change in physical environmental conditions.

planetary boundary layer (PBL)—the air layer that extends from the Earth's surface upward to an altitude that varies from approximately 300 m at night to 1–3 km during the day. Friction between the airflow and the Earth's surface influences atmospheric motions in this layer. The daytime PBL is also called the mixed layer.

plan view—a representation made to scale of the top view or horizontal section of an object or scene.

polar easterlies—shallow cold easterly winds located at high latitudes.

polar front—a semipermanent, semicontinuous boundary within the lower troposphere between air masses of tropical and polar origin.

polar jet stream—a zone of strong winds in the middle troposphere usually located on the downstream side of a trough in a Rossby wave and above the polar front.

pollen—grains or microspores containing male gametophytes of seed plants. Fertilization in seed plants is the transfer of pollen, often by insects, wind, and birds, from the anthers of flowers to the pistils of flowers on the same plants or on plants of the same species far away. Wind-blown pollens of ragweed and many other plants are human allergens.

population—organisms of the same species that inhabit a specific area.

population dynamics (*see* dynamics of populations).

population outbreak—a rapid increase in the density of an organism in an ecosystem in response to environmental conditions that are favorable for reproduction. Aerial movements of populations from other habitats can also contribute to a population outbreak in an area.

positive feedback—a flow of information, mass, or energy that causes an enhancing effect on the stimulus. Contrast negative feedback.

precipitation scavenging—the washing of particulates, including biota, out of the atmosphere by precipitation. Often called rainout or washout.

preconditioning—conditions in source areas that are crucial to the dynamics of a population, initiation of movement, and the status of biota during transport. Also the initial stage in the aerobiology process model.

predator—an animal that kills other animals for its food or pleasure.

prediction—a statement about the future. Same as forecast.

pressure gradient—the variation of atmospheric pressure over distance that drives the wind. Winds are fast when the pressure gradient is steep.

prevailing wind—a large-scale wind that usually flows from a single direction (e.g., the mid-latitude westerlies).

preventive management—a paradigm for manipulating an ecosystem with the objective of maintaining the long-term stability of its diverse populations. The approach focuses on predicting events that destabilize relationships among organisms and between organisms and their environments.

profile view—a representation made to scale of the vertical cross section of an object or scene.

radar—acronym for radio detection and ranging. An electronic instrument for detecting distant objects that reflect microwaves. *See* Doppler weather radar and insect monitoring radar.

radiation—energy emitted in a wave form by substances that possess heat.

radiative cooling—the cooling of the Earth's surface and adjacent air that occurs at night when terrestrial longwave radiation exceeds atmospheric longwave radiation and thus the Earth's surface suffers a net loss of heat.

radiative heating—the warming of the Earth's surface and adjacent air that occurs primarily during the day whenever the shortwave and longwave radiation absorbed by the surface exceeds the longwave radiation emitted by it.

radiosonde—an instrument package borne by a helium-filled balloon for measuring air temperature, pressure, humidity, wind speed, and wind direction. As the balloon floats slowly upward in the atmosphere, a radio transmitter conveys the measurements to a receiver below.

range—the geographic distribution of a species.

receptor ecosystem—a plant and animal community into which aerobiota land and impact. Same as destination area in aerobiology process model.

region—a general term for a geographical unit based on a political entity. A region may comprise ecosystems, landscapes, biotic divisions and/or domains. Contrast ecoregion.

relational database—a flexible data management tool in which objects are stored separately from their attributes. The name of an entity, and its type, attributes, and/or geographical coordinates are linked by a unique identifier.

reproduction—the production of offspring that perpetuates a species.

resistance—a property of an organism that allows it to withstand a poison (e.g., insecticide or fungicide) or other control measure.

Richardson number—a ratio of the mechanical to buoyant forces that provides an indication of atmospheric stability. Positive Richardson numbers denote stable atmospheric conditions, negative values denote unstable conditions, and near-zero Richardson numbers indicate neutral conditions.

ridge—an elongated area of high atmospheric pressure. Contrast trough.

risk—the hazard or chance of loss from a pest or pathogen.

rotor—a turbulent eddy that can form on the leeward side of an obstruction such as a mountain.

Rossby wave—a mid- to upper-tropospheric wave in the westerly flow above the middle and high latitudes characterized by long wavelength (≈ 6,000 km), large amplitude (≈ 3,000 km), and slow movement.

roughness boundary layer—a zone in the atmosphere's surface layer occurring among roughness elements on the Earth's surface. Within this layer, atmospheric constituents are primarily transferred by turbulent eddies moving along constantly changing pathways through the spaces among roughness elements.

roughness element—an object on the Earth's surface that obstructs or distorts airflow (e.g., plants and buildings).

sample location—a geographic site where biological observations are made.

satellite—a device containing scientific instruments in orbit around the Earth that transmits images of the atmosphere and ground surface to Earth.

scale—a graduated classification system of temporal or spatial measurement.

sea (or lake) and land breeze circulation—a system of local winds that commonly develops during calm, clear weather in coastal areas and is driven by temperature differences between the land and water surfaces. The flow is from sea to land during the day and from land to sea during the night.

sea breeze front—the leading edge of a sea breeze. The passage of this front is often associated with showers, a wind shift from seaward to landward, and a sudden drop in temperature.

seed—the mature embryo of a plant from which a new individual grows.

seedbed—a plot of ground where young plants are grown before transplanting.

senescence—an aging process in mature individuals. In most organisms, senescence occurs at the end of the life cycle while in deciduous plants, senescence also occurs before the shedding of leaves in autumn.

sensible heat flux—the transfer of heat that organisms can feel from the Earth's surface to the atmosphere or vice versa. Sensible heat is measured with a thermometer. Contrast latent heat flux.

small scale—used in a qualitative and relative manner to indicate short distances and/or time periods. Contrast large scale.

sounding—a measurement of atmospheric conditions aloft.

source area—the ecoregion from which aerobiota initiate aerial movement. Extrinsic ecology and the terrestrial environment impact preconditioning and the initiation of movement by organisms in the source area. The starting point in the aerobiology process model.

spore—a reproductive body capable of developing into a new organism asexually. Bacteria, fungi, some protozoa, and a few plants produce spores that can be carried long distances by winds.

sporangiospore—an asexual spore produced in a sporangium in fungi. Sporangiospores of many fungi are adapted to disperse in the atmosphere.

sporangiophore—thread-like stalks in fungi that bear sporangia which produce spores.

spore cloud—a parcel of air containing spores.

sporulation—the process of multiple division resulting in the production of spores.

squall line—a narrow band of convective storms.

stable atmospheric conditions—a state of the atmosphere with respect to stability that suppresses vertical motions. A stable condition exists whenever the decrease in temperature is less than the dry adiabatic lapse rate. Radiative cooling of the Earth's surface promotes stable conditions, and a temperature inversion represents an extremely stable atmosphere.

stationary front—a boundary between two air masses with winds blowing nearly parallel but in opposite directions on each side of the front. Extratropical cyclones can form along a stationary polar front when conditions aloft are conducive to uplift of air from the Earth's surface.

storm—an atmospheric disturbance that is characterized by strong winds and precipitation.

strain—a subpopulation that displays behavioral and/or morphological characteristics that differ from the larger population.

strategic biological sampling landscape—a landscape where a set of biological observations are made at appropriate locations within component ecosystems.

stratosphere—the stable atmospheric layer above the troposphere.

stratus cloud—a gray layered cloud often associated with a warm front and drizzle.

subsidence—the slow sinking of air that usually is associated with areas of high pressure at the Earth's surface.

subtropical high pressure (STHP) cells—semi-permanent areas of high pressure located near 30° N and S latitudes over the oceans during the equinoxes. These large surface features are beneath the subsiding wing of the Hadley cell circulations and shift seasonally.

surface layer—the lowest few meters of the planetary boundary layer whose depth is highly variable depending upon the wind speed above, the roughness of the surface below, and the temperature gradient between the ground and air. The surface layer is divided into laminar, roughness, and turbulent surface boundary layers.

susceptibility—the degree to which an organism is prone to infection by a particular disease.

sustainable development—activities that do not deplete or degrade the environmental resources upon which present and future activities depend.

synoptic scale—the scale typical of daily weather maps that show high and low pressure areas and fronts spanning continents.

systemic fungicide—a fungi-killing compound that is absorbed by leaves or roots and transported throughout the plant.

systems approach—a philosophy and a methodology for acquiring knowledge and learning about a set of interconnected phenomena and processes in order to understand and make decisions about difficult problems. As a philosophy, it is a holistic approach to thinking about complex phenomena and processes. As a methodology, it is a logical procedure for building models to analyze the structure and predict the behavior of systems.

systems models—tools for increasing our knowledge about the complex interactions among organisms and their environments and for predicting the behavior of systems in the future and/or under different circumstances.

takeoff—the process of commencing aerial movement. Takeoff is considered active when it involves the use of energy from within the organism and passive when the energy comes from the environment. Liberation is often used for the initiation of aerial movement by spores and other microorganisms.

target organism—the organism a natural enemy or pesticide is intended to kill.

temperature—the degree of coldness or hotness of an object as measured by a thermometer. More technically, a measure of the average kinetic energy of the molecules in a substance.

terrestrial environment—the soil, water, biota, and air near the Earth's surface in which an organism lives. Used with extrinsic ecology to denote factors that govern the vitality of organisms and influence the takeoff & ascent and descent & landing stages in the aerobiology process model.

theory—a group of scientific propositions for explaining related phenomena and processes.

thermal (*see* convection current).

thermal turbulence—movements of air characterized by small-scale changes in direction and speed of flow caused by buoyant forces. Contrast mechanical turbulence.

thunder—the sound emitted by rapidly expanding air along the path of lightning.

thunderstorm—a local convective storm produced by cumulonimbus clouds and accompanied by lightning and thunder.

topography—the physical characteristics of a landscape with particular reference to elevation.

tornado—a funnel-shaped vortex that extends downward from a cumulonimbus cloud. Winds in a tornado are typically between 100 to 400 km/hr and can cause severe damage.

trade winds—the belts of prevailing easterly winds which are surface features of the Hadley cell circulations. These winds blow from the subtropical high pressure cells into the intertropical convergence zone from both sides of the equator.

trajectory analysis—a method to describe the mean displacement of atmospheric constituents, including aerobiota, between successive measurements and to predict movement over subsequent time intervals. A trajectory analysis run backward in time from an end point is called a back trajectory.

trajectory confidence—a measure of the reliability of a forecast trajectory in the atmosphere. Atmospheric trajectory confidence depends on the spatial and temporal variations (complexity) of the wind field during the forecast interval.

translation (*see* horizontal transport).

transport event—a singular occurrence of aerial movement of biota from source to destination areas.

tropical—a geographic reference to the area on Earth between approximately 23.5° N and 23.5° S latitude.

tropical cyclone—the general term for a large storm without weather fronts that originates within the trade wind belt in low latitudes. Tropical cyclones develop from bands of convective cells and are called tropical disturbances, tropical depressions, tropical storms, or hurricanes (typhoons) depending on the speed of their winds.

tropical disturbance—a tropical cyclone exhibiting wind speeds less than $\approx$10 m s^{-1}.

tropical depression—a tropical cyclone exhibiting wind speeds between $\approx$10 and 17 m s^{-1}.

tropical storms—a tropical cyclone exhibiting wind speeds between 17 and 33 m s^{-1}.

tropopause—the boundary between the troposphere and stratosphere characterized by an abrupt change from a decrease to an increase of temperature with altitude. The temperature inversion inhibits mixing and can be considered a "lid" to the weather.

troposphere—the layer of atmosphere that extends from the Earth's surface upward to between 10 and 20 km and is characterized by decreasing air temperature with altitude, vertical wind motion, and appreciable water vapor content.

trough—an elongated area of low atmospheric pressure. Contrast ridge.

turbulence—airflow characterized by small-scale random changes in speed and direction.

turbulent eddy—an irregular swirl of motion or microscale circulation that behaves differently than the larger flow in which it exists. Eddies are set into turbulent motion by buoyant and/or mechanical forces in the atmosphere.

turbulent surface boundary layer—a zone in the atmosphere's surface layer above roughness elements where atmospheric constituents are primarily transferred by turbulent eddies.

typhoon—a hurricane that forms in the western Pacific Ocean.

ultraviolet (UV) radiation—radiation with a wavelength shorter than visible light and longer than x-rays. Exposure to ultraviolet radiation can be lethal to microorganisms during horizontal transport in the atmosphere.

unstable atmospheric conditions—a state of the atmosphere with respect to stability that enhances vertical motions. An unstable condition exists whenever the decrease in temperature is greater than the dry adiabatic lapse rate. The heating of air near the Earth's surface during sunny days promotes unstable conditions.

updraft—an upward rising current of air that may contribute to the genesis of a convective storm.

valley breeze (*see* mountain-valley circulation).

vapor pressure—the pressure exerted by water vapor molecules in a given volume of air. A commonly used measure of atmospheric humidity.

vector—an organism that carries a disease among plants and animals.

velocity—a measure of the speed and direction of air movement.

viability—the ability to live and grow.

virus—the simplest living microorganism. Viruses must reproduce within a host cell and are often able to manipulate the host to create conditions favorable for their own reproduction. They cause many important infectious diseases of plants, humans, and other animals.

vitality—a term used in conjunction with the aerobiology process model to denote the current condition or health of individuals within a population. Vitality is governed by factors related to extrinsic ecology and the terrestrial environment, highly variable both in time and space, and can potentially be manipulated by humans.

volume density—the number of something (aerobiota) in a given volume.

vortex—a spiral flow of air.

warm front—the leading edge of a warm air mass as it rises over colder air within an extratropical cyclone. The warm front slopes gently upward through the lower troposphere and is usually characterized by stratiform clouds and gentle precipitation.

warm sector—the area of warm air within an extratropical cyclone between the cold and warm fronts. Often there is large-scale poleward advection of biota within the warm sector of an extratropical cyclone.

wavelength—the distance covered in one complete cycle of a wave such as radiation.

weather—the condition of the atmosphere (including temperature, humidity, precipitation, wind speed, and cloud cover) in a specific locality at a particular time.

westerlies—the prevailing west to east winds that blow in the middle latitudes on the poleward sides of the subtropical high pressure cells.

wet deposition—the transport of atmospheric constituents, including aerobiota, from the atmosphere to the Earth's surface by precipitation. Contrast dry deposition.

wildlife—organisms living independent of humans.

wind—air in motion relative to the Earth's surface.

wind profile—the change in wind speed and/or direction with altitude.

wind shear—the rate of change in the wind direction or speed in either the vertical or horizontal direction.

windward—facing into the wind or being upwind of an obstruction to airflow. Contrast leeward.

yellow fever—a viral tropical disease of humans that results in high fever, acute hepatitis, jaundice, and hemorrhages in the skin and from the stomach and bowels. Mosquitoes serve as vectors for the yellow fever virus.

zonal flow—a large-scale pattern of circulation that is primarily west to east above the middle latitudes. Contrast meridional flow.

References

Achtemeier, G. L. 1992. "Grasshopper response to rapid vertical displacements within a 'clear air' boundary layer as observed by doppler radar." *Environmental Entomology* 21: 921–38.

Alerstram, T. 1990. Bird migration. Cambridge University: Cambridge, UK.

Alexander, G. 1964. "Occurrence of grasshoppers as accidentals in the Rocky Mountains of northern Colorado." *Ecology* 45: 77–86.

Allen, T. F. H., and T. W. Hoekstra. 1992. Toward a unified ecology. Columbia: New York, N.Y.

Altieri, M. A. (ed.) 1995. Agroecology: the science of sustainable agriculture. Westview: Boulder, Colo.

Angermeier, P. L., and J. R. Karr. 1994. "Biological integrity versus biological diversity as policy directives." *BioScience* 44: 690–97.

Antor, R. J. 1994. "Arthropod fallout on high alpine snow patches of the central Pyrenees, northeastern Spain." *Arctic and Alpine Research* 26: 72–76.

Atkinson, B. W. 1989. Meso-scale atmospheric circulations. Academic: New York, NY.

Avery, D. J. 1966. "The supply of air to leaves in assimilation chambers." *Journal of Experimental Botany* 17: 655–77.

Aylor, D. E. 1978. "Dispersal in time and space: aerial pathogens." In Horsfall, J. G. and E. B. Cowling (eds.), Plant disease: an advanced treatise, vol. 2. Academic, New York, N.Y., pp. 159–80.

Aylor, D. E. 1986. "A framework for examining inter-regional aerial transport of fungal spores." *Agricultural and Forest Meteorology* 38: 263–88.

Aylor, D. E. and G. S. Taylor. 1983. "Escape of *Peronospora tabacina* spores from a field of diseased tobacco plants." *Phytopathology* 73: 525–29.

Aylor, D. E., G. S. Taylor, and G. S. Raynor. 1982. "Long-range transport of tobacco blue mold spores." *Agricultural Meteorology* 27: 217–32.

Bailey, R. G. 1983. "Delineation of ecosystem regions." *Environmental Management* 7: 365–73.

———. 1995. Ecosystem geography. Springer-Verlag: New York, N.Y.

———. 1998. Ecoregions: the ecosystem geography of the oceans and continents. Springer: New York, N.Y.

Barbosa, P. (ed.) 1998. Conservation biological control. Academic: New York, N.Y.

Barbosa, P., and J. C. Schultz (eds.) 1987. Insects outbreaks. Academic: New York, N.Y.

Barry, R. G., and R. J. Chorley. 1987. Atmospheric weather and climate. Methuen: New York, N.Y.

Benninghoff, W. S., and R. L. Edmonds (eds.) 1972. "Ecological systems approaches to aerobiology I. Identification of component elements and their functional relationships." US/IBP Aerobiology Program Handbook Number 2. University of Michigan: Ann Arbor, Mich.

Berry, R. E., and L. R. Taylor. 1968. "High-altitude migration of aphids in maritime and continental climates." *Journal of Animal Ecology* 37: 713–22.

Bertalanffy, L. von. 1968. General systems theory: foundations, development, applications. Braziller: New York, N.Y.

Berthold, P. 1993. Bird migration: a general survey. Oxford University: Oxford, UK.

Blackadar, A. K. 1957. "Boundary layer wind maxima and their significance for the growth of nocturnal inversions." *Bulletin of the American Meteorological Society* 38: 283–90.

Bonner, W. D. 1968. "Climatology of the low level jet." *Monthly Weather Review* 96: 833–50.

Brodie, H. J. 1951. "The splash-cup dispersal mechanism in plants." *Canadian Journal of Botany,* 29: 224–34.

Brown, G. C. 1991. "Research and extension roles in development of computer-based technologies in integrated pest management." *Environmental Entomology* 20: 1236–40.

Brown, G. C., A. R. Lutgardo, and S. H. Gage. 1980. "Data base management systems in IPM programs." *Environmental Entomology* 9: 475–82.

Brown, A. W. A. 1983. "Insecticide resistance as a factor in the integrated control of Culicidae." In Laird, M. and J. W. Miles (eds.) Integrated mosquito control methodologies, Vol. 1. Academic, London, UK, pp. 161–235.

Burge, H. A. (ed.) 1995. Bioaerosols. Lewis Publishers: Boca Raton, La.

Capra, F. 1982. The turning point: science, society, and the rising culture. Bantam: New York, N.Y.

Checkland, P. B. 1981. "Rethinking a systems approach." *Journal of Applied Systems Analysis* 8: 3–14.

Chiang, H. C. 1973. "Bionomics of the northern and western corn rootworms." *Annual Review of Entomology* 18: 47–72.

Churchman, C. W. 1968. The systems approach. Bantam: New York, N.Y.

Coulson, R. N. 1992. "Intelligent geographic information systems and integrated pest management." *Crop Protection* 11: 507–16.

Coulson, R. N., and M. C. Sanders. 1987. "Computer-assisted decision-making as applied to entomology." *Annual Review of Entomology* 32: 415–37.

Coulson, R. N., M. C. Saunders, D. K. Loh, F. L. Oliveria, D. Drummond, P. J. Barry, and K. M. Swain. 1989. "Knowledge system environment for integrated pest management in forest landscapes: the southern pine beetle (Coleoptera: Scolytidae)." *Bulletin of the Entomological Society of America* 34: 26–33.

Cox, C. S. 1987. The aerobiological pathway of microorganisms. Wiley: New York, N.Y.

Danthanarayana, W. (ed.) 1986. Insect flight: dispersal and migration. Springer-Verlag: New York, N.Y.

Davis, J. M. 1987. "Modeling the long-range transport of plant pathogens in the atmosphere." *Annual Review of Phytopathology* 25: 169–88.

Davis, J. M., and C. E. Main. 1984. "A regional analysis of the meteorological aspects of the spread and development of blue mold on tobacco." *Boundary-Layer Meteorology* 28: 271–304.

———. 1986. "Applying atmospheric trajectory analysis to problems in epidemiology." *Plant Disease* 70: 490–97.

———. 1989. "The aerobiology of the sporangiospore of *Peronospora tabacina.*" Proceedings of the Nineteenth Conference on Agricultural and Forest Meteorology. American Meteorological Society: Boston, Mass., pp. 264–67.

Davis, J. M., C. E. Main, and W. C. Nesmith. 1985. "The biometeorology of blue mold of tobacco. Part I: the evidence of long-range sporangiospore transport." In MacKenzie, D. R., C. S. Barfield, G. G. Kennedy and R. D. Berger with D. J. Taranto (eds.), The movement and dispersal of agriculturally important biotic agents. Claitor: Baton Rouge, La., pp. 473–98.

———. 1990. "The aerobiological aspects of the occurrence of blue mold in Kentucky in 1985." In Main, C. E. and H. W. Spurr, Jr. (eds.), Blue mold disease of tobacco. Delmar: Charlotte, N.C., pp. 55–71.

Dimmick, R. L., and A. B. Akers (eds.) 1969. An introduction to experimental aerobiology. Wiley-Interscience: New York, N.Y.

Dingle, H., 1996. Migration: the biology of life on the move. Oxford University: Oxford, UK.

Dorst, J. 1962. The migration of birds. Houghton Mifflin: Boston, Mass.

Drake, V. A. 1982. "Insects in a sea-breeze front at Canberra a radar study." *Weather* 37: 134–43.

———. 1984. "The vertical distribution of macro-insects migrating in the nocturnal boundary layer: a radar study." *Boundary-Layer Meteorology* 28: 353–74.

———. 1985. "Radar observations of moths migrating in a nocturnal low-level jet." *Ecological Entomology* 10: 259–65.

———. 1993. "Insect monitoring radar: a new source of information for migration research and operational pest forecasting." In Corey, S. A., D. J. Dall, and W. M. Milne (eds.), Pest control and sustainable agriculture. Commonwealth Scientific and Industrial Research Organisation Publication: Melbourne, Australia, 452–55.

Drake, V. A., and R. A. Farrow. 1988. "The influence of atmospheric structure and motions on insect migration." *Annual Review of Entomology* 33: 183–210.

Drake, V. A., and A. G. Gatehouse, (eds.) 1995. Insect migration: tracking resources through space and time. Cambridge University: Cambridge, UK.

Drake, V. A., A. G. Gatehouse, and R. A. Farrow. 1995. "Insect migration: a holistic conceptual model." In Drake, V. A. and A. G. Gatehouse (eds.), Insect migration: tracking resources through space and time. Cambridge University: Cambridge, UK, pp. 427–57

Drake, V. A., I. T. Harman, J. K. Westbrook, and W. W. Wolf. 1998. "Corn earworm migration trajectories in the lower Rio Grande valley observed with an insect monitoring radar." In Proceedings of the Thirteenth Conference on Biometeorology and Aerobiology. American Meteorology Society: Boston, Mass., 352–53.

Edmonds, R. L. (ed.) 1979. Aerobiology: the ecological systems approach. Dowden, Hutchinson & Ross: Straudsburg, Pa.

Edwards-Jones, G. 1993. "Knowledge-based systems for crop protection: theory and practice." *Crop Protection* 12: 565–78.

Eidershink, J. C. 1992. "The 1990 conterminous U.S. AVHRR data set." *Photogrammetric Engineering and Remote Sensing* 58: 809–13.

Farrow, R. A. 1990. "Flight and migration in Acridoids." In Chapman, R. F. and A. Joern (eds.), Biology of grasshoppers. Wiley: New York, N.Y., pp. 227–314.

Fraser, D. W., T. R. Tsai, W. Orenstein, W. E. Parkin, H. J. Beecham, R. G. Sharrar, J. Harris, G. F. Mallison, S. M. Martin, J. E. McDade, C. C. Shepard, P. S. Brachman, and The Field Investigation Team. 1977. "Legionnaires' disease: a description of an epidemic of pneumonia." *New England Journal of Medicine* 297: 1189–97.

Gage, S. H., and M. E. Mispagel. 1979. "Design and development of a cooperative crop monitoring system." In Dover, M. and G. N. Hersh (eds.), Automated data processing for integrated pest management. United States Department of Agriculture: Washington, D.C., pp. 157–78.

Gage, S. H., M. E. Whalon, and D. J. Miller. 1982. "Pest event scheduling system for biological monitoring and pest management." *Environmental Entomology* 11: 1127–33.

Gage, S. H., and H. L. Russell. 1987. "Pest surveillance systems in the USA—a case study using the Michigan crop monitoring system (CCMS)." In Teng, P. S. (ed.), Crop loss assessment and pest management. American Phytopathological Society: St. Paul, Minn., pp. 209–24.

Gage, S. H., T. M. Wirth, and G. A. Simmons. 1990. "Predicting regional gypsy moth (Lymantriidae) population trends in an expanding population using pheromone trap catch and spatial analysis." *Environmental Entomology* 19: 370–77.

Garratt, J. R. 1992. The atmospheric boundary layer. Cambridge University: Cambridge, UK.

Gauthreaux, S. A., Jr. 1991. "The flight behavior of migrating birds in changing wind fields: radar and visual analyses." *American Zoologist* 31: 187–204.

Geiger, R., R. H. Aron, and P. Todhunter. 1995. The climate near the ground. Vieweg, Braunschweis\Wiesbaden: Germany.

Georghiou, G. P. 1986. "The magnitude of the resistance problem." In National Research Council, Pesticide resistance: strategies and tactics for management. National Academy of Sciences: Washington, D.C., pp. 14–43.

Georghiou, G. P., and A. Lagunes-Tejeda. 1991. The occurrence of resistance to pesticides in arthropods: an index of cases reported through 1989. Food and Agriculture Organization of the United Nations: Rome, Italy.

Glass, E. H. 1986. Preface. In National Research Council, Pesticide resistance: strategies and tactics for management. National Academy of Sciences: Washington, D.C., pp. ix–xi.

Glick, P. A. 1939. "The distribution of insects, spiders, and mites in the air." United States Department of Agriculture Technical Bulletin 673. Washington, D.C.

Goldsworthy, G. J., and C. H. Wheeler (eds.) 1989. Insect flight. CRC: Boca Raton, Fla.

Gray, W. M. 1979. "Hurricanes: their formation, structure, and likely role in the tropical circulation." In Shaw, D. B. (ed.), Meteorology over the tropical oceans. Royal Meteorological Society, Billings: Guildford, UK, pp. 155–218.

Greer, J. E., S. Falk, K. J. Greer, and M. J. Bentham. 1994. "Explaining and justifying recommendations in an agriculture decision support system." *Computers and Electronics in Agriculture* 11: 195–214.

Gregory, P. H. 1949. "The operation of the puff-ball mechanism of *Lycoperdon perlatum* by raindrops shown by ultra-high-speed schlieren cinematography." *Transaction of the British Mycological Society* 32: 11–15.

———. 1973. The microbiology of the atmosphere. Wiley: New York, N.Y.

Gregory, P. H. and J. L. Monteith (eds.) 1967. Airborne microbes. Cambridge University: Cambridge, UK.

Griffin, D. R. 1964. Bird migration. Anchor: Garden City, N.J.

Grumbine, R. E. (ed.) 1994. Environmental policy and biodiversity. Island: Washington, D.C.

Gunderson, L. H., C. S. Hollings, and S. S. Light (eds.) 1995. Barriers and bridges to the renewal of ecosystems and institutions. Columbia: New York, N.Y.

Gwinner, E. (ed.) 1990. Bird migration: physiology and ecophysiology. Springer-Verlag:, New York, N.Y.

Haynes, D. L., R. K. Brandenburg, and P. D. Fisher., 1973. "Environmental monitoring network for pest management systems." *Environmental Entomology* 2: 889–99.

Haynes, D. L., R. L. Tummala, and T. L. Ellis. 1980. "Ecosystem management for pest control." *BioScience* 30: 690–96.

Heffter, J. L. 1983. Branching Atmospheric Trajectory (BAT) Model. NOAA Technical Memorandum ERL ARL-121. Air Resources Laboratory: Rockville, Md.

Hidore, J. J. and J. E. Oliver. 1993. Climatology: an atmospheric science. MacMillan: New York, N.Y.

Hirst, J. M. and O. J. Stedman. 1963. "Dry liberation of fungus spores by raindrops." *Journal of General Microbiology* 33: 335–44.

Hochbaum, H. A. 1955. Travels and traditions of waterfowl. Banford: Newton, Mass.

Hollings, C. S. 1973. "Resilience and stability of ecological systems." *Annual Review of Ecology and Systematics* 4: 1–23.

———, (ed.) 1978. Adaptive environmental assessment and management. Wiley: New York, N.Y.

Horsfall, J. B. 1972. Preface. In Horsfall J. B. (ed.), Genetic vulnerability of major crops. National Academy of Science: Washington, D.C.

Huffaker, C. B. and R. L. Rabb (eds.) 1984. Ecological entomology. Wiley: New York, N.Y.

Ingold, C. T. 1967. "Liberation mechanisms of fungi." In Gregory, P. H., and J. L. Monteith (eds.), Airborne microbes. Cambridge University: Cambridge, UK, pp. 102–15.

———. 1971. Fungal spores: their liberation and dispersal. Clarendon: Oxford, UK.

Irwin, M. E., and J. M. Thresh. 1988. "Long-range aerial dispersal of cereal aphids as virus vectors in North America." *Philosophical Transactions of the Royal Society of London, Series* B 321:421–46.

Isard, S. A. (ed.) 1993. Alliance for aerobiology research workshop report. Alliance for Aerobiology Research Workshop Writing Committee: Champaign, Ill.

Isard, S. A., and M. E. Irwin. 1993. "A strategy for studying the long-distance aerial movement of insects." *Journal of Agricultural Entomology* 10: 283–97.

———. 1996. "Formulating and evaluating hypotheses on the ascent phase of aphid movement and dispersal." In Proceedings of the Twelfth Conference on Biometeorology and Aerobiology. American Meteorological Society: Boston, Mass., pp. 430–33.

Isard, S. A., M. E. Irwin, and S. E. Hollinger. 1990. "Vertical distribution of aphids (Homoptera, Aphididae) in the planetary boundary layer." *Environmental Entomology* 19: 1473–84.

Isard, S. A., M. A. Nasser, J. L. Spencer, and E. Levine. 1999. "The influence of weather on western corn rootworm flight activity at the borders of a soybean field in east-central Illinois." *Aerobiologia* 14: 95–104.

Jain, M., M. Eilts, and K. Hondl. 1993. "Observed differences of the horizontal wind derived from Doppler radar and a balloon-borne atmospheric sounding system." Proceedings of the Eighth Symposium on Meteorological Observations and Instrumentation. American Meteorological Society: Boston, Mass.

Johnson, C. G. 1954. "Aphid migration in relation to weather." *Biological Review* 29: 87–118

———. 1969. Migration and dispersal of insects by flight. Methuen: London, UK.

Johnson, C. G., and L. R. Taylor. 1957. "Periodism and energy summation with special reference to flight rhythms in aphids." *Journal of Experimental Biology* 34: 209–21.

Johnson, S. J. 1995. "Insect migration in North America: synoptic-scale tranpsort in a highly seasonal environment." In Drake, V. A., and A. G. Gatehouse (eds.), Insect migration: tracking resources through space and time. Cambridge University: Cambridge, UK, pp. 31–66.

Karlen, A. 1995. Man and microbes: disease plagues in history and modern times. Putman: New York, N.Y.

Kendrick, B. 1992. The fifth kingdom. Focus Information Group: Newburyport, Mass.

Kennedy, J. S. 1985. "Migration, behavioral and ecological." In Rankin, M. A. (ed.), Migration: mechanisms and adaptive significance. Contributions in Marine Science, vol. 27. Marine Science Institute, University of Texas at Austin: Port Aransas, Tex., pp. 5–26.

Kerlinger, P. 1995. How birds migrate. Stackpole: Mechanicsburg, Pa.

Klazura, G. E., and D. A. Imy. 1993. "A description of the initial set of analysis products available from the NEXRAD WSR-88D system." *Bulletin of the American Meteorological Society* 74: 1293–1311.

Knight, J. D., and M. E. Cammell. 1994. "A decision support system for forecasting infestations of the black bean aphid, *Aphis fabae* Scop., on spring-sown field beans, *Vicia faba.*" *Computers and Electronics in Agriculture* 10: 269–79.

Knight, J. D., G. M. Tatchell, G. A. Norton, and R. Harrington. 1992. "FLYPAST: an information management system for the Rothamsted aphid database to aid pest control research and advice." *Crop Protection* 11: 419–26.

Krajick, K. 1997. "The floating zoo." *Discover* 18: 66–73.

Krysan, J. L. 1986. "Introduction: biology, distribution, and identification of pest *Diabrotica.*"

In Krysan, J. L., and T. A. Miller (eds.), Methods for the study of pest *Diabrotica.* Springer-Verlag: New York, N.Y, pp. 1–24.

Kuchler, A. W. 1964. “Potential natural vegetation of the conterminous United States.” Special Publication #36, American Geographical Society: New York, N.Y.

———. 1970. Potential natural vegetation. Map (scale 1: 7,500,000). In The national atlas of the United States of America. United State Geological Survey: Washington, D.C., pp. 89–91.

Larkin, R. P. 1991. “Sensitivity of NEXRAD algorithms to echoes from birds and insects.” In Proceedings of the Twenty Fifth International Conference Radar Meteorology. American Meteorological Society: Boston, Mass., pp. 203–5

Lele, S., M. Taper, and S. H. Gage. 1998. “Statistical analysis of population dynamics in time and space.” *Ecology* 79:1489–1502.

Lemke, D. E., and C. E. Main., 1990. “Distribution of *Nicotiana repanda* and *Peronospora tabacina* in southern and central Texas: a potential source of inoculum.” In Main, C. E., and H. W. Spurr, Jr. (eds.), Blue mold disease of tobacco. Delmar: Charlotte, N.C., pp. 179–82.

Leonard, D. E. 1981. “The bioecology of the gypsy moth.” In Doane, C. C. and M. L. McManus (eds.), The gypsy moth: research toward integrated pest management. United States Department of Agriculture, Forest Service Technical Bulletin 1584. Washington, D.C., pp. 9–29.

Levizzani, V., T. Georgiadis, and S. A. Isard. 1998. “Meteorological aspects of the aerobiological pathway.” In Mandrioli, P., P. Comtois, and V. Levizzani, (eds.), Methods in aerobiology. Pitagora Editrice: Bolgnia, Italy, pp. 113–84.

Liebhold, A. M., R. E. Rossi, and W. P. Kemp. 1993. “Geostatistics and geographic information systems in applied insect ecology.” *Annual Review of Entomology* 38: 303–27.

Lighthart, B., and A. J. Mohr (eds.) 1994. Atmospheric microbial aerosols. Chapman and Hall: New York, N.Y.

Lingren, P. D., J. K. Westbrook, V. M. Bryant, Jr., J. R. Raulston, J. F. Esquivel, and G. D. Jones. 1994. “Origin of corn earworm (Lepidoptera: Noctuidae) migrants as determined by citrus pollen markers and synoptic weather systems.” *Environmental Entomology* 23: 562–70.

Liquido, N. J., and M. E. Irwin. 1986. “Longevity, fecundity, change in degree of gravidity and lipid content with adult age, and lipid utilisation during tethered flight of alates of the corn leaf aphid.” Rhopalosiphum maidis. *Annals of Applied Biology* 108: 449–59.

Litovitz, T. L., B. F. Schmitz, and K. M. Bailey. 1990. “1989 annual report of the American Association of Poison Control Centers National Data Collection System.” *American Journal of Emergency Medicine* 8: 394–442.

Loh, D. K., and E. J. Rykiel, Jr. 1992. “Integrated resource management systems: coupling expert systems with data-base management and geographic information systems.” *Environmental Management* 16: 167–77.

Lowry, W. P., and P. P. Lowry II. 1989. Fundamentals of biometeorology: interactions of organisms and the atmosphere: the physical environment. Peavine: McMinnville, Ore.

Lucas, G. B. 1975. Diseases of tobacco. Biological Consulting Associates: Raleigh, N.C.

———. 1980. "The war against blue mold." *Science* 210: 147–53.

Lutgens, F. K., and E. J. Tarbuck. 1998. The atmosphere: an introduction to meteorology. Prentice Hall: Englewood Cliffs, N.J.

MacKenzie, D. R., C. S. Barfield, G. G. Kennedy, and R. D. Berger with D. J. Toranto (eds.) 1985. The movement and dispersal of agriculturally important biotic agents. Claitor: Baton Rouge, La.

Madden, L. V. 1992. "Rainfall and the dispersal of fungal spores." *Advances in Plant Pathology* 8: 39–79.

Main, C. E., and J. M. Davis. 1989. "Epidemiology and biometeorology of tobacco blue mold." In McKean, W. E. (ed.), Blue mold of tobacco. American Phytopathology Society: St. Paul, Minn., pp. 201–15

Main, C. E. and H. W. Spurr, Jr. (eds.) 1990. Blue mold disease of tobacco. Delmar: Charlotte, N.C.

Main, C. E., J. M. Davis, and M. A. Moss. 1985. "The biometeorology of blue mold of tobacco. Part I: a case study in the epidemiology of the disease." In MacKenzie, D. R., C. S. Barfield, G. G. Kennedy and R. D. Berger with D. J. Taranto (eds.), The movement and dispersal of agriculturally important biotic agents. Claitor: Baton Rouge, La., pp. 451–69

Mani, M. S. 1962. Introduction to high altitude entomology. Methuen: London, UK.

Mandrioli, P., P. Comtois, and V. Levizzani, (eds.) 1998. Methods in aerobiology. Pitagora Editrice: Bolgnia, Italy.

Mason, C. J., and M. L. McManus. 1981. "Larval dispersal of the gypsy moth." In Doane, C. C. and M. L. McManus (eds.), The gypsy moth: research toward integrated pest management. United States Department of Agriculture, Forest Service Technical Bulletin 1584. Washington, D.C., pp. 161–202.

McKeen, W. E. (ed.), 1989. Blue mold of tobacco. American Phyotpathology Society: St. Paul, Minn.

Merriam, C. H. 1898. "Life zones and crop zones in the United States." Biological Survey Publication 10. Washington, D.C.

Metcalf, R. L. 1983. "Implications and prognosis of resistance to insecticides." In Georghiou, G. P., and T. Saito (eds.), Pest resistance to pesticides. Plenum: New York, N.Y., pp. 703–34.

Mitchell, M. J., R. W. Arritt, and K. Labas. 1995. "A climatology of the warm season Great Plains low-level jet using wind profiler observations." *Weather and Forecasting* 10: 576–91.

Monteith, J. L., and M. H. Unsworth. 1990. Principles of environmental physics. Edward Arnold: London. U.K.

Moser, M. R., T. R. Bender, H. S. Margois, G. R. Noble, A. P. Kendal, and D. G. Ritter. 1979. "An outbreak of influenza aboard a commercial airliner." *American Journal of Epidemiology* 110: 1–6.

Nagarajan, S., and D. V. Singh. 1990. "Long-distance dispersion of rust pathogens." *Annual Review of Phytopathology* 28: 139–53.

Nesmith, W. C. 1984. "The North American blue mold warning system." *Plant Disease* 68: 933–36.

Newton, C. W. 1967. "Severe convective storms." In Landsberg, H. E., and J. van Mieghen (eds.), Advances in geophysics, vol. 12. Academic: New York, N.Y., pp. 257–308.

Norton, G. A. 1991. "Formulating models for practical purposes." *Aspects of Applied Biology* 26: 69–80.

Oke, T. R. 1990. Boundary layer climates. Routledge: London, UK.

Omernik, J. M., and G. E. Griffith. 1991. "Ecological regions versus hydrologic units: framework for managing water quality." *Journal of Soil and Water Conservation* 46: 334–40.

Onstad, D. W., M. G. Joselyn, S. A. Isard, E. Levine, J. L. Spencer, L. W. Bledsoe, C. R. Edwards, C. D. Di Fonzo, and H. Willson. 1999. "Modeling the spread of western corn rootworm (Coleoptera: Chrysomelidae) populations adapting to soybean-corn rotation." *Environmental Entomology* 28: 188–94.

Pair, S. E., J. R. Raulston, J. K. Westbrook, W. W. Wolf, and S. D. Adams. 1991. "Fall armyworm (Lepidoptera: Noctuidae) outbreak originating in the Lower Rio Grande valley, 1989." *Florida Entomologist* 74: 200–13.

Patz, J. A., P. R. Epstein, T. A. Burke, and J. M. Balbus. 1996. "Global climate change and emerging infectious diseases." *Journal of the American Medical Association* 275: 217–23.

Pedgley, D. E. 1982. Windborne pests and diseases: meteorology of airborne organisms. Ellis Horwood: Chichester, UK.

Pedgley, D. E. 1986. "Long distance transport of spores." In Leonard, K. J. and W. E. Fry, (eds.), Plant disease epidemiology: population dynamics and management, vol. 1. Macmillan: New York, NY, pp. 346–65.

———. 1990. "Concentration of flying insects by the wind." *Philosophical Transaction of the Royal Society of London, Series B* 328: 631–53.

Pickering, J., W. W. Hargrove, J. D. Dutcher, and H. C. Ellis. 1990. "RAIN: a novel approach to computer-aided decision making in agriculture and forestry." *Computers and Electronics in Agriculture* 4: 275–85.

Pike, D. R. and M. E. Gray. 1992. "A history of pesticide use in Illinois." In Gray, M. E., and M. E. Ostroski, (eds.), Proceedings of the 1992 Eighteenth Annual Illinois Crop Protection Workshop. Cooperative Extension Service, University of Illinois: Champaign, Ill., pp. 43–52.

Pimentel, D. (ed.) 1997. Techniques for reducing pesticide use: economic and environmental benefits. John Wiley & Sons: New York, N.Y.

Pimentel, D., L. McLaughlin, A. Zepp, B. Lakitan, T. Kraus, P. Kleinman, F. Vancini, W. J. Roach, E. Graap, W. S. Keeton, and G. Selig. 1990. "Environmental and economic impacts of reducing U.S. agricultural pesticide use." In Pimentel, D. (ed.), CRC Handbook on pest management in agriculture, vol. 1. CRC: Boca Raton, Fla., pp. 679–718.

Pimentel, D., H. Acquay, M. Biltonen, P. Rice, M. Silva, J. Neson, V. Lipner, S. Giordano, A. Horowitz, and M. D'Amore. 1993. "Assessment of environmental and economic impacts of pesticide use." In Pimentel, D. and H. Lehman (eds.), The pesticide question: environment, economics, and ethics. Chapman and Hall: New York, N.Y., pp. 47–84.

Plant, R. E., and R. S. Loomis. 1991. "Model-based reasoning for agricultural expert systems." *AI Application* 5: 17–28.

Plant, R. E., and N. E. Stone. 1991. Knowledge-based systems in agriculture. McGraw-Hill: New York, N.Y.

Power, J. M. 1993. "Object-oriented design of decision support systems in natural resource management." *Computers and Electronics in Agriculture* 8: 301–24.

Purdy, L. H., S. V. Krupa, and J. L. Dean. 1985. "Introduction of sugarcane rust into the Americas and its spread to Florida." *Plant Disease* 69: 689–93.

Rabb, R. L. 1985. "Conceptual bases to develop and use information on the movement and dispersal of biotic agents in agriculture." In MacKenzie, D. R., C. S. Barfield, G. G. Kennedy and R. D. Berger with D. J. Taranto (eds.), The movement and dispersal of agriculturally important biotic agents. Claitor: Baton Rouge, La., pp. 5–34.

Rabb, R. L. and F. E. Guthrie (eds.) 1970. Concepts of pest management. North Carolina State University: Raleigh, N.C.

Rabb, R. L. and G. G. Kennedy (eds.) 1979. Movement of highly mobile insects: concepts and methodology in research. North Carolina State University: Raleigh, N.C.

Rainey, R. C. 1951. "Weather and the movements of locust swarms: a new hypothesis." *Nature (Lond)* 168: 1057–60.

Rainey, R. C. 1958. "Atmospheric movements and the biology of the desert locust (*Schistocerca gregaria* Forskal)." *Proceedings of the Linnaen Society of London* 169: 73–74.

———, (ed.) 1976. "Insect flight." Symposium of the Royal Entomological Society of London, no. 7. Blackwell Scientific: Oxford, UK.

———. 1989. Migration and meteorology: flight behaviour and the atmospheric environment of locusts and other migrant pests. Clarendon: Oxford, UK.

Rainey, R. C., K. A. Browning, R. A. Cheke, and M. J. Haggis (eds.) 1990. "Migrant pests: problems, potentialities and progress." *Philosophical Transactions of the Royal Society of London, Series B* 328: 515–764.

Rankin, M. A. (ed.) 1985. "Migration: mechanisms and adaptive significance." Contributions in Marine Science, vol. 27. Marine Science Institute, University of Texas at Austin: Port Arkansas, Tex.

Raynor, G. S., J. V. Hayes, and E. C. Ogden. 1974. "Mesoscale transport and dispersion of airborne pollens." *Journal of Applied Meteorology* 13: 87–95.

Real, L. A. 1996. "Sustainability and the ecology of infectious disease." BioScience 46: 88–97.

Reynolds, D. R., and J. R. Riley. 1997. "Flight behaviour and migration of insect pests." National Resource Institute Bulletin 71. University of Greenwich: Chatham, UK.

Richardson, W. J. 1978. "Timing and amount of bird migration in relation to weather: a review." *Oikos* 30: 224–72.

Riley, J. R. 1992. "A millimetric radar to study the flight of small insects." *Electronics & Communication Engineering Journal* 4: 43–48.

———. 1993. "Long-term monitoring of insect migration." In Isard, S. A. (ed.), Alliance for Aerobiology Research Workshop Report. Alliance for Aerobiology Research Workshop Writing Committee: Champaign, Ill., pp. 28–32.

Riley, J. R., A. D. Smith, D. R. Reynolds, A. S. Edwards, J. L. Osborne, I. H. Williams, N. L. Carreck, and G. M. Poppy. 1996. "Tracking bees with harmonic radar." *Nature* 379: 29–30.

Ritchie, M., and D. E. Pedgley. 1989. "Desert locusts cross Atlantic." *Antenna* 13: 10–12.

Rotem, J., and D. E. Aylor. 1984. "Development and inoculum potential of *Peronospora tabacina* in the fall season." *Phyotpathology* 74: 309–13.

Ruesink, W. G. 1976. "Status of the systems approach to pest management." *Annual Review of Entomology* 21: 27–44.

Russell, R. W., and J. W. Wilson. 1997. "Radar-observed "fine lines" in the optically clear boundary layer: reflectivity contributions from aerial plankton and its predators." *Boundary-Layer Meteorology* 82: 235–62.

Schaefer, G. W. 1976. "Radar observations of insect flight." In Rainey, R. C. (ed.), Insect flight. Symposium of the Royal Entomological Society of London, no 7. Blackwell Scientific: Oxford, UK, pp. 157–97.

Schaefer, G. W. 1979. "An airborne radar technique for the investigation and control of migrating pest insects." *Philosophical Transactions of the Royal Society of London, Series B* 287: 459–65.

Schumann, G. L. 1991. Plant diseases: their biology and social impact. American Phytopathology Society: St. Paul, Minn.

Scott, R. W., and G. L. Achtemeier. 1987. "Estimating pathways of migrating insects carried in atmospheric winds." *Environmental Entomology* 16: 1244–54.

Sellers, R. F., and A. R. Maarouf. 1988. "Impact of climate on western equine encephalitis in Manitoba, Minnesota and North Dakota, 1980–1983." *Epidemiology and Infection* 101: 511–35.

Shelford, V. E. 1913. Animal communities in temperate America. University of Chicago: Chicago, Ill.

Showers, W. B., F. Whitford, R. B. Smelser, A. J. Keaster, J. F. Robinson, J. D. Lopez, and S. E. Taylor. 1989. "Direct evidence for meteorologically driven long-range dispersals of an economically important moth." *Ecology* 70: 987–92.

Showers, W. B., A. J. Keaster, J. R. Raulston, W. H. Hendrix III, M. E. Derrick, M. E. McCorcle, J. R. Tobinson, M. O. Way, M. J. Wallendorf, J. L. Goodenough. 1993. "Mechanism of southward migration of a noctuid moth *Agrotis ipsilon* (Hufnagel): a complete migrant." *Ecology* 74: 2303–14.

Showers, W. B. 1997. "Migratory ecology of the black cutworm." *Annual Review of Entomology* 42: 393–425.

Simpson, J. E. 1967. "Swifts in sea-breeze fronts." *British Birds* 60: 225–39.

———. 1994. Sea breeze and local wind. Cambridge University: Cambridge, UK.

Sisterson, D. L., and R. Frenzen. 1978. "Nocturnal boundary-layer wind maxima and the problem of wind power assessment." *Environmental Science and Technology* 12: 218–21.

Slocombe, D. S. 1993. "Implementing ecosystem-based management: development of theory, practice, and research for planning and managing a region." *BioScience* 43: 612–22.

Smith, A. D., J. R. Riley, and R. D. Gregory. 1993. "A method for routine monitoring of the aerial migration of insects by using a vertical-looking radar." *Philosophical Transactions of the Royal Society of London, Series B* 340: 393–404.

Sparks, A. N. (ed.) 1986. "Long-range migration of moths of agronomic importance to the United States and Canada: specific examples of occurrence and synoptic weather patterns conducive to migration." United States Department of Agriculture, Agricultural Research Service, no. 43: Washington, D.C.

Spencer, D. M. (ed.) 1981. The downy mildews. Academic Press: New York, N.Y.

Stakman, E. C. 1942. "The field of extramural aerobiology." In Moulton, F. R. (ed.), Aerobiology. American Association for the Advancement of Science, no 17. Washington, D.C., pp. 1–7.

Stakman, E. C., and J. C. Harrar. 1957. Principles of plant pathology. Ronald: New York, NY.

Starfield, A. M., D. H. M. Cumming, R. D. Taylor, and M. S. Quadling. 1993. "A frame-based paradigm for dynamic ecosystem models." *AI Applications* 7: 1–13.

Stensrud, D. J. 1996. "Importance of low-level jets to climate: a review." *Journal of Climate* 9: 1698–1711.

Stone, N. D. and L. P. Schaub. 1990: "A hybrid expert system/simulation model for the analysis of pest management strategies." *AI Applications* 4: 17–26.

Stone, N. D. 1994. "Knowledge-based systems as a unifying paradigm for IPM." In Proceedings of the Second National Integrated Pest Management Symposium/Workshop. North Carolina State University: Raleigh, N.C., pp. 13–24.

Stover, R. H. 1966. "Intercontinental spread of banana leaf spot (*Mycosphaerella musicola* Leach)." *Tropical Agriculture* 39: 327–38.

Strahler, A. N. 1966. Introduction to physical geography. Wiley: New York, N.Y.

Stull, R. B. 1989. An introduction to boundary layer meteorology. Kluwer: Dordrecht, Netherlands.

Tatchell, G. M. 1991. "Monitoring and forecasting aphid problems." In Proceedings of the Conference on aphid-plant interactions: populations to molecules. United States Department of Agriculture, Agricultural Research Service, Oklahoma State University: Stillwater, Okla., pp. 215–31.

Tang, J. Y., J. A. Cheng, and G. A. Norton. 1994. "HOPPER—an expert system for forecasting the risk of white-backed planthopper attack in the first crop season in China." *Crop Protection* 13: 463–73.

Tatum, L. A. 1971. "The southern corn leaf blight epidemic." *Science* 171: 1113–16.

Taylor, L. R. 1958. "Aphid dispersal and diurnal periodicity." *Proceedings of the Linnaen Society of London* 169: 67–73.

———. 1973. "Monitor surveying for migrant insect pests." *Outlook on Agriculture* 7: 109–16.

———. 1974. "Insect migration, flight periodicity and the boundary layer." *Journal of Animal Ecology* 43: 225–38.

Teng, P. S. 1985. "A comparison of simulation approaches to epidemic modeling." *Annual Review of Phytopathology* 23: 351–79.

Teng, P. S., and S. Savary. 1992. "Implementing the systems approach in pest management." *Agricultural Systems* 40: 237–64.

Tummala, R. L., and D. L. Haynes. 1977. "On-line pest management systems." *Environmental Entomology* 6: 339–49.

Viennot-Bourgin, G. 1981. "History and importance of downy mildews." In Spencer, D. M. (ed.), The downy mildews. Academic Press: New York, N.Y., pp. 1–15.

Visher, S. S. 1925. "Tropical cyclones and the dispersal of life from island to island in the Pacific." *American Naturalist* 59: 70–78.

Wallace, A. R. 1876. The geographic distribution of animals, with a study of the relations of living and extinct faunas as elucidating the past changes of the earth's surface, vol. 1 and 2. MacMillian: London, UK.

Warwick, C. J., J. D. Mumford, and G. A. Norton. 1993. "Environmental management expert systems." *Journal of Environmental Management* 39: 251–70.

Watt, K. E. F. 1961. "Use of a computer to evaluate alternative insecticidal programs." *Science* 133: 706–7.

——— 1966. Systems analysis in ecology. Academic: New York, N.Y.

Wellington, W. B. 1945a. "Conditions governing the distribution of insects in the free atmosphere, I. Atmospheric pressure, temperature and humidity." *Canadian Entomologist* 77: 7–15.

——— 1945b. "Conditions governing the distribution of insects in the free atmosphere, II. Surface and upper winds." *Canadian Entomologist* 77: 21–28.

——— 1945c. "Conditions governing the distribution of insects in the free atmosphere, III. Thermal convection." *Canadian Entomologist* 77: 44–49.

Wellington, W. B., and R. M. Trimble. 1984. "Weather." In Haffaker, C. B., and R. L. Rabb (eds.), Ecological entomology. Wiley-Interscience: New York, N.Y., pp. 399–425.

Weltzien, H. C. 1981. "Geographical distribution of downy mildews." In Spencer, D. M. (ed.), The downy mildews. Academic Press: New York, N.Y., pp. 31–43.

Westbrook, J. K., R. S. Eyster, W. W. Wolf, P. D. Lingren, and J. R. Raulston. 1995a. "Migration pathways of corn earworm (Lepidoptera: Noctuidae) indicated by tetroon trajectories." *Agricultural and Forest Meteorology* 73: 67–87.

Westbrook, J. K., J. R. Raulston, W. W. Wolf, S. D. Pair, R. S. Eyster, and P. D. Lingren. 1995b. "Field observations and simulations of atmospheric transport of noctuids from northeastern Mexico and south-central U.S." *Southwestern Entomologist, Supplement* 18: 25–44.

Westbrook, J. K., and W. W. Wolf. 1998. "Migratory flights of bollworms, *Helicoverpa zea* (Boddie), indicated by doppler weather radar." In Proceedings of the Thirteenth Conference on Biometeorology and Aerobiology. American Meteorology Society: Boston, Mass., pp. 354–55.

White, W. B., and N. F. Schneeberger. 1981. "Socioeconomic impacts." In Doane, C. C., and M. L. McManus (eds.), The gypsy moth: research toward integrated pest management. United States Department of Agriculture, Forest Service Technical Bulletin 1584: Washington, D.C., pp. 681–94.

WHO/UNEP. 1990. Public health impact of pesticides used in agriculture. World Health Organization/United Nations Environment Programme: Geneva, Switzerland.

Williams, C. B. 1958. Insect migration. Collins: London, UK.

Wilson, L. T. 1989. "Changing perspectives on the use of simulation models in IPM programs." In National Integrated Pest Management Symposium, National Integrated Pest Management Coordinating Committee, New York State Agricultural Experiment Station. Cornell University: Ithaca, N.Y., pp. 129–36.

Wolf, W. W., J. K. Westbrook, J. Raulston, S. D. Pair, and S. E. Hobbs. 1990. "Recent airborne radar observations of migrant pests in the United States." Philosophical Transactions of the Royal Society of London, B 328: 619–30.

Worner, S. P. 1991. "Use of models in applied entomology: the need for perspective." *Environmental Entomology* 20: 768–73.

Index

A

B

C

D

E

F

G

H

R

S

T

V